Lecture Notes in Social Networks

For further volumes:
http://www.springer.com/series/8768

Anthony J. Masys
Editor

Networks and Network Analysis for Defence and Security

Springer

Editor
Anthony J. Masys
University of Leicester
Leicester
UK

ISSN 2190-5428 ISSN 2190-5436 (electronic)
ISBN 978-3-319-04146-9 ISBN 978-3-319-04147-6 (eBook)
DOI 10.1007/978-3-319-04147-6
Springer Cham Heidelberg New York Dordrecht London

Library of Congress Control Number: 2013957876

Printed on acid-free paper

Springer is part of Springer Science+Business Media (www.springer.com)

Preface

Introduction

Shocks to regional, national, and global systems stemming from natural hazards, acts of armed violence, terrorism, and transnational crime have significant defence and security implications. Today, nations face an uncertain and complex security landscape in which threats impact/target the physical, social, economic, and cyber domains. For example, acts of terrorism and organized crime are considered one of the greatest threats to national security. In the UK alone, the social and economic costs associated with organized crime are estimated between £20 and £40 billion per year [1]. Threats to national security, such as that against critical infrastructures not only stem from man-made acts but also from natural hazards. Katrina (2005), Fukushima (2011), and Hurricane Sandy (2012) are examples that highlight the vulnerability of critical infrastructures to natural hazards and the crippling effect they have on the social and economic well-being of a community and a nation.

With this dynamic and complex threat landscape, network analysis has emerged as a key enabler in supporting defence and security. With the advent of "big data" and increasing processing power, and innovative interpretive approaches, network analysis can reveal insights with regard to structural and dynamic properties thereby facilitating greater understanding of complex networks, their entities, interdependencies, and vulnerabilities.

Network "Mindset"

Network thinking, as described in Barabasi [2] opens up novel perspectives to understanding complex systems such as markets and economic system, socio-technical systems, and criminal and terrorist networks. The foremost challenge offered by complex networks resides in their interconnectedness (networks of networks) and multiscale nature [3]. The network thinking mindset leverages a topological analysis which is based upon classical graph theory through which interesting properties of structure and dynamics of a network system can be revealed. The inherent complexity of the defence and security domain characterized by interdependencies, interconnectivity, dynamic, nonlinear behavior therefore calls for this "network" mindset.

The "network" mindset is very much about having a systems approach with regard to defence and security. Link analysis for example can be used to map associations among people, places, and commodities, after a crime or terrorist attack has occurred in order to identify the perpetrators. Further, tools associated with social network analysis (SNA) can enrich the interpretive power of network analysis. Insights derived from analysis of dark networks or critical infrastructures can position law enforcement and disaster management professionals with the requisite information for informed decision making.

As part of the Springer book series: *Lecture Notes in Social Networks*, this edited volume: **Networks and Network Analysis for Defence and Security**, focuses on the contribution of network science to the following areas:

- Defence and security risk analysis
- Criminal intelligence
- Cybercrime
- Cognitive analysis
- Counter-terrorism and Social Network Analysis
- Transnational Crime
- Critical infrastructure analysis
- Support to defence and security intelligence

This book comprises 12 chapters from leading researchers engaged in network analysis within the defence, security, and intelligence domains. The chapters present state-of-the-art research on network analysis tools, techniques, and applications supported by case studies and computational simulation.

Content

The chapter "Network Analysis in Criminal Intelligence" by Steven Strang provides an overview of network analysis beginning with the application of link analysis and social network analysis (SNA) techniques that support criminal intelligence in order to understand and act against serious crimes, criminal groups, and criminal markets. Strang highlights how the effective visualization of both quantitative and qualitative data support intelligence assessments. It is through the application of network analysis in criminal intelligence that the nature and extent of relationships between data points (individuals, locations, organizations, objects, and events) emerge. Strang further illustrates how SNA measures have utility in producing targeting recommendations for intelligence collection and operational disruption.

Overall, this chapter will provide the readers with an understanding of:

- Network concepts in the analysis of serious crime, organized crime, and terrorism.
- Applications of network analysis methods and technique in criminal intelligence analysis.
- Operational uses of network analyses to support the investigation of crimes and the disruption of criminal groups.
- Strengths and limitations of network analysis in criminal intelligence.

The chapter "Identifying Mafia Bosses from Meeting Attendance" by Francesco Calderoni describes how the application of simple network analysis methods to surveillance data can identify the main players in a large mafia network. The analysis relies on data from a large investigation on the presence of the 'Ndrangheta (a mafia from the Southern Italian region of Calabria) around Milan, Italy's second largest city and economic capital. Operation Infinito, lasted for more than 2 years, identified several 'Ndrangheta families and tracked the main mafia meetings. Network analysis of an individuals' participation in mafia meetings enabled identification of the different roles and positions within the mafia network. Results show that the most central subjects also had leading positions within the mafia families. This suggests that the use of network analysis applied to meetings may provide useful information for law enforcement agencies to identify high-status criminals.

The chapter "Macrosocial Network Analysis: The Case of Transnational Drug Trafficking" by Rémi Boivin aims to put criminal activity in the larger context in which it takes place. More precisely, it is argued that country-level features may complicate or facilitate legitimate business as well as criminal activities. Transnational crime is a collaboration to avoid law enforcement; it is expected that, as far as possible, countries with higher risks of seizure and arrest will be avoided in favor of more lenient countries. Macrosocial network analysis may then be used to understand the impact of law enforcement efforts on criminal activities. It is also argued that some criminal activities must be understood in a relational perspective. Drug trafficking is presented as a case study to illustrate the contributions of macrosocial network analysis. The general conclusion of the chapter is that, despite the fact that macrosocial features are largely neglected in most analyses, network analysis is a relevant tool in understanding global criminal markets. Transnational crime is supported by a set of relations between individuals and groups operating in larger networks of countries.

The chapter "Policing the Hackers by hacking Them: Studying Online Deviants in IRC Chat Rooms" by David Décary-Hétu, Benoit Dupont and Francis Fortin develops a framework that allows criminologists to easily tap into this pristine source of data on cybercriminals. Past research has shown that the Internet Relay Chat (IRC), a worldwide instant messaging system, has been and is still one of the favorite meeting grounds for cybercriminals. Unfortunately, criminologists have yet to fully take advantage of this network of chat rooms where millions of individuals interact every day. This chapter begins with a review of how criminologists have addressed the question of cybercrime to date. The authors present the inherent limits to the current tools that are used and demonstrate that gathering data on hackers with IRC has many advantages including access to unbiased sources of information and increased efficiency. This chapter then showcases the versatility of IRC logs in criminological research and details both the solutions and the new challenges that such a methodology brings forward. This framework will enable law-enforcement agencies to go beyond traditional social platforms and to track delinquents in underground chat rooms where they hide in security through obscurity. If law-enforcement agencies are to gather the best intelligence, they will therefore need to monitor IRC networks just as some of them have been doing with social networks like Facebook and Twitter.

The chapter "Why Terror Networks are Dissimilar: How Structure Relates to Function" by Christian Leuprecht compares Al Shabab (AS) networks as they relate to recruitment, fundraising, and attacks across the United States and Australia with corroborating evidence from Canada, the United Kingdom, The Netherlands, and Denmark. Although networks differ markedly across these attributes, unrelated networks performing similar functions are consistent in their nature and structure. These findings suggest that networks are functionally differentiated insofar as they serve as strategic repertoires. This is a significant finding. Knowing how a network's function is related strategically to its structure means being able to infer a network's function if only its structure is known and, conversely, being able to infer a network's structure if only its function is known. Not only does SNA thereby facilitate detection and dismantling of networks, it also suggests that recruitment, fundraising, and attack networks require differentiated approaches by defence and security agencies insofar as SNA shows them to be distinct phenomena.

The chapter "Social Network Analysis Applied to Criminal Networks: Recent Developments in Dutch Law Enforcement" by P. A. C. Duijn and P. P. H. M. Klerks ask the questions: What can we learn from developments within The Netherlands concerning the application of Network Analysis in controlling Organized Crime? What are the practical implications of applying network analysis within targeting criminal networks and strategic intelligence gathering? What does dynamical network analysis research tell us about the effectiveness of criminal network disruption? How does the network paradigm connect with law enforcement decision making? Addressing these questions the authors' objective is:

1. To offer insight into the recent developments of the application of network analysis in controlling Organized Crime within The Netherlands.
2. To offer insight into the practical application of network analysis within targeting criminal networks and gathering intelligence about criminal cooperation.

In the chapter "The Networked Mind: Collective Identities and the Cognitive-Affective Nature of Conflict," Manjana Milkoreit and Steven Mock apply a network-based analysis to an area that suffers from even greater empirical challenges than social network analysis: the human mind. Working with the assumption that all human behavior has cognitive origins, this chapter has three aims. First, the authors make the case for the relevance of a cognitive approach to defence and security studies, exploring areas of application and potential insights. Second, they use a theory of the mind—emotional coherence—and a complex systems approach to explore the role of specific cognitive elements—collective identities—for the emergence and resolution of inter-group conflict. Third, the authors introduce cognitive-affective mapping as a tool to apply a theoretical framework to specific empirical cases and walk the reader through two specific case analysis: the Israeli-Palestinian dispute over the Western Wall as a long-term low-violence conflict and—maybe surprisingly—the international climate negotiations as an example of non-violent political conflict at the global level.

The chapter "Conflict Cessation and the Emergence of Weapons Supermarkets " by Gisela Bichler and Juan Franquez present the results of a study to uncover the underlying network funneling small arms to actors that use them as fodder for conflict. This study uses a dynamic actor-based simulation modeling software, called SIENA to capture changes in the gray market of gun trade following the termination of armed conflict. The simulation models permit the detection of evolutionary change in networks. The key advantage to using SIENA is that it permits the inclusion of structural variables (e.g., changes in reciprocity, transitivity, and degree centrality metrics) as independent variables, in addition to dynamic covariates (changing values over time), static covariates, and other dynamic and static networks (dyadic variables). This flexibility has been shown to be useful in examining international gun trade [4].

In the chapter "A Conspiracy of Bastards?", Simon Bennett presents a unique perspective on security. With reference to two case studies—the 1983 "Holloway Road incident" and the 1989 Hillsborough cover-up—he seeks first, to understand why public servants responsible for civil safety and security can sometimes elevate their own interests above those of the public. Secondly, to show how system theory, specifically organization theory [5, 6], actor-network theory [7, 8], and theories of isomorphic learning [9] can be used to understand how such deviant behavior is organized and why it is so difficult to identify and correct.

The chapter "Decision Support Through Strongest Path Method Risk Analysis" by Philip O'Neill presents a technique, known as the *strongest path method*, for performing risk analysis of multiple systems of systems that are highly interconnected. The *connectedness* entails complex dependency relationships throughout the system of systems and these propagate risk. Users, managers, and regulators of such systems who want to understand the impact and vulnerability of its component parts require a risk analysis method that deals explicitly with chains of dependency relationships. The strongest path method provides such capability. Not only are the strongest path or paths taken into consideration when estimating the potential impact of one entity on another but the compound effects of all pathways are included. The results of the path analysis can be used to prioritize risk and to prepare risk mitigation plans and contingency plans. The chapter focuses on developing the risk analysis technique and showing its value to decision makers. Some areas where the method is useful will be put forward with particular emphasis on infrastructure risk analysis.

The chapter "Critical Infrastructure and Vulnerability: A Relational Analysis Through Actor Network Theory" by Anthony J. Masys describes how threats to national security, such as that against critical infrastructures not only stem from man-made acts but also from natural hazards. Katrina (2005), Blackout Canada-US (2003), Fukushima (2011), Hurricane Sandy (2012), and Alberta Floods (2013) are examples that highlight the vulnerability of critical infrastructures to natural hazards and the crippling effect that failures can have on the social and economic well-being of a community and a nation. Focusing on the initiating event that precipitated the critical infrastructure failure does not capture the root vulnerabilities or "resident pathogens" that are "hard-wired" into the greater networked system. Through the complexity/systems lens of actor network theory, this chapter explores critical infrastructure and how key "actors" within a network can align other actors creating "unseen" vulnerabilities.

The final concluding chapter brings together common themes from across the chapters to highlight the complexity space that characterizes the defence and security domain and applies the "network thinking" paradigm to the comprehensive approach.

Collectively, the chapters present the reader with a broad spectrum yet detailed analysis of the defence and security domain, thereby highlighting the requirement for a "network mindset." It advances our understanding and state of the art regarding network analysis and lays the foundation for continued exploitation and development of network analysis tools and techniques to support the defence and security domains.

References

1. National Crime Agency: a plan for the creation of a national crime-fighting capability (2011). Presented to Parliament by the Secretary of State for the Home Department by Command of Her Majesty (https://www.gov.uk/government/publications/national-crime-agency-a-plan-for-the-creation-of-a-national-crimefighting-capability)
2. Barabasi A-L (2003) Linked. Plume, New York Penguin Group
3. Vespignani A (2009) Predicting the behavior of techno-social systems. Science, 325:425–428
4. Bichler G, Malm A (2013) Small arms, big guns: a dynamic model of illicit market opportunity. Global Crime 14(2–3):261–286
5. Tsoukas H, Knudsen C (2005) The oxford handbook of organization theory. Oxford University Press, Oxford

6. Ravasi D, Schultz M (2006) Responding to organizational identity threats: exploring the role of organizational culture, Acad Manage J, 49(3): 433–458.
7. Latour B (1991) Technology is society made durable. In A sociology of monsters: essays on power, technology and domination, J. Law (ed.) Routledge, London, pp 103–131
8. Woolgar S (1991) Configuring the user: the case of usability trials. In: Law J (ed) A sociology of monsters: essays on power, technology and domination, Routledge, London, pp 57–99
9. Toft B, Reynolds S (1997) Learning from disasters, Perpetuity Press, Leicester
10. Callon M, Law J (1995) Agency and the hybrid collectif. South Atlantic Q, 94(2):481–507
11. Perrow C (1984) Normal accidents: living with high-risk technologies, Basic Books

Glossary

Activity Focus Networks	Represent the complex activity system of an organization. An activity focus is a conceptual or physical entity around which joint activity is organized.
Actor Network Theory (ANT)	An approach to social theory that treats objects as part of social networks.
Betweenness Centrality	The number of geodesics the node is on.
Bridge	A link between two nodes in different networks or network components.
Bonacich Power	The extent to which an actor is connected to other actors that score high in degree centrality.
Bots	software application that runs automated tasks over the internet.
Centrality	Describes the relative position of an individual in a network.
Closeness Centrality	The length of the geodesics from a specific node to all the other nodes.
Cognitive-Affective Mapping	A method for graphically diagramming cognitive systems as networks of mental representations.
Cognitive-Affective Maps	Represent an individual's concepts and beliefs about a particular subject.
Cognitive Analysis	The attempt to identify, describe and understand the content, structure and dynamics of systems of mental representations.
Cognitive Science	The multidisciplinary study of mind and intelligence.
Comprehensive Approach	is based on the assumption and requirement for some level of coherence amongst the actors/stakeholders regarding shared goals and objective and to create a dialogue to address the various dimensions of the problem space (political, security, safety, socio-economic, humanitarian and human rights).

Critical Infrastructure	Assets, systems and networks, whether physical or virtual, so vital…that their incapacitation or destruction would have a debilitating effect on security, national economic security, national public health or safety.
Cutpoint	is a single node connecting two or more components of a network. Removing that node should disconnect those components.
Dark Networks	Groups that do their best to conceal themselves and their activities from the authorities (i.e. Terrorist, organized crime organizations).
Degree centrality	The number of nodes adjacent to a specific node.
Density	The number of links in a network, as a percentage of the number of possible links.
Directed Graph	Defined by a set of nodes and a set of ordered pairs of nodes called edges.
Directed Path	A sequence of nodes with the property that each node is connected to its respective successor by an edge.
Equivalence	is the similarity of connections/roles of two or more nodes (individuals).
Geodesic	Shortest available path between two nodes.
High Reliability Organization	An organization that has avoided a catastrophe in an environment where normal accidents can be expected.
Hybrid Collectif	An actor network theory term that describes actors that are neither purely technical nor purely social [10].
Internet Relay Chat (IRC)	A worldwide instant messaging system.
Isomorphic Learning	Drawing lessons from disasters that have occurred, to prevent future recurrence of the same.
K-core Analysis	Uses degree centrality to identify clusters of actors that are tightly connected.
Link Analysis	is a data-analysis technique used to evaluate relationships (connections) between nodes.
Macrosociology	The analysis of social systems on a large or at high level of abstraction.
Ndrangheta	An Italian mafia from Calabria, Southern Italy.
Normal Accident Theory	Articulates the view that accidents are inevitable or 'normal' [11].
SIENA	A dynamic actor-based simulation modeling software.

Social Network Analysis	The study of social relationships in terms of network theory consisting of nodes and ties (also called edges, links, or connections).
Translation	An actor network theory process in which actors within a network will try to enroll the other actors into positions that suit their purposes.
Transnational Crime	Organized crime coordinated across national borders.

Contents

Network Analysis in Criminal Intelligence 1
Steven J. Strang

Identifying Mafia Bosses from Meeting Attendance 27
Francesco Calderoni

Macrosocial Network Analysis: The Case of Transnational Drug Trafficking 49
Rémi Boivin

Policing the Hackers by Hacking Them: Studying Online Deviants in IRC Chat Rooms 63
David Décary-Hétu, Benoit Dupont and Francis Fortin

Why Terror Networks are Dissimilar: How Structure Relates to Function 83
Christian Leuprecht and Kenneth Hall

Social Network Analysis Applied to Criminal Networks: Recent Developments in Dutch Law Enforcement 121
Paul A. C. Duijn and Peter P. H. M. Klerks

The Networked Mind: Collective Identities and the Cognitive-Affective Nature of Conflict 161
Manjana Milkoreit and Steven Mock

Conflict Cessation and the Emergence of Weapons Supermarkets 189
Gisela Bichler and Juan Franquez

A Conspiracy of Bastards? 217
Simon Bennett

Decision Support Through Strongest Path Method Risk Analysis 247
Philip O'Neill

Critical Infrastructure and Vulnerability: A Relational Analysis Through Actor Network Theory 265
Anthony J. Masys

Dealing with Complexity: Thinking About Networks and the Comprehensive Approach 281
Anthony J. Masys

Network Analysis in Criminal Intelligence

Steven J. Strang

Abstract A set of approaches to network analysis are used in criminal intelligence to understand and act against serious crimes, criminal groups, and criminal markets. These approaches are based in link analysis and increasingly include techniques from social network analysis. Network analysis techniques in criminal intelligence are used to organise data and reveal patterns in the nature and extent of relationships between data points. They also provide effective visualizations of both qualitative and quantitative data which are valuable in presenting intelligence assessments. The link analysis methods used in criminal intelligence are a form of network analysis designed to discover and illustrate patterns in the connections between individuals, locations, organizations, objects and events. Social network analysis has a tighter focus, concentrating on the relationships between people. Some social network analysis measures have utility in producing targeting recommendations for intelligence collection and operational disruption.

Keywords Criminal intelligence · Intelligence analysis · Link analysis · Network analysis · Social network analysis · Organized crime · Terrorism

---

Opinions expressed are those of the author and do not necessarily reflect those of the Royal Canadian Mounted Police or the Government of Canada

---

S. J. Strang (✉)
Royal Canadian Mounted Police, Ottawa, Canada
e-mail: steve.strang@rcmp-grc.gc.ca

A. J. Masys (ed.), *Networks and Network Analysis for Defence and Security*,
Lecture Notes in Social Networks, DOI: 10.1007/978-3-319-04147-6_1,

1 Introduction

1.1 The Criminal Intelligence Context

Network analysis has been a core technique of criminal intelligence analysis since the early 1970s, primarily in the form of link analysis, a specific adaptation of network analysis to criminal intelligence and investigation. More recently, elements of social network analysis have also been used and adapted into intelligence work.

For the purpose of the following discussion, criminal intelligence is defined as knowledge and understanding of current and anticipated criminal threats. The main uses of criminal intelligence in law enforcement are:

- Guiding and supporting ongoing operations to prevent crime, detect crime, and apprehend criminals
- Informing priority-setting for policing resources and operations
- Warning of upcoming threats.

Criminal intelligence has both a tactical and strategic role in policing. To paraphrase Carl von Clausewitz: tactics are used to win fights; strategies are used to win wars.

Tactical intelligence is part of specific criminal investigations or other police operations. The tactical intelligence may be reactive, in support of an ongoing investigation or other operation. It may also be proactive, such as: intelligence probes to prepare for planned operations, or threat assessments for upcoming major events. Tactical intelligence analysis produces assessments and charts, such as link charts, for specific ongoing criminal investigations. The primary client for a tactical intelligence product will often be the lead investigator. Some law enforcement agencies and organizations use the term "Operational" to describe this type of intelligence.

Strategic intelligence is part of the police force's strategic planning process. It is used to identify and warn of emerging crime issues, and to assess the current state of major crime problems. Strategic intelligence analysis produces assessments which support the development and implementation of strategies to counter organized and serious crime. These assessments provide warnings of emerging criminal threats, and enhance our understanding of the nature and extent of current criminal threats in order to improve our ability to act against them. The primary clients for strategic intelligence will often be senior managers who are required to make decisions about operational resource allocation. Strategic assessments should also provide valuable insights to investigators, program managers, and other personnel.

Criminal intelligence analysis is the process of deriving meaning from information about crimes, criminals, and their operating environment in order to produce intelligence. The processes of criminal intelligence analysis are framed by the intelligence cycle (Direction, Planning, Collection, Evaluation, Collation, Analysis, and Dissemination). A wide range of qualitative and quantitative analytic techniques

Table 1 Quadrant of criminal intelligence functions

Emerging crime issues	*Planned operations*
Proactive strategic intelligence provides warning and recommendations on emerging and expected crime issues. For example:	Proactive tactical intelligence prepares for planned investigations into criminal activity, and other upcoming police operations. For example:
• Early warning assessments identifying and explaining emerging and foreseeable threats • Assessments explaining implications of changes in criminal activity and recommending strategic policing responses	• Intelligence probe preparing the foundation for an investigation • Threat assessments in preparation for a major event
Current crime issues	*Current operations*
Reactive strategic intelligence supports initiatives to investigate, suppress, and prevent criminal activity related to current priorities. For example:	Reactive tactical intelligence supports ongoing investigations and other police operations. For example:
• Assessments explaining national/international context, implications, and opportunities to attack criminal activity in a current priority • Assessments of changes in criminal activity in a current priority area in relation to policing strategies	• Assessments of criminals and criminal activity under investigation which may incorporate: time lines; link charts; profiles of suspects or other persons involved in the crimes; profiles of companies or other organizations involved in the crime; statistical analysis of crime occurrences; or, geographic analysis of crime locations

and approaches is used in criminal intelligence analysis, including several specialized applications of network analysis (Table 1).

The information available for analysis comes from sources including: victims, witnesses, informants, suspects, convicts, undercover police, surveillance teams, telephone intercepts and other technical sources, public registries and other open sources. Some of the individuals providing information are reliable, telling the truth to the best of their knowledge. Others will attempt to deceive the police in order to protect themselves or their associates, or to achieve some other goal. The information provided includes known facts which can be used as evidence in court, and also information which is less certain or comes from sources of unknown reliability and needs to be used more carefully. This uncertain information may include facts, partial truths, false information, and lies. Any analytic technique used in criminal intelligence must be robust in the face of missing and uncertain data.

One other significant issue with the information used in intelligence is the problem of determining which information is relevant. This is usually referred to as the problem of signal and noise, in which important information is mixed in with large amounts of irrelevant information. For example, if conducting a network analysis centered on a known organized crime figure, his social network will include other members of his criminal group amongst all his other contacts. It can be extremely difficult and labour intensive to sort out which of his contacts are actually relevant to his criminal activity. Analytic techniques used in intelligence need to be able to cope with large amounts of information and help sort the signal from the noise.

Network analysis techniques in criminal intelligence are applied to investigations and issues related to: serious crime, organized crime, and terrorism. All of these terms have multiple definitions, including legal, scholarly, and popular meanings. In this discussion, serious crimes are major offences which usually involve violence or the threat of violence, such as: homicide, rape, kidnapping, or armed robbery. An organized crime group, as defined in Canadian law, consists of three or more persons, however organized, acting together to commit serious offences for material benefit (Criminal Code of Canada S.467.1). A terrorist group in Canada is essentially defined as an entity or association of entities which commits or facilitates terrorism—acts of violence or threats of violence motivated by ideology and intended to intimidate the public or a segment of the public (Criminal Code of Canada S83.01).

1.2 Network Analysis Approaches

Networks in human society consist of individuals and their relationships with others, so in social network analysis the nodes are always people, and the links are their relationships. Networks of human behaviour consist of individuals, things, places and events; and of the relationships between them. The nodes in a link analysis of a criminal conspiracy may include entities such as: individuals, companies, vehicles, houses, weapons, amounts of money, bank accounts, and so on.

Most criminals motivated by profit or ideology do not operate in isolation. Individual thieves have relationships with purchasers of stolen goods, drug dealers have suppliers and customers, and even "lone wolf" terrorists do not operate in a complete social vacuum. Criminals who are members of structured groups may have supervisors, employees, and colleagues. If they are, instead, participants in criminal networks they have collaborators and associates. When applied to organized criminals, network analysis can reveal patterns of collaboration and relative social positions, identify key individuals and subgroups, and also identify vulnerabilities which could be exploited to disrupt an organization or degrade its capabilities.

Link analysis, also known as association analysis, explores the connections between individuals involved in criminal activity through their links to each other and through their links to organizations, objects, places and events related to the crimes. As a result of this focus on criminal activity, link analysis has evolved specifically to work under conditions of persistent uncertainty, where important information is often hidden and where the accuracy of any given piece of information may be uncertain or unknown.

Other specialized applications of network analysis are related to, and often contribute to a link analysis. For example; analysis of telephone records, known as toll analysis, reveals the patterns of calls between specific telephones, including direction, time, and duration of the call. While this does not reveal the content of the conversations, or necessarily which individual is using each telephone, the traffic patterns can be revealing when collated with other information, and provide leads for further investigation. As another example; analysis of commodity flows reveal the patterns of economic transactions within a criminal network or criminal market, with goods and/or services moving in one direction and payment in the other. The flows of goods and payments are not always symmetrical, that is they do not always follow the same routes or pass through the same individuals. As a general rule, money flows towards individuals holding power within a criminal network.

Social network analysis is the study of patterns of social connections. There are many applications of network analysis; social network analysis has been developed to improve our ability to understand social structures and behaviour. It can be used to reveal patterns of communication, exchange, friendship, trust, cooperation or kinship within an organization or community. It can also show the inverse of these, revealing patterns of secrecy, competition, mistrust, and enmity.

In social network analysis, links are the known relationships between individuals. The links have content, and knowing the content and nature of these links is essential to using social network analysis in criminal intelligence. The link might be a kinship relation through descent or marriage, it might also be role-based (e.g. supervisor/subordinate), based on interactions (e.g. giving advice, sharing gossip, selling drugs), or based on affiliation (e.g. same organized crime group, same prison, same neighbourhood). The link also contains affect; the individuals may like or dislike, trust or mistrust each other. The links can also be asymmetrical, the link between two individuals may have trust in one direction and mistrust in the other, or one might always be the seller and the other the buyer.

A standard organizational chart shows the hierarchical structure of a network by representing the formal relationships of the organization's members. These relations include authority and responsibility, reporting and tasking. Within any organization there are also informal networks based on interpersonal relationships. These informal relations include sharing information and gossip, and are marked by trust and suspicion, respect and derision, cooperation and competition. The informal relations reveal social capital: who is trusted and respected, who has the most needed skills and resources, who has the largest personal network, and who controls key connections within the network. While some organized crime and terrorist groups have a hierarchical structure which can be shown in an organizational chart, many do not. In those groups which function without formal structure or hierarchy, the social network itself is the main organizing principle.

1.3 Visual Representations

Network analysis charts can be highly eloquent visualization tools for both qualitative and quantitative information.

Link Analysis charts organize and present the patterns of connections between entities, which include the individuals, groups, objects, places and events relevant to a criminal intelligence assessment or criminal investigation.

These charts can use:

- Annotations on links to document the content of conversations, the amounts of money transferred, the number of times calls were placed between two telephones, or any other information relevant to the analysis of that link. This can also include the source(s) of our knowledge of that link, our assessment of the source's reliability, and our assessment of the validity of the information
- Different types of lines to indicate whether a link is known or suspected. The standard convention is to use solid lines for proven links and dashed or dotted lines for uncertain or suspected links. The lines can also be marked to show the direction of the link
- Distinct symbols or icons to indicate clearly whether the entity is a person, business, address, vehicle, weapon, or other specific type of thing relevant to the assessment. Some software packages allow the use of individuals' photographs as their icons on a link chart.

Social network analysis charts develop and present the patterns of relationships between individuals within an organization or community.

These charts can use:

- Directed links which use arrows to mark the direction of exchanges relevant to the intelligence question, such as the movement of goods, money, advice, permission and information between the individuals charted
- Weighted or valued links which show the relative strength, intensity, frequency, duration or quantity of the link. The weight can be shown through numbers attached to the link, by varying the thickness or colour of the line, or by other appropriate techniques such as showing close relationships as physically close on the chart
- Weighted nodes which show the relative importance of individuals by varying the size, colour, or placement of the node
- Layout to reveal structural features, for example by clustering kinship groups or business partnerships, or by overlaying the chart on a map to show the correlation of links to geographic locations.

2 Networks of Criminals

Each organized crime group is a network in itself, and operates in a wider network (or meta-network) of suppliers, facilitators, consumers and competitors. Each terrorist group is also a network inside a wider network of supporters, suppliers, audiences, and opponents. While the different motives of organized crime and terrorist groups drive some very important differences between what these people do, there are also some fundamental similarities between all covert groups operating under risk of detection, disruption, or destruction by the police or by competitors. These core similarities relate to the groups' need to maintain secrecy, enable resilience, promote internal discipline, and communicate inside and outside the group.

The shape of a criminal network helps determine its usefulness to the individual members and their roles within the network. For example, small tightly-knit networks can be less useful to their members than networks with many loose connections to individuals outside the main network. This is because more open networks, with many weak ties and social connections, are more likely to introduce new ideas and opportunities to their members than closed networks with many redundant ties. This is one of the effects of the power of weak ties [1].

However, smaller and tighter networks are more secure than a large network with many weak ties, since they are more effective at keeping information inside which could otherwise lead to their detection and prosecution. Organized criminals and terrorists have to balance their need for collaboration with their need for security; the form and density of the networks reveal how they balance those two conflicting requirements. Changes in these aspects of a covert network are often adaptations to changes in pressure on the group from the police, other security forces, or competitors.

An individual involved in organized criminal activity may have connections to a variety of networks rather than many connections within a single network. These individuals can profit and exercise influence by acting as brokers between two or more networks that are not otherwise linked. Carlo Morselli's detailed analysis of the criminal career and network of Howard Marks, an international cannabis broker, examines the network properties specific to criminal brokers [2].

Relations between individuals can be understood as transactions. Through these transactions, individuals develop social capital, which is the value of their network of relationships to others based on who the individual knows and their ability to make useful connections for others. Individuals also bring personal capital to these transactions. Personal capital consists of qualities such as intelligence, self-confidence, charisma, and knowledge.

Within criminal networks, individuals have criminal capital, which is the combined value of their personal attributes and network to other criminals. Elements of criminal capital include:

- Useful knowledge, for example: knowledge about criminal methods, potential targets, or police operations
- Useful skills for criminal operations, for example: money laundering, safe cracking, drug production, or bomb making
- Access to equipment or infrastructure, for example: trucks, dockyards, pill-pressing machines, garages, firearms
- Access to raw materials such as chemical drug precursors
- Links to other individuals with useful knowledge or skills, or to individuals with an established criminal reputation who can provide protection.

Network analysis reveals and illustrates patterns of behaviour indicating social and criminal capital.

2.1 Social Network Analysis Concepts

Social network analysis concepts of particular value to intelligence analysis include: centrality, equivalence, density, strong/weak ties, and cutpoints.

Centrality. Centrality describes the relative position of an individual in a network. Centrality can be measured in several ways, based on the nature and number of the relationships that individual has to the other individuals in the network. A *path* is any route which can be followed from one person to another, linked through any number of intervening people. The *length* of a path is the number of links on it, sometimes referred to as the 'degrees of separation'. The *geodesic* in a network is the shortest available path between two individuals.

Degree Centrality is the number of other people adjacent to the individual. The higher this measure is, the more direct associates the individual has in the known network. The person may be a formal leader, a skilled networker, or poor at keeping his connections secret. This measure can be subdivided into *indegree* and *outdegree*, the number of adjacent people on *directed links* which come into the individual or which go outwards. Directed links are those showing flows, such as of money, commands, or permissions.

Betweeness Centrality is the number of geodesics the individual is on. The higher this measure is, the more indirect associates the individual has. He or she may be a central actor in the communications or exchange network, and may be a key individual in holding the network together.

Closeness Centrality is the length of the geodesics from the individual to all the other people in the network. This measure shows how close the person is to all the other members of the network through both direct and indirect paths. The lower the number, the stronger the closeness is between the individuals.

The extent and type of centrality an individual has in the network will reflect his or her role. For example, a manager in a hierarchical group will probably have higher centrality in that group than a service provider would. In a group with no formal hierarchy, individuals with high centrality play the greatest role in keeping the group together and may be the most influential participants. These individuals would be the most valuable targets for intelligence collection because of the frequency and range of their communications. They may also be valuable targets for investigation to disrupt the group.

Some criminal networks include a few extremely highly connected individuals, people who have direct links to an unusually large number of other network participants. These individuals are hubs and their presence indicates that the network has scale-free properties. A scale-free network has one or more hubs and the distribution of the degree centrality scores of all the participants fits a power law, forming a distinctive curve when plotted on a graph.

Criminal networks without hubs may alternatively have small-world properties. A small-world network has relatively short path lengths between all participants, and any two individuals both linked to a third are fairly likely to know each other directly. The distribution of centrality scores in a small-world network tends to form a bell curve when plotted on a graph.

Equivalence. Equivalence describes the redundancy of an individual in a network. Individuals with low equivalence are harder to replace, so removing them through investigation and arrest can have a greater impact on the ability of the group to operate. Equivalence can be measured in several ways:

Equivalence is the similarity of connections/roles of two or more individuals. The higher the equivalence the more similar the specific links in each individual's *neighbourhood.* What appear to be two individuals with very strong equivalence may actually be one individual using an alias.

Substitutability is when an individual has one or more counterparts in the network who can maintain the same pattern of connections. Substitutable individuals are redundant in the network.

Stochastic Equivalence is when two or more individuals are linked to the same percentage of other actors; they have the same probability of being linked to any other actor in the network. This is a potentially useful measure under conditions of incomplete information, as is common in intelligence analysis.

Role Identification is the number of members of the network who can fill each role. Individuals with unique roles are vulnerabilities for the network since they are more difficult to replace from within.

Strong and Weak Ties. Network charts can show the relative strength of the connections between individuals. Granovetter [1] defined this notion of strength as including a "combination of the amount of time, the emotional intensity, the intimacy (mutual confiding), and the reciprocal services which characterize the tie". When analyzing criminal networks, the judgment of how strong a tie is could represent: the frequency with which the two individuals cooperate in criminal acts; the duration or frequency of their interactions; the value of goods or services exchanged; the relative intensity of their shared experiences; or any other identified factor which reveals the importance and resilience of the relationship.

When assigning weights to links in order to discover and show the relative strong and weak ties, the analyst must use measures relevant to the nature of the group and the purpose of the analysis.

The relative strength or weakness of a link is sometimes estimated simply from the frequency of interaction between the two individuals. This is easily quantified, but frequent interactions are not a certain indicator of the importance of the relationship. Covert networks can deliberately reduce the contacts between co-conspirators in order to maintain operational security and insulate leading individuals from prosecution, so it may be the infrequent contacts which are most important. To pick examples from two ends of the spectrum of groups encountered in intelligence analysis: frequent interaction would indicate a strong tie between individuals in a gang of young drug dealers; it is unlikely to identify the most important tie within an espionage network.

Density. Network density is the number of links in a network, as a percentage of the number of possible links. A network with a density of 100 % is completely interconnected, every participant in the network knows everyone else. The denser a network is, the more difficult it is to break into components by removing individuals. However, highly dense criminal networks are more vulnerable when penetrated by undercover officers or when existing members become informants. Small-world networks tend to have higher densities than scale-free ones do.

Groups with lower densities are more resistant to effective penetration through source development or the insertion of undercover operators. Links between the individuals are more "stove-piped" within the network, so a human source is less likely to have access to a broad range of information useful for intelligence or investigative purposes.

Groups engaged in conspiracies and other criminal acts over time usually adapt to pressure from law enforcement and other security operations by reducing their overall network density. They do this by insulating their leadership, isolating operational cells, and other measures to ensure operational security.

Network Horizon. A social network analysis chart can show indirect connections on paths of any length. However in practical terms there are limits to the usefulness of increasingly indirect relationships. Two degrees of separation describes, at best, a "friend of a friend", someone you may have heard about but have never met. Under normal circumstances, two degrees of separation represent the network horizon which circumscribes the area of an individual's network containing the other individuals they know or could know of.

The nature of the relationships, as well as the length of the paths, is significant in establishing the network horizon. For example, most RCMP intelligence analysts are two degrees from a current or former Prime Minister of Canada, through meeting or working with current or former members of the Prime Minister's protective detail. These same analysts are therefore at three degrees or less from many other world leaders. This does not mean an analyst can make use of these links to ask a favour from the Prime Minister. The existence of a link does not necessarily show the existence of a useful relationship.

Networks engaged in ongoing conspiracies or other covert activities will sometimes use various methods to conceal the most important links. These methods include the use of one or more layers of "go-betweens" to insulate the leaders from the individuals engaged in operations. In these cases the most important operational relationships will deliberately cross the normal network horizon.

Cutpoints and Bridges. A cutpoint is a single individual connecting two or more components of a network. Removing that individual should disconnect those components. A cutset is a set of individuals who connect two or more components of a network. A minimum weight cutset is the smallest number of individuals who must be removed to disconnect components. A bridge is a link between two individuals in different networks or network components, so this relationship is also the connection between the two networks or sub-networks.

Line connectivity is the number of links which must be removed to disconnect components of a network. Node connectivity is the number of nodes which must be removed to disconnect components.

So there are two main approaches to separate components of a network: by removing individuals who are cutpoints or who make up a cutset, or by removing relationships which function as bridges.

2.2 Patterns of Self-Organization and Leadership

In human societies the primary direct driver of organizational structure is group size. As Johnson [3] and others have demonstrated, increasing population in a group causes a decreased ability to reach consensus, increasing dispute frequency, and so causes a greater need for conflict management and resolution within and between groups. The larger the group becomes, the greater its need for internal structures and a formal leadership hierarchy.

There are various conditions which allow large groups to form, the most important of which is access to large quantities of predictable resources. For example, an organized crime network in a location which allows them to control the passage of large drug shipments and in which there is little interference from law enforcement can become a large, hierarchical group. A similar network in a location which is not on a significant trade route or other critical market node, and which is subject to effective pressure from law enforcement, will remain smaller in size and less hierarchical in structure.

For terrorist groups, a key additional driver of organizational structure is the need to maintain operational security against law enforcement and other security forces while communicating their ideological position through acts of violence and, often, public statements. For organized crime groups, the key additional driver is the need to maintain security against law enforcement and competing groups while marketing their goods and services to their clients. These combined drivers relate to the types of organizational forms taken by covert groups engaged in terrorism and organized crime, and therefore to the patterns revealed or indicated in network analyses. Understanding the nature of the organizational structure in a network is important because it has a significant influence on the law enforcement approach taken to disrupt the network's functioning.

2.3 Organizational Structures in Organized Crime

Organized crime groups also develop and modify their organizational structures for cultural, economic, and operational reasons. Groups committing crimes for profit are essentially criminal businesses. Their structures are adapted to maximize profit from their core businesses, balanced by a need to remain resilient to pressure from competitors within their criminal market and from law enforcement. The range of groups includes: collaborating subsistence criminals, family-based groups, networks of criminal entrepreneurs, criminal franchises sharing a brand name, and formal named groups with internal hierarchies.

Some criminal networks are formed around a family group, or some other core group of individuals linked through mutual trust and interest. Other networks take the shape of the production, shipment, and distribution paths of a commodity such as illegal drugs, or of the management and provision of an illegal service such as money laundering. Some covert networks have leaders who occupy a formal position and who can command and control members of his or her group. However many networks function with leading individuals, who have prestige and respect based on their abilities and reputation, but whose leadership role is situational and who cannot issue commands to the other network participants. A law enforcement strategy of "taking out the leader" does not work on a network which has no fixed leader.

The range of organized crime economic activities is also very wide, some groups only engage in crime for profit, others in a mix of criminal and licit activities. Some groups focus on a single criminal activity while others are highly diversified.

There are essentially three main categories of criminal economic activities:

- Trading in illicit goods, such as: illegal drugs, counterfeit goods, smuggled goods, or human trafficking
- Trading in illicit services, such as: loan sharking, prostitution, money laundering, or smuggling
- Taking property by crime, such as: robbery, theft, extortion, or fraud.

The first two categories consist of crimes which are most often consensual, the client wants to purchase heroin so he buys it from a drug trafficker, or the client wants to conceal the origin of his income so he uses a money launderer. These are the criminal mirror image of the manufacturing, retail, and service sectors in the licit economy. The third category consists of taking money or other private property from victims, so that economic network has both willing and unwilling participants. The core business of a criminal group will influence the shape of its network.

Organized crime groups can be divided into three broad categories of network structure:

- Formal hierarchical groups
- Informal stable groups
- Informal fluid groups.

The first three of these categories are roughly equivalent to the standard forms in which humans self-organize. Using the categories outlined by Elman Service, these are: chiefdoms, tribes, and bands [4]. These ethnographic categories are valuable for the understanding of organized crime, because they explain the context and functions of the different types of leadership roles found in each form of social organization.

Formal Hierarchical Groups. Formal hierarchical groups have named ranks in a structure which persists over time despite changes in which person occupies a position. Formal leadership structures are reported in the Italian and Italian-based mafias, outlaw motorcycle gangs, the Japanese boryokudan, and in a number of other types of organized crime groups. For example, Cosa Nostra Families have a quasi-feudal structure. At the top is the Boss, advised by a Consigliere and assisted by an Underboss. Below this level are several mid-level managers called Caporegime, below each Caporegime are a number of Soldiers [5]. As another example, the Black Kings gang described by Venkatesh [6, p. 34] had a structure in which there was a "board of directors" running the gang, below them were "captains" and "lieutenants" who each managed several gang factions, and each faction had an appointed leader.

These hierarchical groups operate much like chiefdoms. The chief lives in part or wholly off of taxes, just as the heads of hierarchical criminal groups receive a portion of his subordinates' profits. A chief is able to give orders and make rules, but his power is opportunistic and personal rather than institutional or bureaucratic. Unlike headmen or band leaders, the leadership roles in tribes and bands, chiefs come to power inside an existing social structure. Earle [7] observes that chieftaincies tend to form in the illegal economy. In developed countries these criminal chiefdoms are constrained by the justice system. But in countries with weak or failing justice systems these chiefs have the opportunities to establish themselves as warlords or drug lords with substantial political and economic power.

A network analysis of a formal hierarchical group needs to take into account the rank structure as well as the informal web of connections. However, the official rank does not always fully correspond to an individual's influence and importance to the functioning of the group, since a low-ranking or marginal individual may have non-redundant and essential skills or contacts. This is where social network analysis and link analysis techniques are essential to discover aspects of the group which are not revealed by the bare bones of an organizational chart.

Informal Stable Groups. Informal stable groups do not have formal leadership structures, but have stable membership and sources of income, and some individuals emerge as leaders based on their criminal capital. These groups often do not name themselves, but they have a predictable resource base, such as a steady market for their range of illegal services.

These groups are similar in leadership roles to tribes as described by Service. The leading individuals are personally respected, a respect which may be based in part on his personal wealth and generosity as well as his abilities. An individual becomes important through his own efforts as a "political entrepreneur" who creates his own power by amassing reputation [7]. However, these individuals lead only by example and persuasion, they cannot give orders.

Informal Fluid Groups. Informal fluid groups are small networks of criminals. They may be opportunistic in the crimes they commit, or they may have a core source of criminal income. These small groupings are sometimes ad-hoc collaborations of career criminals who are part of a larger network of criminals who know each other but who do not work together on an ongoing or organized basis. Leadership is also fluid, an individual might take the lead in one situation but not in others. These groupings may be based primarily on specific offender convergence settings, locations where individuals have the opportunity to establish and make use of relationships with other criminals [8].

These fluid groups are similar to foraging bands; they have an unstable membership (although sometimes with a small stable core of individuals) and may fragment seasonally to exploit changing resources. An individual may move between bands easily, they are participants in a network rather than exclusive members of one grouping. Bands have little if any specialization, individuals all have essentially the same set of skills. Groups with little specialization have very high redundancy, so while their operating capacity can be reduced by removing individuals, their operating capabilities will remain largely intact.

Decision making in bands is primarily by consensus. There are leading individuals in bands who are given respect based on personal skill, often rhetoric and hunting. They may be asked for advice, and are sometimes asked to make decisions. This leadership is often situational and limited, a particularly good caribou hunter may be asked for advice or suggestions on hunting, but might not be asked for advice on fishing.

While small organized crime groups can function on the basis of consensus and situational leadership, that does not mean they always do. Some will include an individual with a strong and controlling personality who can run the group as his own business. In other cases the group may be a criminal family business in which the patriarch or matriarch has firm control and the younger members are clearly subordinate.

In groups without any formal structure or leadership roles the social network is the foundation and the context, they operate entirely through personal connections. The boundaries of these groups will be unclear, since they draw from a pool of participants who change over time.

2.4 Organizational Structures in Terrorism

The structural organization of terrorist/criminal extremist groups can be described in terms of four main categories. In some cases, a specific terrorist group will exhibit elements of more than one of these categories as a result of historical or ongoing changes in organization structure.

Terrorist groups develop their specific organizational forms for various cultural, operational or ideological reasons. Some terrorist groups deliberately attempt to change their organizational structure as an adaptation to pressure from security forces; in an attempt to improve internal security and to improve internal redundancy and resilience to counter-terrorism operations. Whether the structure is static or in flux, the group is striking a compromise between two opposing forces. The first is the need for efficient internal communications and control to maintain ideological consistency and ensure that the members' activities stay "on message". The opposing force is the need for tight internal security which requires dividing the group into components with minimal interaction to maximize resilience to disruption when the group is compromised by informers or undercover agents of the security forces.

There are numerous approaches taken to categorize types of terrorist groups. For the purpose of this discussion the four main categories of network structures are:

- Paramilitary
- Centralized Cells
- Decentralized Cells
- Ad-Hoc Cells/Independent Action.

Paramilitary. Paramilitary groups have a hierarchical structure with clearly defined roles and lines of command and control within an essentially pyramidal organizational chart. These structures provide the strongest command and control functions, clearest demarcation of authority and responsibility, and clearest lines of communication. This structure may have minimal insulation of the command levels from the operational personnel, and may have minimal insulation between operational and support business lines. This structure is arguably the most vulnerable to disruption through infiltration by the security forces.

Terrorist groups which are established in imitation of regular military forces, and groups engaged in guerrilla/insurgent warfare, are most likely to show this kind of structure. For example, the Provisional Irish Republican Army (PIRA) originally had a paramilitary structure, and persisted in using role titles from that structure after moving to a centralized cell structure.

While the hierarchical structure of the group mimics a military or corporate organizational chart, the group will also be similar to armies and corporations in that it will tend to have networks of social connections which cross components and ranks. Also, when terrorist groups establish themselves as insurgent groups, they take on more features of a state.

Centralized Cells. Centralized cell networks have a hierarchical structure with command and control from a headquarters cell linked to specialized operational and support cells. In its ideal form, the specialized operational and support cells have limited or no direct communication with each other, in order to insulate them in case one is penetrated by the security forces. Communication links with the HQ cell are also restricted in order to maximize the degree of insulation of the group leadership from the functional cells. A network analysis of this type of group will most likely show distinct components or cliques linked by bridges or cutpoints.

Decentralized Cells. Decentralized cell networks have a less hierarchical structure, in which leadership takes the form of suggestion and guidance rather than command and control. In its ideal form, the operational cells carry out their own support functions (e.g. fundraising and weapons procurement) as well as planning and executing their own attacks. Targets and methods are selected based on the group's ideological and cultural norms, guided by statements by the leading figures in the group. The group's leadership may operate in the open, by working within the legal boundaries of their jurisdiction and not directly engaging in or counseling criminal acts. In a network analysis, this type of group may appear to be several unconnected networks, since there may be no apparent person-to-person relationship between two cells.

Ad-Hoc Cells. Ad-hoc cells and independent action are the organizational modes with the lowest density, although the social networks from which these ad-hoc cells or individual actors emerge may have a high density. A well-known example of an ad-hoc cell is the two men who carried out the 1995 bombing of the Murragh Building in Oklahoma City. The two brothers who allegedly carried out the bombing at the 2013 Boston Marathon may also turn out to have been an ad-hoc cell. This is a situational structure within the network based on joint participation in an attack, rather than group membership. In a small temporary group of like-minded individuals there is no need for one person to take on overall leadership, although there may be a leading individual and a division of labour based on individual skills and knowledge. Ad-hoc cells and independent actors cannot be controlled by a central authority, and so run the greatest risk of going "off message" and committing attacks which disrupt support for the terrorist group or movement. However, this approach to terrorism is also the most difficult for the security forces to predict and infiltrate.

A network analysis of an ad-hoc cell or independent terrorist may show no direct connections to other terrorists, although the individuals are likely to be in direct or indirect contact with other participants in the terrorist group's or movements' network of supporters.

3 Networks of the Criminal Economy

3.1 Criminal Markets

A criminal market, such as the market for synthetic drugs, is a network of economic activities including production, distribution, and sale. A network analysis of the market will examine the relationships facilitating the supply of precursor chemicals and equipment, the stages of production, packaging, shipment, wholesale and retail.

The approach to mapping a criminal market for illicit goods could take the form of a commodity flow analysis, tracking the production and distribution chains. The operational goal of this analysis is to find the choke points where law enforcement or regulatory action would cause the greatest disruption to the market. The choke points may involve critical precursors or equipment. They might also be in the shipment from production locations to the distribution network, or in the money laundering operations.

Mapping the supply chain network for illicit services is somewhat different in content, since the clients are purchasing actions rather than goods. Illicit services may include: money laundering, prostitution, or smuggling in cases where the smuggler simply provides transportation, and has not purchased the goods for resale.

Vulnerability analyses have been conducted on supply chain networks for some major licit businesses for the purpose of increasing resilience. These can be mirror-imaged to provide models and suggestions for supply chain disruption. Brintrup et al. [9] analysed Toyota's supply chain network and found that it is exponentially scaled (close to scale free) so has hubs which are vulnerabilities for disruption. Their detailed vulnerability analysis showed that the average failure threshold from node removal varied based on two factors, network structure and rarity of the products supplied by the nodes. Network malfunction occurred after removing an average of: 6.22 random nodes, 3.78 hubs, or 3.19 nodes producing rare components. This suggests that functional redundancy can be more important than centrality in supply chain network resilience.

A network analysis of a criminal market must also consider the financial flows. The flows of profits may also have vulnerable choke points, such as in the personnel and organizations used for money laundering. Disconnecting the network components which transfer and handle the profits of crime may be more disruptive than targeting the commodity, service, or the distribution network.

3.2 *Terrorist Supply Chains*

The commodity flow approach also has value for investigating the networks supplying finances and materiel for terrorist groups. Some of these networks are elements of the terrorist network, others are organized groups or networks which knowingly or unknowingly do business with terrorists. As above, mapping the supply chains should reveal choke points and vulnerabilities which can be exploited.

3.3 *Critical Paths*

Another network-based approach to finding failure points in criminal activities uses critical path analysis. This can be applied to the planning and preparations for any complex criminal act, from major drug shipments to major terrorist attacks. This approach has been used to identify indicators of attack preparations, and to identify critical elements of the conspiracy where the police or other security forces can intervene.

4 Disconnecting the Dots

Networks can be structurally disrupted by removing nodes and/or by removing links. Criminal networks can be functionally disrupted by disrupting their structures and/or their other operational requirements, such as the supply chain of the goods they sell or the political conflict they commit terrorism in furtherance of.

Only by understanding the structure and functioning of a network in its operational context can we plan enforcement actions to disrupt it to maximum effect. The history of organized crime in Canada and elsewhere has shown that most medium to large criminal groups are very resilient to law enforcement efforts to disrupt and destroy them.

Network analysis can reveal vulnerabilities in criminal networks which could be exploited for intelligence collection and for disruption of the network's structures. Vulnerabilities to intelligence collection relate to information holdings and communication patterns within the network. Analysis can suggest where the security forces can collect the most important information in the most timely and efficient way. Vulnerabilities to structural disruption relate to the internal cohesion of the group. Analysis can suggest which individuals and relationships are most vulnerable, and most important to the operations of the network. These analyses will always be subject to a degree of uncertainty, since it is rare to have the comprehensive and accurate data needed when studying covert groups.

4.1 Targeting Nodes

Studies done on social network analysis and network disruption [10–12] have consistently recommended targeting the individuals with the highest centrality. Since these individuals arguably have the greatest role in connecting the individuals making up the network, their removal should do the most to disrupt the pattern of connections within the network. In scale-free networks centrality measures are extremely important, since these networks are vulnerable to hub removal but highly resilient when nodes are removed randomly.

There is also suggestion that the form of the network determines whether the individuals with high centrality are the appropriate focus for disruption. Everton [13] suggests that differences in the density and in the extent to which the network has a hierarchical structure determine the effectiveness of such tactics. He argues that a covert network needs to maintain a balance on these two dimensions to operate effectively in its environment, and that the disruption strategy should focus on making changes that will push the network away from that balance point. In effect, making it too hierarchical or too egalitarian, and/or making it too dense or too sparse to function effectively.

One of the key concerns with applying social network analysis to intelligence problems is the technique's sensitivity to missing data. There is qualified assurance from experimentation that centrality measures remain robust under small amounts of error—up to 10 % randomly generated error in random networks [14]. However, centrality calculations may not be as robust when the errors are not random, such as when a covert group is successful in maintaining operational security on its most important elements. These calculations may also be more vulnerable when the network itself is not random.

Both Sparrow and Carley et al. also recommend targeting the individuals with the most unique roles. These individuals are the most difficult to replace, so their removal should have the longest-lasting effects. This approach is supported by Brintrup et al's findings on the importance of unique roles in supply chain network resilience cited above.

Carley et al. also discuss other personal factors, or criminal capital, to look for in individuals who are current or emergent leaders, which include:

- Highest task accuracy
- Highest amount of unique knowledge
- Highest cognitive load (number and complexity of tasks).

These factors identify the highest performing individuals in the network. The removal of these individuals will do the most to degrade the network's operational capabilities.

A more recent study on structural vulnerability in illegal drug markets [15] identifies different network properties in separate phases of the commodity chain. The conclusions suggest the police would have the greatest impact on the commodity chain by focusing on individuals in the middle of the chain (smuggling, supply, and finance) and individuals involved in complex activities. These individuals are more likely to be both highly connected and difficult to replace. The suggestion to target the individuals involved in complex activities (multiple roles in the drug trade) seems to correspond with Carley et al.'s suggestion to target the individuals with the highest cognitive load.

Malm and Bichler recommend the police focus on three possible network vulnerabilities in the illegal drug market, depending on market niche:

- Bridge—individuals with a bridging role in smuggling or supply in addition to other market roles
- Hub and bridge—individuals with a bridging role and very high degree centrality, operating in production in addition to other market roles
- Repeated hub—individuals with a very high degree centrality, operating in production or transport.

The initial published work on network disruption focused on counter-terrorist operations, and relates most strongly to military operations to disrupt the networks. If a military operation kills an individual, he is removed from the network. In law enforcement, the methods used to deal with an individual are very different. Arresting an organized crime member or terrorist, even when he is convicted and sentenced to imprisonment, removes him from some of the network activities but not from all. The individual may still be an active part of the network of communications, command, advice and permission within the group, though no longer personally committing crimes on the street. In some cases, the individual will continue their criminal career and networking while inside the correctional system.

While law enforcement can degrade the communication links around a specific individual or individuals, it will not necessarily break those links. The individual in question may decide not to invest the increased effort and risk in maintaining existing links post arrest/conviction:

- Out of a desire to rehabilitate
- As part of an agreement to cooperate with the Crown
- Because he or she has lost trust in those criminal colleagues.

However, the individual in question may make the effort to maintain existing links:

- In order to remain in the group
- To demonstrate that they have not cooperated with the Crown
- For status and protection while imprisoned.

The individual in question may also use their term of imprisonment to develop new links:

- Which will increase their criminal capital through access to other networks, knowledge, and expertise
- Which will also increase their status in their network.

4.2 Targeting Links

Actions which degrade or destroy the trust between individuals in criminal networks may prove to be an effective additional approach to network disruption. This is a method to attack the links rather than the nodes by changing the nature of the relationships. Effects of this can be seen in criminal groups after they discover that informants, agents, or undercover investigators have been used against them, and in which the consequent mutual suspicion starts to interfere with their collaborations.

It should be noted that any deliberate effort to disrupt or destroy the trust between individuals in an organized crime or terrorist group could have severe and ethically unjustifiable consequences. When a member or associate of one of these groups is suspected of being an informant or agent of the police it is not unusual for the group to murder him.

The trust between individuals in a criminal network can have two faces. There is the level of confidence a criminal has, based on past experience or on reputation, that the other will not cheat him, rob him, or inform on him to the authorities; that he is an "honest crook". The other face is the level of confidence a criminal has that the other will choose not to cheat, rob or inform on him out of fear of reprisal. These two faces of trust also apply to relations between criminal groups operating in the same market.

This issue of trust as an element of the links in criminal networks has been categorized into four types by Lampe and Johansen [16]. Each type of trust has specific applications within criminal networks, and can form the basis for a bridge or other critical link in that network. Exploring these types of trust further reveals differences in the impacts that law enforcement operation have on network links.

Individualized trust is based in a history of reliable interactions between the individuals involved, perhaps through friendship, kinship, shared experience on the streets or in prison. When two individuals do not have a shared history with each other, a third party trusted by both may act as a guarantor of the transaction. The implication for law enforcement is that if any of the parties are revealed as untrustworthy, for example if they turn out to have been an informant or police agent, this reduces trust in the connections they and their associates helped establish.

Trust based on reputation depends on communication through the criminal network, media, or through other means, of the personal reputation of reliability and skill acquired by individual criminals. A strong personal reputation allows business opportunities beyond the individual's circle of direct contacts, and is the basis for leadership roles in informally-structured networks. Loss of reputation, such as through losing shipments of goods or money, reduces the individual's status and ability to establish and maintain their criminal network.

Trust based on generalizations depends on the reputation of a group to which the individual belongs. This is the power of the 'brand names' in organized crime, since an individual will be expected (rightly or wrongly) to adhere to the standards of fair business dealings or violent retribution set by his group. Damage to a group's reputation reduces the members' ability to maintain and establish trusted links based on their group allegiance.

Abstract trust is the trust in institutions, such as the monetary system or the government, to function in a predictable way. This type of trust plays a role in networks of the criminal markets where banks and other licit institutions are used to unwittingly move illicit money and goods. Decreasing the trust criminals have to move money through licit financial firms or goods through licit transportation services without detection reduces their efficiency and forces them to invest in establishing and maintaining their own institutions.

Trust is, of course, not the only element in the relationships between collaborating criminals. The criminal network links are also based on greed, and in pre-existing social relations including: kinship, ethnicity, neighbourhood ties, institutional experiences, or other social factors [17]. These pre-existing social links are arguably the basis for sufficient trust for the individuals to commit crimes together. Disrupting greed as a shared bond in organized crime depends on disrupting the economic functions of the network. While greed plays a role for some terrorists, the shared desire driving a terrorist group is the desire for power to make political, religious, and/or social change. Disrupting this bond between terrorists also involves disrupting the functions of the network.

4.3 Approaches to Functional Disruption of Networks

There is a range of possible objectives, or desired tactical or strategic effects for operational initiatives against organized crime and terrorist groups. Some approaches to countering terrorism or organized crime focus on disrupting a network's structure, all of them ultimately attempt to disrupt network functions. Network analysis can help identify which approaches will be most effective and efficient when the analysis incorporates descriptions and assessments of network functioning as well as of the network structure.

Tactical objectives are the goals of a specific investigation or project. The objectives of a tactical initiative could include, depending on circumstance:

- Preventing a terrorist attack by disrupting an ongoing conspiracy
- Detecting money laundering through a seemingly legitimate business
- Apprehending members of a drug trafficking network.

Strategic objective are the goals of a major project or other strategic initiative. The objectives of a strategic initiative against a specific organized crime or terrorist group/network could include:

- Destroying the group
- Reducing the current size of the group/network through arrests and defections of members/participants
- Reducing the future size of the group/network by impeding recruiting
- Reducing the efficiency of the group's/network's activities
- Mitigating the impacts of the group's/network's activities by strengthening their intended victims
- Mitigating the impacts of the group's/network's activities by reducing their access to needed materiel, merchandise, or service providers.

The objectives of a strategic initiative against a criminal market, such as human trafficking, chemical drug production, or insurance fraud, could include:

- Reducing the scale of the market by reducing supply and/or demand for the criminal goods or services, or by reducing criminal opportunities for fraud, theft, or robbery
- Reducing the severity of the market by reducing the harmful impacts on consumers and victims.

Strategic success against a criminal market may be achieved by coordinating operations with policy initiatives for regulations, and legislation.

4.4 Targeting Critical Capabilities and Critical Requirements

Operations intended to contribute to a strategic success need to disrupt or destroy the critical capabilities of the target, and deny the critical requirements the target needs to operate. The role of intelligence is to identify those critical capabilities and requirements in order to recommend operational goals. At the strategic level, these operational objectives sometimes also require policy solutions to provide needed legislative and collaborative frameworks.

Critical Capabilities. Critical capabilities are the capabilities essential to the network's functions. If the network cannot accomplish these activities it will cease to operate. These vary depending on the motives and operating methods of the target but, for example,

Critical capabilities of a criminal network may include the ability to:

- Corrupt public officials
- Intimidate victims or competitors
- Launder funds
- Cross borders with contraband
- Maintain operational security
- Recruit new members.

When law enforcement succeeds in efforts to degrade, disrupt, or destroy critical capabilities, the group's ability to operate is removed or reduced. Strategic approaches to degrade these capabilities include:

- Investigation and arrest of individuals with key skills
- Criminalization or increased regulation of essential activities
- Strengthening security and resilience of victims and unwitting accomplices.

Critical Requirements. Critical requirements are the essential elements of the network's processes or market. If the network does not have access to these things it will cease to operate. These requirements vary depending on the motives, markets, and operating methods of the network.

The critical requirements for an organized crime group may include:

- Chemicals used to manufacture synthetic drugs
- Essential equipment, such as pill pressing machines
- Customers
- Potential victims vulnerable to a type of fraud.

The critical requirements for a terrorist group may include:

- Potential recruits open to the ideological message
- Information about potential targets
- Weapons (improvised or acquired).

Network analysis of commodity chains, critical paths, and other business processes identify the potential failure points in providing critical requirements. When law enforcement succeeds in denying access to critical requirements there is the potential to disrupt entire markets, not just specific target groups or networks. Strategic approaches to deny or disrupt access critical requirements may include:

- Criminalization or increased regulation of essential raw materials or equipment
- Public information to reduce the number of potential fraud victims
- Community outreach initiatives to reduce the available pool of potential recruits.

4.5 Exploiting Critical Vulnerabilities

A group or market will have weak points which can be exploited. The most important vulnerabilities are those which can lead to network failure. Intelligence needs to identify these critical vulnerabilities where they exist in order to recommend operational approaches. These may provide routes to disrupt critical capabilities and requirements directly, or they may prepare the ground for such disruptions. Successful operations will often create or expose new vulnerabilities which can be exploited in follow-up operations.

The critical vulnerabilities in a target group or network could include:

- Flawed vetting of new members
- Internal conflict
- Lack of redundancy in key roles
- Flawed operational security practices.

Successful investigations can create or amplify network vulnerabilities. For example, demonstrated use of human sources or agents can reduce trust among group members, harming the integrity of their internal code of silence and making it psychologically easier for others to become sources. Arrest of key individuals can create a power vacuum, and increase internal conflict.

4.6 Centres of Gravity

The notion of a target's centre of gravity comes from von Clausewitz's theory of warfare [18]. The centre of gravity of a nation, army, or other group is the force that ties it together. According to von Clausewitz the purpose of strategy in war is to damage the target's centre of gravity, which removes the target's ability to operate. Organized crime and terrorist groups may have centres of gravity.

- For an organized crime group, the centre of gravity might be: their code of silence, profit sharing, or family relationships
- For a terrorist group this might be ideology, or social bonds between members
- For a violent cult this might be a charismatic leader.

An organized crime network might not have a centre of gravity. It has collaborating career criminals who participate in a network, rather than belonging to a group. However, an organized crime network may still have critical capabilities and requirements.

4.7 Effects of Strategic Success

Effects of success against a group or network functions can include:

- Reduced operating capacity of group/network
- Reduced internal cohesion/trust in the group or network (creating additional vulnerabilities which can be exploited in follow-up investigations)
- Reduced size of group/network (fewer members or associates for group, fewer participants for network)
- Reduced size of customer base—fewer people purchasing the criminal goods or services.
- Reduced number of victims—fewer people being attacked, defrauded, extorted, or stolen from.
- Group/network shifts to less harmful activities.

5 Conclusions

The goal of criminal intelligence is to enable the police to disrupt the structure and functions of networks of individuals engaged in criminal activity. In order to discover and map the critical elements of these networks the range of analytic approaches need to be holistic. The various applications of network analysis are a set of different views into a complex problem, and the problem cannot be understood and effectively addressed without seeing it from many angles.

The core questions for the intelligence professional applying network analyses are deceptively simple:

- How does the network function?
- How could we break it?

The developments in social network analysis over the past two decades have given us far better insights and tools to discover which individuals are most important to a network, and so we have some guidance on identifying the critical people. Other applications of network analysis, to processes such as supply chains, production chains, attack preparations and other criminal conspiracies, provide insights into core activities of organized crime and terrorist networks and help identify the critical requirements and capabilities of these networks.

In order to be effective in disrupting organized crime and terrorism, the goal has to be disrupting the functions of those networks, disrupting the structures are merely one means to that end. To use an unlikely example: if the Cosa Nostra retained its structure but turned into a social club instead of an economic network based on robbery, extortion, and other crimes, it would cease to be a problem.

At its most basic, the minimal network is a dyad: two nodes with a link between them. The network fails if either a node or a link is removed. In the discussions on disrupting criminal networks to date the focus has so far been on removing nodes, assuming that the links will be removed as well. But even a highly effective justice system cannot always remove the nodes from the network, so lessons also need to be learned and applied from the ways in which law enforcement has succeeded in disrupting the crucial links.

The network analysis approaches which have had greatest success in criminal intelligence have been adapted to work under specific conditions: short time-lines for analysis, unknown amounts of missing data, and persistent uncertainty about the quality of some of the data at hand. Increasingly sophisticated techniques and applications of network theory create exciting new opportunities for intelligence analysis. However, these techniques will need to be tested and adapted to function reliably when used in attempts to understand covert groups and activities.

References

1. Granovetter MS (1973) The strength of weak ties. Am J Sociol, 1360–1380
2. Morselli C (2005) Contacts, opportunities, and criminal enterprise. University of Toronto Press, Toronto
3. Johnson, Gregory A (1982) Organizational structure and scalar stress. In: Renfrew Colin, Rowlands Michael, Barbara A, Segraves-Whallon (eds) Theory and explanation in archaeology. Academic Press, New York, pp 389–421
4. Service ER (1971) Primitive social organization: an evolutionary perspective, 2nd edn. Random House, New York
5. de Champlain P (2004) Mobsters, gangsters and men of honour: cracking the mafia code. Harper Collins, Toronto
6. Venkatesh Sudhir (2008) Gang leader for a day: a Rogue sociologist takes to the streets. Penguin Press, New York
7. Earle T (2011) Chiefs, chieftancies, chiefdoms, and chiefly confederacies: power in the evolution of political systems. Soc Evol Hist 10(1):27–54
8. Felson M (2006) The ecosystem for organized crime. The European Institute for Crime Prevention and Control (HEUNI Paper No. 26) http://www.heuni.fi
9. Brintrup A, Kito T, Lopez E, New S, Reed-Tsochas F (2011) The structure of the Toyota supply network: the emergence of resilience, CABDyN working paper
10. Sparrow MK (1991) Network vulnerabilities and strategic intelligence in law enforcement. Int J Intell CounterIntell 5(3):255–272
11. Carley KM, Lee JS, Krackhardt D (2002) Destabilizing networks. Connections 24(3):79–92
12. Schwartz, Daniel M., and Tony Rouselle (2008) Using Social Network Analysis to Target Criminal Networks. Trends in Organized Crime, 188–207
13. Everton SF (2009) Network topography key players and terrorist networks. Connections, 12–19
14. Borgatti SP, Carley KM, Krackhardt D (ND) On the Robustness of Centrality Measures under Conditions of Imperfect Data, Dynamic Networks project, Carnegie Mellon University
15. Malm A, Bichler G (2011) Networks of collaborating criminals: assessing the structural vulnerability of drug markets. J Res Crime Delinq 48(2):271–297
16. von Lampe K, Ole Johansen P (2004) Organized Crime and Trust: on the conceptualization and empirical relevance of trust in the context of criminal networks. Glob Crime 6(2):159–184
17. Klerks P (2001) The network paradigm applied to criminal organizations: theoretical nitpicking or a relevant doctrine for investigators? recent developments in the Netherlands. Connections 24(3):53–65
18. von Clausewitz Carl (1968) On war. J.J. Graham trans. Penguin Books, UK

Identifying Mafia Bosses from Meeting Attendance

Francesco Calderoni

Abstract Law enforcement agencies have frequently shown skepticism toward the results of the application of social network analysis to organized crime. Indeed, most studies to date have analyzed data (e.g. telephone intercepts) whose content was already well-known to the practitioners. Shifting the focus to data on mafia meetings, this chapter explores whether network analysis can identify the bosses in a large mafia network. The analysis relies on data from a large-scale investigation on the presence of the 'Ndrangheta, a mafia from the Southern Italian region of Calabria. Operation *Infinito* identified several mafia families and tracked a number of mafia meetings. The results show that betweenness centrality is the most significant predictor of leadership in the mafia. A logistic regression model, using network measures as predictors, is able successfully to predict the position (boss or other) of 92 % of the individuals in the network. If supported by further studies, network analysis of meetings may provide law enforcement agencies with information useful for identifying the bosses of criminal organizations.

Keywords Mafia · Organized crime · Social network analysis · 'Ndrangheta · Statistical analysis

1 Introduction

Since its first appearance, social network analysis of organized crime has been associated with the idea of its operational exploitation for law enforcement purposes [1, 2].

On the research side, the idea that organized crime may be better understood as a network rather than a hierarchical and structured organization is not new in the

---

F. Calderoni (✉)
Università Cattolica del Sacro Cuore and Transcrime, Milan, Italy
e-mail: francesco.calderoni@unicatt.it

A. J. Masys (ed.), *Networks and Network Analysis for Defence and Security*,
Lecture Notes in Social Networks, DOI: 10.1007/978-3-319-04147-6_2,

literature. The reaction against the alien conspiracy approach, which suggested analyzing organized crime as a bureaucratic organization, with a formal hierarchy and with detailed rules for its functioning, soon led to alternative hermeneutic perspectives leaning towards more flexible and informal mechanisms. For example, Albini argued that organized crime is "a system of loosely structured relationships" mainly based on patron-client relations [3]; the works of Ianni suggested that mafia-type organizations should be better understood as social systems based on shared social cultural and ethnic relations. He explicitly mentioned network analysis, as "an anthropological tool that is used to chart social interactions" [4]. Ianni analyzed a number of Puerto Rican and African American criminal groups as networks, although the application was quite distant from current use. Overall, although the concepts of network and network analysis were recurrently evoked to describe the functioning and structure of organized crime, there were very limited empirical applications using network analysis methods [4].

On the law enforcement side, since the mid-1970s there has been growing attention to the processing and analysis of intelligence data on organized crime. In this regard, link analysis (visual representation of the structure of a criminal group performed through manual or computer-assisted drawings) was increasingly applied by law enforcement agencies, and a private industry offering methodologies and training courses quickly developed [2, 5, 6]. Link analysis made it possible to "establish the relationships that exist among individuals and organizations from bits and pieces of available evidence" [5]. It became increasingly popular thanks to the development of intelligence software. The step from link analysis to social network analysis was relatively short [2, 7, 8].

Academic interest in organized crime networks gradually met with attempts at network analysis by law enforcement agencies. From the first half of the 1980s onwards, the first applications of network analysis to organized crime already suggested possible applications to law enforcement action [6, 7, 9]. Lupsha argued that effectively combating organized crime required a shift to the strategic analysis of groups and operations, including network analysis as an "essential and necessary step" (1980). One year later, FBI special agent Davis provided a first hypothetical example of the application of network methods to a criminal organization adopting some basic network concepts, such as density and centrality, with a fictional criminal organization [9]. In the following years, some studies advocated the application of network methods to criminal organizations [1, 2, 8, 10–13]. However, most of these now "classic" contributions, albeit with some exceptions (e.g. [6]), did not engage in the empirical analysis of criminal groups. Not surprisingly, only 12 years ago Coles [14] still complained about the "failure by criminologists to adopt Social Network Analysis techniques and concepts in the investigation of criminal networks, particularly of organized crime".

Only in the last decade has the use of network analysis in the study of organized crime seen significant developments. Since the beginning of the 2000s, academic interest in this specific field has greatly increased and contributed to opening new research directions in the study of criminal organizations [15–34].

Despite the increase in empirical applications, and despite the fact that some law enforcement agencies have started to apply network analysis in their criminal investigations, the method is not yet established as a standard instrument in law enforcement tactical and strategic analysis. Many law enforcement operators are still skeptical in regard to the application of network methods in organized crime cases. According to most of these operators, the current applications of network analysis are not likely to provide everyday investigations with critical advantages. This is because, in long-lasting investigations, the police and the prosecution normally have extremely detailed knowledge about the case. Each suspect is constantly monitored, her/his background is scanned, and telephone calls and emails are frequently intercepted. Network analysis could hardly offer any additional knowledge to operators working daily on the case. Network analysis can assist a researcher in reconstructing a network and in making sense of the structure and functioning of a criminal organization. While these analyses have provided unprecedented academic insights into the structure and functioning of criminal organizations, the risk is that current applications may be of limited interest to law enforcers. More often than not, scholars identify as the most important individuals, the best targets for network disruption, the "usual suspects" that law enforcement has already identified.

The limited operational implications of current applications of network analysis to organized crime should not lead to the conclusion that the method is useless. In fact, network analysis of organized crime is still in its infancy, and new developments may be generated by the advancement of the methods and increasing familiarity with them of scholars and practitioners [35, 36].

The reasons for the limited interest in network analysis by law enforcement may be due to two different reasons.

First, most network analysis studies have focused on individual judicial cases, with the result of an overlap with the information already gathered by law enforcement. From this perspective, one possible direction of evolution is the application of social network analysis across different investigations [37]. This development may have a geographical and a chronological dimension. From the geographical perspective, network analysis may merge investigative data from different investigations or jurisdictions and provide an overall picture which otherwise might be too complex or difficult to achieve. One such application has demonstrated that network structure and positioning may significantly change when different cases are merged [38]. From the chronological perspective, some criminal organizations such as mafias are particularly resilient, and last for years notwithstanding constant law enforcement action. The application of network analysis to different investigations on the same group may provide strategic information about the development and structure of a single criminal organization through time.

Second, most existing studies have analyzed data from telephone interceptions. While these data enable scholars to examine the communication flows within a criminal network, they offer much more additional information to the law enforcement agencies and the prosecution. Wiretapped conversations provide important evidence for both the investigation and the trial phase, particularly in organized crime cases, where most criminal activities are without victims, or the victims may fear retaliation if they report to the police. Academic exploration of network dynamics on the basis of intercepted communications is unlikely to add any value to the information already gathered by the authorities. Furthermore, in most jurisdictions, the possibility to intercept the personal communications of suspects requires a number of conditions. Typically, it is restricted to serious offences for a limited period of time. Lastly, in order to proceed, the police and/or the prosecution may need a court order, which usually requires them to provide substantial evidence about the criminal activities.

In this regard, one possible development could be the application of network analysis approaches to different types of data more easily accessible even in the preliminary stages of a criminal investigation.

The analysis of the meetings of the members of a criminal network may be an example of such an approach. Frequently, law enforcement agencies can more easily control individuals' movements and meetings than their communications. For example, procedural requirements to shadow suspects or monitor public places (e.g. bars, restaurants, parks) are less constraining. The application of social network analysis to meeting data may provide insights into the functioning of a criminal network, with a limited investment not only in terms of procedural requirements but also in financial terms.

The analysis of meeting attendance may prove particularly effective in combating traditional organized crime groups, such as mafias, owing to their characteristics. First, the mafias have a formal structure with an established hierarchy. Formal affiliation is an important passage which provides the initiates with a new social status [39, 40]. Attendance at mafia meetings may be indicative of an individual's affiliation. Furthermore, some meetings may be reserved for only a few affiliates, especially when criminal activities are discussed. Second, mafias have a social structure which is embedded in family and kinship connections [41, 42]. Law enforcement may easily extend the analysis to social events such as religious ceremonies (e.g. baptisms, weddings, funerals) and family celebrations (e.g. birthday parties). Third, the most active and visible members of mafias and other organized crime networks may take precautions against over-exposure in telephone communications in order to reduce the risk of interception. They may manage the criminal activities through their status or more indirect forms of control [22, 31]. Individuals that make most use of the telephone may be middle-level criminals, who then report to the bosses. The analysis of meetings may penetrate these countermeasures.

Despite the possible advantages of the analysis of meetings, to date no study has applied network analysis methods to organized crime meeting patterns.

This chapter addresses this gap by exploring individuals' attendance at meetings within a large mafia organization. The aim is to show how the network analysis of meetings can successfully predict which individuals are the bosses of the mafia. The analysis is based on a single case study of a law enforcement operation (Operation *Infinito*) against the 'Ndrangheta, an Italian mafia from Calabria, Southern Italy.

Section 2 of this chapter provides background information on the 'Ndrangheta, focusing on its internal organization, which is a necessary basis for the prediction exercise. Subsequently, Sect. 3 describes the case study and the methodology. Section 4 presents and discusses the results. The Sect. 5 concludes.

2 The 'Ndrangheta

Historically, the 'Ndrangheta originated in the nineteenth century, or even earlier according to some sources [43, 44]. Until recent years, in Italy and abroad, the 'Ndrangheta has received less attention than Cosa Nostra, which is widely considered to be the archetypical mafia organization [45]. Yet some recent events have brought the Calabrian organization into the world media spotlight.[1] Since then, foreign law enforcement agencies and institutions have increasingly paid attention to the 'Ndrangheta.[2]

The 'Ndrangheta has shown a remarkable capacity to establish its presence outside Calabria. Despite years of denial and minimization, its presence in the North of Italy is by now acknowledged as a fact [58, 59]. Moreover, a number of sources report the presence of this mafia in various foreign countries (e.g. Australia, Canada, Germany and the Netherlands) [57, 60–62].

Numerous factors have favored the spread of the 'Ndrangheta outside Calabria. They include the migration of Calabrians in search of better living conditions; the establishment of profitable connections for criminal trade; the demand for criminal protection by some legal entrepreneurs; the need to evade prosecution; and mafia wars [63, 64].

Another important factor in the 'Ndrangheta's success may have been its complex social and organizational structure. Blood and community ties, rituals, affiliation ceremonies, mythology and a hierarchy comprising a number of formal

---

[1] The first event was the so-called 'Duisburg massacre' of 15 August 2007, when six people of Calabrian origin were murdered in Duisburg, Germany, in connection with a decade-long blood feud between two *'ndrine* [46–49]. Investigators found evidence of an affiliation ritual in the pocket of one of the victims (the remnants of a burned holy image), which further confirmed the tight relation between the massacre and the 'Ndrangheta [50]. Second, on 30 May 2008 the President of the United States designated the 'Ndrangheta a significant foreign narcotics trafficker (to date, the only Italian mafia) and included it on a special list, thus prompting a number of sanctions by U.S. authorities [51, 52]. Third, two exceptional law enforcement operations mounted on 15 July 2010 led to the arrest of more than 300 people and brought the 'Ndrangheta to international media attention [53–56]. The first investigation (Operation *Il Crimine*) focused on Calabria and for the first time recorded the secret annual 'Ndrangheta meeting at the Sanctuary of Our Lady of Polsi. The second operation (Operation *Infinito*, the case study of this chapter) highlighted the stable presence of the 'Ndrangheta in Milan (the capital of Lombardy, Italy's second largest city and its main economic and financial center).

[2] For example, in 2011, the Dutch Police issued a report on the 'Ndrangheta, arguing that "no reliable statements can be made about the organizational structure in the Netherlands. It is likely that the 'Ndrangheta also applies the structure it uses in Calabria and elsewhere in the world in the Netherlands, which means that there are one or more *locales* that organize criminal activities" [57].

ranks (*doti*) and offices (*cariche*) strengthen trust among the members, even at a great distance from the motherland [40, 57].[3]

The organization of the 'Ndrangheta revolves around blood kinship [40, 63]. It is very common for men of the same family to join the organization, unlike original Cosa Nostra, where the "family" was in fact a group of individuals who were rarely consanguineal kin (there was even a rule against accepting more than two blood relatives into a Cosa Nostra family) [40, 65]. One or several 'Ndrangheta families, frequently connected by marriage, godfatherhood and similar social ties, form a *'ndrina*. The *'ndrine* from the same area may form a *locale*, which controls a specific territory [66]. This social structure reinforces the cohesion of the criminal group, reducing the risk of betrayal.[4]

The 'Ndrangheta also has a complex internal hierarchy made of ranks (*doti*) and offices (*cariche*) [40]. Competition for advancement in the hierarchy is fierce, and concern about his criminal career is a constant element in the life of a *'ndranghetista*. Affiliates frequently discuss matters relating to ranks and positions. Information on the formal organization is fragmented, despite the discovery of written "regulations" and the accounts of collaborators with justice describing the rules of the organization [40, 44, 68]. The 'Ndrangheta continues to evolve, and the number of ranks and offices has increased, possibly due to internal pressure to climb the mafia career ladder.

In general, every affiliate holds a specific rank (*dote*) within the organization. The ranks are a sort of internal career advancements for the members. Indeed, the *dote* indicates each affiliate's status within the 'Ndrangheta.[5] While the lower levels have remained constant over the years, recent decades have seen the creation of new higher ranks [40]. As a result, members are today divided into two main tiers: the higher society (*società maggiore*) and lower society (*società minore*), with multiple ranks within each society [40]. Before associates formally join the 'Ndrangheta, they are called *contrasto onorato*. After their initiation, they start within the lower society, which includes affiliates with lower ranks (*picciotto, camorrista, sgarrista*). The higher society comprises members who have reached higher ranks (*santa*, *vangelo*, *quartino*, *tre quartino*, *quintino, associazione*, and other ranks discovered in recent investigations).[6]

---

[3] Indeed, neither Cosa Nostra nor the Camorra have established a stable organized presence outside their regions. They have instead adopted more flexible structures. During the "golden age" of the twentieth century, only a few Cosa Nostra families were based outside Sicily [65]. A study on the Camorra *La Torre* clan highlighted that its presence in the Netherlands and Scotland was associated with money laundering and participation in illegal markets. However, the core business of the group (criminal protection) remained in the region of origin [27].

[4] As proof of this, very few *pentiti* (collaborators with justice) come from the 'Ndrangheta, as opposed to the Sicilian Mafia or the Neapolitan Camorra [45, 67].

[5] The award of a new rank is a very important moment for a *'ndranghetista*. It is usually celebrated with a dinner or a party with the closest affiliates. Higher ranks can only be awarded by top-ranked individuals, with the agreement of the most prominent figures in Calabria.

Each locale has a number of formal offices (*cariche*) with specific functions. Offices are assigned to higher-ranked affiliates. The boss of the *locale* is the *capobastone* or *capolocale*. The *contabile* (accountant) is responsible for the common fund of the *locale*, the *crimine* (crime) oversees criminal, and particularly violent, actions, and the *mastro di giornata* (literally "master of the day") attends to communication flows within the *locale*. Information is incomplete as to the ranks, with investigations uncovering new offices and functions. In general, however, the number of offices is small, possibly to avoid overlaps among competences. For example, one internal rule requires every affiliate to state the names of the individuals holding the three main offices (the *copiata*) in his *locale* (*capobastone*, *crimine* and *contabile*) whenever he presents himself to members of other groups.

3 Methodology

Attendance at mafia meetings has been largely unexplored in the literature. This chapter focuses on the relation between meeting attendance and hierarchy in the 'Ndrangheta, exploring whether network positioning can predict who the leaders of the mafia are. The objective is important for a number or reasons. First, the identification of the bosses of a criminal organization is one of the main goals of both scholars and law enforcement agencies [31, 38]. Second, a number of sources report that the 'Ndrangheta replicates its complex internal organization also outside its region of origin [40, 57, 61, 62, 69]. The results of this study may be of help in analyzing the structure of the 'Ndrangheta in other areas.

As argued in the next sub-sections, the case study conducted in this chapter is especially helpful for the objective of the analysis: it concerns the presence of the 'Ndrangheta in Lombardy, a Northern Italian region far from Calabria. Further, the specific focus of the investigation reported was to provide reliable information about the internal hierarchy, independently from the patterns of participation at the meetings.

3.1 Case Study

This study focused on one large judicial case, Operation *Infinito*, named after one of the new ranks in the 'Ndrangheta discovered during the investigation. The case concerned the establishment of several 'Ndrangheta *locali* in the area around Milan, the capital city of the Lombardy region, Italy's second largest city and economic

[6] The evolution of the rank of *santa* clearly illustrates the impact of internal competition on the formal hierarchy. The rank defines what is called *La Santa*, an inner circle of affiliates created during the 1970s. Initially, the *santa* was the highest rank in the 'Ndrangheta, accessible to no more than thirty-thee individuals. A *santista* could enter into contact with freemasonry, politics and entrepreneurs and even betray lower affiliates if this was in the organization's higher interest. Nevertheless, in subsequent years, new and higher ranks emerged, and today the *santa* is only the first rank of the higher society [40].

capital. The data sources included a number of judicial documents, from law enforcement agency reports, through pretrial detention orders, to first and second grade judgments.

The history of *Infinito* dates back to 2006 and the launching of several criminal investigations of the *'Ndrangheta* in Lombardy. These were subsequently joined and concluded in 2010 by the Antimafia Prosecutor's Office of Milan and the Carabinieri (a national law enforcement agency). On 13 July 2010, the authorities arrested more than 160 individuals. At the time of writing, the trials are still ongoing.[7]

The main concern of *Infinito* was the evolution of a special 'Ndrangheta structure called *la Lombardia* (Lombardy in Italian) and of the person charged with its management, the *Mastro generale della Lombardia* (general master of Lombardy). Although information about the functioning of this body is still unclear, it seems that *la Lombardia* was a sort of coordination chamber among the *locali* based in Lombardy, but also with a role in relations between these and the main *locali* in Calabria. At the beginning of the investigation, N161 was the leader of *la Lombardia*.[8] However, he showed excessive autonomy from the powerful Calabrian *locali* by pressing for greater independence for the Northern groups in the management of criminal activities and the award of ranks and offices. As a reaction, the *locali* from Calabria ordered his murder, which took place on 15 July 2008, in a bar in San Vittore Olona, a town about 30 km from Milan. After N161's murder, the dominant Calabrian families decided to suspend the office of *Mastro generale della Lombardia* and appointed a temporary manager, N099. Eventually, after machinations concerning occupancy of the vacant office, N157 was appointed as *Mastro generale* on 31 October 2009.

One of the salient characteristics of Operation *Infinito* is that the main charge is the offence of mafia-type association (Article 416-bis of the Italian Criminal Code). This offence criminalizes participation in any association, whatever its origin, which uses the so-called mafia method.[9] The key elements of the mafia method are: (a) the intimidatory power deriving from the strength of the associative bond, (b) the condition of subjection and (c) of *omertà* or silence, which are both consequences of the mentioned power. Due to this specific criminal charge,

---

[7] The Court of Milan issued two first-grade judgments on 11 November 2011 and 6 December 2012, convicting most of the suspects. On 23 April 2013 the Court of Appeal of Milan confirmed the convictions of the first judgment. Further appeals and the third-grade trials at the Cassation Court of Rome will follow in the next months.

[8] The study coded each individual in the court order as N001, N002, etc. to prevent identification.

[9] Article 416-bis para 3 of the Italian Criminal Code: "The association is of a mafia-type when the participants use the power of intimidation, which arises from the association, and the system of subordination and *omertà* (code of silence) that arises from it, to commit crimes, or to obtain—directly or indirectly—control over economic activities, public procurement or concession contracts, or to obtain unfair profits for themselves or for other people, or for the purpose of impeding or jeopardizing the free exercise of the right to vote or to gain votes for themselves or others upon elections".

Infinito sought to demonstrate the stable presence of a mafia-type association in the area around Milan. The account of the struggle for the office of *Mastro generale della Lombardia* and reconstruction of the hierarchy of the *locali* in the area were crucial in proving the existence of a mafia-type association. Consequently, most of the investigative activities, from background checks to wiretaps and surveillance, focused on describing the organizational structure of the 'Ndrangheta, with a particular concern to chart the hierarchy of *la Lombardia* and the different *locali* present in the region.[10]

The most important source for this study was the pretrial detention order issued by the preliminary investigation judge (*Giudice per le indagini preliminari*, henceforth GIP) of Milan upon request by the prosecution [70]. This is a judicial document with the function of restricting the suspects' freedom, and in particular of remanding some of them in custody or pretrial detention. In Italy, detention orders are issued towards the end of criminal investigations, when the prosecution has gathered sufficient evidence to request the remand in custody or other pretrial restrictions against the suspects.[11] From a comparative perspective, these court orders are broadly similar to arrest warrants in other jurisdictions. The Italian law requires the judge to motivate his/her decision. For this reason, the documents provide a wealth of information for the analysis of criminal groups and their criminal activities. They report personal details of the suspects (e.g. name, birth date, residence, citizenship) and of the crimes (for every crime, e.g. extortion, they record the date, the place, the participants and the *modi operandi*). They also provide details on the structure of the organization, the roles played by each member, and the evolution of the group during the investigation. Furthermore, the documents report excerpts from telephone conversations intercepted during the investigation and detailed information about meetings among the suspects observed or reconstructed by the investigators.

The court order of Operation *Infinito* consisted in more than 700 pages providing abundant information for the analysis of the organizational structure of the 'Ndrangheta in Lombardy. The wealth of information and the focus of the investigation made *Infinito* useful for analyzing the relation between the hierarchy and individuals' positions within a meeting network. The use of judicial documents as secondary sources for the analysis of criminal organizations is common. Indeed, in recent years an increasing number of studies have used similar sources to study criminal organizations and mafias, applying network analysis, content analysis and other quantitative methods [15–23, 25, 27–34, 71–73].

---

[10] The investigation also uncovered a number of other crimes, ranging from extortion to usury, from corruption to the infiltration of public procurement.

[11] In the Italian criminal justice system, criminal investigations actually finish sometime after the arrest of the most dangerous suspects, once the prosecution has wrapped up the evidence and formulated the indictment (*richiesta di rinvio a giudizio*).

3.2 Data Gathering

3.2.1 The Meeting Network

The court order of Operation *Infinito* provided information on a large number of meetings among members of the criminal network. Some of these meetings occurred in houses, private premises (e.g. warehouses) or cars, others in public places (e.g. bars, restaurants or public parks).

The study analyzed and coded each meeting mentioned in the court order.[12] In total, the document provided information on 308 individuals participating in 574 of such meetings. The number of participants per meeting varied considerably, with a minimum of 2 and a maximum of 25 identified individuals. In most cases, the law enforcement agencies were able to identify all or most of the participants in any meeting.

Given the specific focus of the investigation, most of the meetings reported in the court order were true 'Ndrangheta meetings. These had the purpose of discussing criminal activities or matters relating to the life of the various *locali* (e.g. affiliations, attribution of ranks and offices) or of the structure *Lombardia* (including the vicissitudes of N161, his murder, the suspension of the office, and the appointment of N157 as the new *Mastro generale della Lombardia*). For example, on 31 October 2009, the law enforcement managed to film a 'Ndrangheta meeting in Paderno Dugnano, a town approximately 10 km from the center of Milan. The meeting took place in a social club for elderly people dedicated to the names of Giovanni Falcone and Paolo Borsellino, two judges killed by Cosa Nostra in 1992. The purpose of the meeting was to appoint N157 as the new *Mastro generale*. Twenty-three men sat at a horseshoe-shaped table, having dinner and conversing. The temporary manager N099 gave a speech and proposed N157. Thereafter, all the bosses of the *locali* approved the candidate, whose appointment was celebrated with a toast.

In general, the vast majority of the participants were 'made members' of the 'Ndrangheta, further proof of the mafia-related nature of the meetings. In some cases, however, the authorities were unable to verify the actual affiliation of some individuals owing to a lack of information. These men may have been actually affiliated with the organization or associates not yet admitted into the 'Ndrangheta, or external individuals who might participate or otherwise in the criminal activities of the mafia. Moreover, the law enforcement classified some individuals as not affiliated to the organization. Also in these cases, the persons concerned may have had personal (e.g. family, friendship) or business (either legal or illegal) connections with the 'Ndrangheta.

The analysis was conducted on meetings with at least four participants. The selection was designed to focus on the most important interactions among the

[12] The study coded each meeting as E001, E002, etc., recording the meeting's time, place and participants.

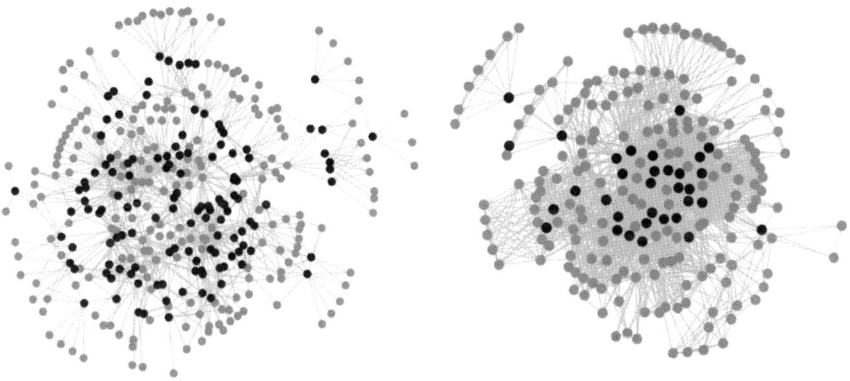

Fig. 1 The infinito 1-mode and 2-mode networks. (In the *left* graph, *black* nodes are meetings, *grey* nodes are participants. In the *right* graph, *black* nodes are the bosses, *grey* nodes are other individuals)

members of the 'Ndrangheta, with removal from the sample of the smallest meetings. Indeed, the objective of the study was to verify if and how attendance at mafia meetings could reveal the 'Ndrangheta's internal hierarchy. The inclusion of the smallest mafia meetings might have biased the sample towards those individuals more intensely controlled by law enforcement, e.g. because their cars were wiretapped. This kind of selection is frequent in studies taking a social network approach to criminal groups [15, 18, 22].

The sample of selected meetings comprised a total of 215 individuals participating in 129 meetings. Interestingly, while the sample included only 22 % of the total 574 recorded meetings, it also comprised nearly 70 % of the total participants. The number of participants per meeting ranged from four to 25 identified individuals. The majority of meetings had four ($n = 47$), five ($n = 19$) or six ($n = 22$) participants, while 41 meetings had seven or more. Of the 215 individuals in the sample, 99 attended only one meeting, 29 two meetings, 65 3–10 meetings and 22 more than 10 and up to a maximum of 38 meetings (Table 1).

Participants and meetings were entered into a two-mode, affiliation matrix (left graph in Fig. 1). From this, a one-mode, valued and undirected adjacency matrix was computed (right graph in Fig. 1), recording each individual's co-participation in meetings with any other individual in the network. The affiliation matrix enabled the calculation of a number of social network measures (all based on a binary matrix, except for valued degree centrality). All network analysis operations were performed using the Ucinet 6 [74].

Compared to previous social network analysis studies, mostly focused on telephone communication flows, the analysis of coparticipation in meetings has different implications. Telephone conversations usually occur between two subjects, and only exceptionally among more than two. Moreover, telephone conversations imply a direct exchange of information between the participants. By contrast, meetings generally involve several people, in this case at least four due to

Table 1 Meetings per number of participants and individuals per meetings attended

Participants per meeting	No of meetings	Meetings attended	No of individuals
4	47	1	99
5	19	2	29
6	22	3–10	65
7+	41	11+	22
Total participants	129	Total meetings	215

the sample selection. Coparticipation in a meeting does not necessarily imply a direct exchange of information among all those present. Rather, in the context of a hierarchical criminal organization such as the 'Ndrangheta, meeting attendance may relate to the status of each subject, implying that each member has a position granting him access to other affiliates and, in some cases, to the bosses of the organization.

Given the lack of previous literature on meeting attendance networks, the analysis adopted an exploratory approach and calculated the centrality measures most frequently used in the literature. These include degree (calculated on both binary and valued matrix), betweenness and closeness centrality, three classic measures of centrality developed by Freeman and frequently used in the literature on criminal networks [38, 75, 76].[13] Other measures are eigenvector centrality (calculated on the binary matrix), an adaptation of degree taking into account the degree of a node's contacts (i.e. neighborhood) [38, 77] and the clustering coefficient, a measure of the level of interconnection among a node's neighborhood [38, 78, 79].[14] Table 2 summarizes the minimum, maximum, mean, and standard deviation for each variable.

3.2.2 The Hierarchy

Given the focus of the investigation on proving the existence and the organizational structure of the 'Ndrangheta in Lombardy, law enforcement dedicated particular attention to the identification of the bosses.

The investigation identified the individuals who led the coordination chamber *la Lombardia,* with the office of *Mastro generale della Lombardia.* The first in office during the observed period was N161, followed by N099 as a temporary manager, until the appointment of N099.

[13] Closeness was measured so that the most central nodes had the highest scores.

[14] The clustering coefficient is strongly influenced by the size of a node's neighborhood. With a high number of direct contacts, it is less likely that they will be densely connected. For this reason, the clustering coefficient is always presented along with the number of pairs (i.e. the number of possible combinations among a node's direct contacts).

Table 2 Descriptive statistics of the variables

	Min	Max	Mean	St.dev.
Leader	0.0	1.0	0.2	0.4
Meetings attended	1.0	38.0	4.0	5.0
Degree (normalised)	1.4	38.8	8.8	7.5
Valued degree	3.0	221.0	31.1	38.4
Betweenness (normalised)	0.0	15.2	0.7	2.0
Closeness (normalised)	24.0	59.0	39.8	6.8
Eigenvector (normalised)	0.0	32.0	6.6	7.1
Clustering coefficient	0.3	1.0	0.8	0.2
Number of pairs	3.0	3,403.0	294.3	520.8
Mafia charge	0.0	1.0	0.5	0.5

The law enforcement also carefully charted the hierarchy of each *locale*. The court order report gave detailed information on 17 *locali* in Lombardy and the membership of each of them. The investigation identified the heads (*capolocale* or *capobastone*) of each *locale* and, in the majority of cases, also the other main officers (*caposocietà*, *contabile*, *crimine*). The law enforcement and the prosecution based their assessment of the offices in each *locale* on different sources. These included evidence from previous investigations and trials, but also and more frequently wiretaps of conversations and meetings among members of the 'Ndrangheta. The offices within each *locale* were frequently discussed by the wiretapped suspects, also due to the particular period of tension which followed the murder of N161 and the competition for appointment as new *Mastro generale della Lombardia.*

In Operation *Infinito*, identification of the 'Ndrangheta bosses was based on direct evidence concerning the mafia's internal hierarchy. Two grades of trial further confirmed the evidence on the organization's internal hierarchy in Lombardy. For these reasons, the assessment of the hierarchical positions within each *locale* is not based on individuals' activities and appears to be the most reliable depiction of the organization's formal hierarchy. In other studies, reconstruction of the internal hierarchy has relied on the evaluations of the law enforcement agency or the court, or on individual tasks or roles in criminal activities, or on content analysis of the conversations [15, 17, 18, 31, 38, 79]. By contrast, in this study the independence between identification of the bosses by law enforcement and their positioning within the network of the meetings offers an interesting opportunity to explore how social network measures may contribute to the identification of mafia bosses.

The analysis coded each individual's office within the 'Ndrangheta. It created a dichotomous variable assessing whether each subject held one of the main offices (either head of the *locali* or the other main offices) in *la Lombardia* or within the *locali* of the 'Ndrangheta in Lombardy or not. As a result, among the 215 individuals in the sample of meetings, 33 had a leadership office and 182 were simple members of the 'Ndrangheta (n = 135), unidentified (n = 43) or not formally affiliated (n = 4).

3.3 Statistical Analysis

The statistical analysis consisted of two steps.

The first step involved preliminary bivariate analyses to discover possible network patterns specific to 'Ndrangheta bosses. It compared the means of the network variables across the groups of the bosses and the other individuals and the Pearson's correlation coefficient among the variables. Given the non-normal distribution of network measures, the comparison between the mean scores of bosses and other individuals used the Mann–Whitney U test, a non-parametric test which allows comparison of the mean values of non-normal variables across two independent samples.

The second step was to conduct a logistic regression model to verify whether network measures were able to identify a criminal leader in the *Infinito* network. The dependent dichotomous variable was the hierarchy (with leader = 1 and other = 0). As already mentioned, the law enforcement agencies carefully identified the criminal leaders by various means, including previous investigations and judicial cases and wiretaps of conversations minutely describing the offices within each *locale* in Lombardy and the overall structure of the *Lombardia.* This assessment was independent from the observation of meeting attendance. No individual was classified as a leader due to attendance at a specific meaning, since the court would have dismissed any such argument for lack of evidence.

Independent variables were individual network measures and two controls (Table 2). The first was the number of meetings attended for each individual. To control for a possible prosecution bias, a dummy was included (mafia charge) in order to assess whether each node was charged with the offence of mafia-type association (data extracted from the court order). Given the exploratory nature of the analysis, the model used a stepwise selection method.

3.4 Limitations

Studies based on judicial sources are inevitably biased by the strategies of law enforcement. In fact, the special focus of Operation *Infinito* provides a valuable opportunity to analyze the formal hierarchy of the 'Ndrangheta and its relation with meeting attendance. As discussed above, various elements suggest that the classification of the offices was particularly accurate and, more importantly, independent from the observation of the meetings.

The judicial investigations may have partially covered the criminal group and this may have affected the reliability of the network measures. Even in a large operation like *Infinito*, the authorities identified only a limited number of members for a few *locali*. This problem is common to most network studies on criminal organizations, and it requires particular care in handling the data. Nevertheless, there are reasons to believe that this has not affected the results of this study. Some studies have found that criminal network measures are strong even if randomly tested for missing data [38, 80]. Furthermore, *Infinito* was a long and thorough investigation specifically aimed at identifying the most important individuals in the 'Ndrangheta in Lombardy, and the first- and second-grade judgments confirmed the charges brought against the suspects. The risk that possible missed or excluded individuals may have radically changed the structure of the network is relatively low. Indeed, Berlusconi tested the robustness of degree and betweenness centrality across different phases of the judicial process (wiretap records, arrest warrants, and judgments) and argued that "arrest warrants and judgments seem to be reliable data sources to identify key players regardless of whether a large proportion of peripheral nodes is missing" [32].

4 Results and Discussion

The bivariate analysis of the hierarchy and individual network measures highlighted a common pattern. The small set of 'Ndrangheta bosses participated in the meetings more frequently (Table 3).

On average, the bosses of the 'Ndrangheta participated in meetings approximately 4.5 times more than other individuals (11.7 vs. 2.6 respectively). Other network measures generally replicate this relation, although to a different extent. The measure with the least difference was closeness centrality (ratio of 1.3 times), while the most remarkable disparity was in betweenness centrality, where bosses had a 14.7 times higher mean.

These results reveal a specific pattern in the meeting network. Bosses obviously take part in a higher number of gatherings. This comes as no surprise, since their role requires them to be present at events important for the organization. Also network measures show more intense participation by the leaders. The measures of direct connectivity (degree, valued degree and eigenvector centrality) are strongly correlated (all coefficients above 0.85, Table 4). There is not much difference between the total number of a node's coparticipants in meetings (degree) and the number of coparticipations (valued degree) or the centrality of the coparticipants (eigenvector). In fact, the ratio between the mean scores of degree and eigenvector centrality is lower than the ratio between meetings attended. This indicates that the two measures are less able to capture the bosses' behavior in the network than the simple observation of the number of meetings. Indeed, the more meetings that one individual attends in a given group, the less likely it is that he will meet new people. Valued degree centrality (i.e. the number of a node's coparticipations) has the same ratio as the participation in meetings. Also this figure is rather intuitive, since as the number of meetings attended increases, so will the total number of encounters with any other subject (correlation coefficient of 0.94, Table 4). Bosses have less clustered neighborhoods than other nodes (0.5 vs. 0.9 respectively), suggesting that they may play a connecting function. However, this may be due to the sheer difference of the sizes in the neighborhoods of the two groups (an average of 1,081 pairs for bosses and 151 for others), since the clustering

Table 3 Mean number of meetings and centrality measures by hierarchy

	Bosses	Others	Ratio bosses/others
Meetings attended	11.7	2.6	4.5
Degree (normalised)	20.0	6.7	3.0
Valued degree	90.7	20.3	4.5
Betweenness (normalised)	3.5	0.2	14.7
Closeness (normalised)	48.1	38.3	1.3
Eigenvector (normalised)	16.2	4.8	3.4
Clustering coefficient	0.5	0.9	0.6
Number of pairs	1081.5	151.6	7.1

The differences of the means are statistically significant at $p < 0.001$ (Mann–Whitney U test)

Table 4 Pearson's correlation coefficients among the main variables

	Variable	1	2	3	4	5	6	7	8
1	Meetings attended	–	0.822	0.940	0.789	0.688	0.681	−0.742	0.862
2	Degree (normalised)		–	0.940	0.635	0.869	0.946	−0.725	0.940
3	Valued degree			–	0.699	0.789	0.854	−0.710	0.951
4	Betweenness (normalised)				–	0.522	0.480	−0.624	0.714
5	Closeness (normalised)					–	0.866	−0.658	0.754
6	Eigenvector (normalised)						–	−0.570	0.863
7	Clustering coefficient							–	−0.642
8	Number of pairs								–

All correlations are statistically significant at $p < 0.01$ level

coefficient generally decreases as the number of pairs increases (the higher the number of possible dyads among a node's contacts, the less dense its neighborhood is likely to be).

The scores of betweenness centrality are the most interesting. They show a remarkable difference between the two groups, which is not immediately attributable to participation in more meetings. Rather, the distance in the scores of betweenness may be a signal that the bosses participated in the network with the purpose of pursuing a particularly strategy in their meeting attendance.

The logistic regression model successfully predicted the hierarchical position for 92.6 % of individuals (66.7 % for bosses and 97.3 % for others). The Hosmer-Lemeshow test for goodness of fit indicated that the model fitted the data acceptably (X^2 15.343, $p < 0.53$, df 8).[15] Nagelkerke's R^2 of 0.641 revealed a good capacity of the predictors to identify the leaders.

[15] The Hosmer and Lemeshow's (H-L) goodness of fit test divides subjects into deciles based on predicted probabilities, and then computes a X^2 from observed and expected frequencies. Differently from the standard X^2 test, the null hypothesis for the H-L test is that the two distributions are equal. Non-significant results accept the null hypothesis and suggest that the model correctly fits the data, since the predicted and observed frequency are statistically similar.

Table 5 Results of the logistic regression (dependent variable: bosses/others)

Predictor	B	SE	p	Odds ratio	95 % CI	
Valued degree	0.036	0.009	0.000	1.037	1.018	1.056
Betweenness (normalised)	0.942	0.280	0.001	2.565	1.482	4.439
Constant	−4.261	0.544	0.000	0.014		

The stepwise selection retained only valued degree and betweenness as independent variables (Table 5). Betweenness centrality had an odds ratio of 2.565 (95 % confidence interval 1.482–4.439) with $p < 0.001$, indicating that a higher value of this measure increases the probability of identifying a leader (approximately +150 % per unit increase of betweenness, keeping the variable valued degree constant). Valued degree also had a positive influence on the leadership (OR 1.037, 95 % CI 1.018–1.056, equal to an increased probability of being a boss of approximately +4 % per unit of valued degree, irrespective of the value of betweenness).

The results demonstrate that the analysis of meetings within an organized crime group can successfully identify the ranks of the participants. In particular, betweenness centrality is the measure which best captures the leadership positions within the network. In this case, betweenness measures the capacity of bosses to meet individuals who do not meet each other on other occasions. This is consistent with the specific role that individuals with important offices within the 'Ndrangheta may fulfill. The mafia leaders are responsible for important decisions about the activities of their *locale,* or the organization as a whole, and for a number of formal or "ceremonial" activities, including attendance at special events. For example, 'Ndrangheta bosses may participate in high-level meetings reserved for leaders, or they may visit other *locali* as a sign of respect. The inclusion of valued degree centrality in the model is more intuitive. Bosses tend to meet other participants more frequently. Valued degree is strongly correlated with other direct connectivity variables (number of meetings attended, binary degree centrality, and the number of pairs). This is a natural consequence of the fact that, in the network, the 'Ndrangheta leaders participated in more meetings.[16]

The important role of betweenness is partially in line with the existing literature on criminal networks. Betweenness centrality reflects a brokering role within a criminal network, i.e. the capacity to bring people, resources and information together. Brokering skills are extremely important for a successful criminal career in a number of illegal markets and activities [38, 81, 82]. While brokering is frequently an element crucial for success, it does not necessarily mirror status leadership within a criminal group. In larger and more structured criminal groups, like mafias, the internal organization may have several layers. Some evidence of such a differentiation is provided by a study on two 'Ndrangheta-related networks

[16] Test logistic regressions including betweenness centrality and the other direct connectivity measures yielded results very similar to those of the model presented above.

involved in drug trafficking. In the two networks, high betweenness centrality was associated with individuals directly coordinating drug smuggling operations. These nodes were not the bosses of the two groups, but middle-status traffickers actively involved in the operational management of the drug trade. The criminal leaders (identified through content analysis of the communications intercepted) remained distant from everyday activities to reduce the risks of detection and arrest [31].

The difference between the results of the two drug trafficking networks related to the 'Ndrangheta and Operation *Infinito* may be due to the types of crimes and network. When engaging in criminal markets, mafias adapt to market rules, adopting more flexible organizational structures [39]. The formal organization usually plays a marginal role, and management of the traffic is delegated to middle-status members, who may see this as a career opportunity within the group. These characteristics of the traffic also emerge in the network of communications among the criminals [31]. *Infinito* focused on the establishment of the 'Ndrangheta in the area around Milan, and demonstrating the organization's structure was the main objective of the prosecution. The meetings reported in the court order rarely addressed drug trafficking issues, but rather internal organizational matters. Most of the other offences were extortion, corruption and loansharking, crimes associated with a stable and structured criminal organization [27, 83]. The difference between drug trafficking and operations associated with a mafia's organizational management may explain why, in the latter, betweenness centrality better identifies the bosses. The meeting network mirrors the "institutional" life of a branch of the 'Ndrangheta, where hierarchy (ranks and offices) is an important issue, object of frequent discussions and occasional disputes. Mafia bosses cannot delegate to other middle-level players some of the most important tasks relating to the organization's internal functioning, and this includes attending certain meetings. For this reason, the meetings in *Infinito* may reflect the fundamental tasks of the bosses more closely than do studies analyzing telephone communications and drug trafficking.

To return to the problem of research, i.e. that most results of network analysis of organized crime are useless for operational purposes, exploration of mafia meetings has some practical implications. The examination of meetings within a mafia may contribute to the identification of potential bosses. The prediction may be useful for selecting the individuals on which to concentrate law enforcement resources, which are frequently limited by definition. It may prompt increased surveillance, background checks, and other actions to verify the prediction. In *Infinito*, the authorities invested a significant amount of (human and financial) resources to: collect preliminary evidence sufficient for judicial authorization to wiretap the suspects; successfully intercept and maintain surveillance on telephone lines and place bugs; gather, listen to and interpret (members of the 'Ndrangheta normally speak Calabrian dialect) thousands of audio recordings. While it is true that these activities were necessary to record some mafia meetings, in some cases it may be possible to monitor known members, record who they encounter, and build a sort of snowball-like network of meetings. This may eventually provide predictions on the leadership positions within the network.

5 Conclusions

This chapter has demonstrated that the analysis of meeting attendance within a criminal network can provide interesting insights into its internal organization and hierarchy. Particularly, social network analysis measures can predict the rank of the participants and who the bosses of the group are. This approach may have useful practical implications for law enforcement activity, which may analyze meeting attendance under a new light. However, the study is only a first exploration of organized crime meetings, and it is too early to generalize the results. Additional studies should focus on meeting patterns to verify whether the results are confirmed in the case of both mafias and criminal groups of other types.

Research might also focus on how the position within a meeting network may influence the judicial outcome of a case. The literature, mostly based on communication networks, argues that high betweenness may make it possible to maintain control over the criminal activities but also to minimize direct involvement and risks. Some studies have pointed out that high betweenness centrality gives individuals better chances of avoiding detection and arrest [25, 38, 76]. By contrast, in some other cases, betweenness does not effectively protect against law enforcement action, although this may be due to high correlation with direct connectivity measures such as degree centrality [34, 79]. Exploration of the impact of network positioning in meetings might clarify this point from a different perspective. This may be important for two reasons. First, because individuals with high betweenness control criminal activities more indirectly, this condition may protect them against law enforcement action. Second, since brokers are key players in criminal organizations, it is important to verify whether they receive an adequate punishment.

To verify whether network analysis of meetings can have operational implications, it may be necessary to test the capacity of network measures to identify the ranks within a mafia during an investigation. Alternatively, it may be possible to order the meetings chronologically and verify how many weeks and meetings yield significant successful predictions. This may show whether it is possible to make reliable and successful predictions with a relatively limited sample of meetings. If possible, the network analysis of meetings may equip law enforcement agencies with an operational intelligence tool for identification of the bosses of a criminal organization.

References

1. Ianni FAJ, Reuss-Ianni E (1990) Network analysis. In: Andrews PP, Peterson MB (eds) Crim. Intell. Anal. Palmer Enterprises, Loomis, pp 67–84
2. Sparrow MK (1991) The application of network analysis to criminal intelligence: an assessment of the prospects. Soc Netw 13:251–274. doi:10.1016/0378-8733(91)90008-H
3. Albini JL (1971) The American Mafia; genesis of a legend. Appleton-Century-Crofts, New York
4. Ianni FAJ (1973) Ethnic succession in organized crime: summary report. Government Printing Office, Washington
5. Harper WR, Harris DH (1975) The application of link analysis to police intelligence. Hum Factors J Hum Factors Ergon Soc 17:157–164
6. Lupsha P (1983) Networks vs. networking: analysis of an organized crime group. Career Crim
7. Lupsha P (1980) Steps toward a strategic analysis of organized crime. Police Chief 36–38
8. Sparrow MK (1991) Network vulnerabilities and strategic intelligence in law enforcement. Int J Intell Count Intell 5:255–274
9. Davis RH (1981) Social network analysis: an aid in conspiracy investigations. FBI Law Enforc Bull 50:11–19
10. Jackson JL, Herbrink JCM, Jansen RWJ (1996) Examining criminal organizations: Possible methodologies. Transnatl Organ Crime 2:83–105
11. Jackson JL, Herbrink JCM (1996) Profiling organised crime: the current state of the art. Netherlands Institute for the Study of Criminality and Law Enforcement (NISCALE), Leiden
12. McAndrew D (1999) The structural analysis of criminal networks. Soc Psychol Crime Groups Teams Netw
13. McIllwain J (1999) Organized crime: A social network approach. Crime Law Soc Change 32:301–323–323
14. Coles N (2001) It's not what you know—it's who you know that counts. Analysing serious crime groups as social networks. Br J Criminol 41:580–594
15. Natarajan M (2000) Understanding the structure of a drug trafficking organization: a conversational analysis. In: Natarajan M, Hough M (eds) Illegal Drug Markets Research Prevention Policy. Criminal Justice Press/Willow Tree Press, Monsey, pp 273–298
16. Morselli C, Giguere C (2006) Legitimate strengths in criminal networks. Crime Law Soc Change 43:185–200
17. Varese F (2006a) The structure of a criminal network examined: the Russian Mafia in Rome. Oxf Leg Stud Res Pap 21
18. Natarajan M (2006) Understanding the structure of a large Heroin distribution network: a quantitative analysis of qualitative data. J Quant Criminol 22:171–192
19. Morselli C, Petit K (2007) Law-enforcement disruption of a drug importation network. Glob Crime 8:109–130
20. Morselli C, Giguère C, Petit K (2007) The efficiency/security trade-off in criminal networks. Soc Netw 29:143–153
21. Malm AE, Kinney JB, Pollard NR (2008) Social network and distance correlates of criminal associates involved in illicit drug production. Secur J 21:77–94
22. Morselli C (2009) Hells Angels in springtime. Trends Organ Crime 12:145–158
23. Malm AE, Bichler G, Van De Walle S (2009) Comparing the ties that bind criminal networks: is blood thicker than water? Secur J 23:52–74
24. Bouchard M, Nguyen H (2010) Is it who you know, or how many that counts? Criminal networks and cost avoidance in a sample of young offenders. Justice Q 27:130–158. doi:10.1080/07418820802593386
25. Morselli C (2010) Assessing vulnerable and strategic positions in a criminal network. J Contemp Crim Justice 26:382–392. doi:10.1177/1043986210377105

26. Bouchard M, Ouellet F (2011) Is small beautiful? The link between risks and size in illegal drug markets. Glob Crime 12:70–86
27. Campana P (2011) Eavesdropping on the Mob: the functional diversification of Mafia activities across territories. Eur J Criminol 8:213–228
28. Scaglione A (2011) Reti Mafiose: Cosa Nostra e Camorra: organizzazioni criminali a confronto. FrancoAngeli, Milano
29. Bright DA, Hughes CE, Chalmers J (2012) Illuminating dark networks: a social network analysis of an Australian drug trafficking syndicate. Crime Law Soc Change 57:151–176. doi:10.1007/s10611-011-9336-z
30. Varese F (2012) How Mafias take advantage of globalization the Russian Mafia in Italy. Br J Criminol 52:235–253
31. Calderoni F (2012) The structure of drug trafficking mafias: the 'Ndrangheta and cocaine. Crime Law Soc Change 58:321–349. doi:10.1007/s10611-012-9387-9
32. Berlusconi G (2013) Do all the pieces matter? Assessing the reliability of law enforcement data sources for the network analysis of wire taps. Glob Crime 14:61–81. doi:10.1080/17440572.2012.746940
33. Mancuso M (2013) Not all madams have a central role: analysis of a Nigerian sex trafficking network. Trends Organ Crime 1–23. doi:10.1007/s12117-013-9199-z
34. Morselli C, Masias VH, Crespo F, Laengle S (2013) Predicting sentencing outcomes with centrality measures. Secur Inform 2:4. doi:10.1186/2190-8532-2-4
35. McGloin JM, Kirk DS (2010) Social network analysis. Handb. Quant. Criminol
36. Carrington PJ (2011) Crime and social network analysis. In: Scott JP, Carrington PJ (eds) SAGE Handbook of Social Network Analysis. SAGE Publications, London, pp 236–255
37. Calderoni F (2014) Social network analysis of organized criminal groups. Encycl Criminol Crim Justice
38. Morselli C (2009) Inside criminal networks. Springer, New York
39. Paoli L (2002) The paradoxes of organized crime. Crime Law Soc Change 37:51–97
40. Paoli L (2003) Mafia brotherhoods: organized crime, Italian style. Oxford University Press, Oxford
41. Van de Bunt H, Kleemans E (1999) The social embeddedness of organized crime. Transnatl Organ Crime 5:19–36
42. Papachristos AV, Smith CM (2011) The small world of Al Capone: the embedded nature of criminal and legitimate social networks. Third Annu. Illicit Networks Work
43. Ciconte E (1992) 'Ndrangheta dall'Unità a oggi. Laterza, Roma Bari
44. Gratteri N, Nicaso A (2009) Fratelli di sangue, 2nd edn. Mondadori, Milano
45. Paoli L (1994) An underestimated criminal phenomenon: the Calabrian 'Ndrangheta. Eur J Crime Crim Law Crim Justice 2:212–238
46. D'Emilio F (2007) Mysterious mob eclipses Cosa Nostra. Wash, Post
47. Landler M, Fisher I (2007) 6 Italians in Germany killed as organized crime feud crosses border. New York
48. Spiegel Online International (2007) The "Vendetta of San Luca" in Duisburg: a deadly Mafia export from Italy. Spieg. Online Int
49. Williamson H, Bompard P (2007) Mafia feud blamed for "executions" in Germany. Finan Times
50. McKenna J (2001) Codes of dishonour. The age
51. OFAC (2012) Narcotics: what you need to know about U.S. sanctions against drug traffickers
52. U.S. Department of State (2011) International narcotics control strategy report: volume I drug and chemical control. U.S. Department of State, Washington
53. BBC (2010) Italian "Ndrangheta steps out of Mafia"s shadow. BBC News
54. De Cristofaro M, Chu H (2010) Italy arrests 300 in sweep, including alleged mafia boss. Los Angeles
55. Dinmore G (2010) Italian police arrest 300 in anti-Mafia raids. Finan Times
56. Panigiani G (2010) Italy arrests hundreds in Mob Sweep. New York A4

57. KLPD (2011) The 'Ndrangheta in the Netherlands: the nature, criminal activities and modi operandi on Dutch territory. National Crime Squad National—Police Agency of the Netherlands, Amsterdam
58. DNA (2010) Relazione annuale sulle attività svolte dal Procuratore nazionale antimafia e dalla Direzione nazionale antimafia nonché sulle dinamiche e strategie della criminalità organizzata di tipo mafioso nel periodo 1° luglio 2009–30 giugno 2010. Direzione Nazionale Antimafia, Roma
59. DNA (2011) Relazione annuale sulle attività svolte dal Procuratore nazionale antimafia e dalla Direzione nazionale antimafia nonché sulle dinamiche e strategie della criminalità organizzata di tipo mafioso nel periodo 1° luglio 2010–30 giugno 2011. Direzione Nazionale Antimafia, Roma
60. Ciconte E, Macrì V (2009) Australian 'Ndrnagheta. Rubbettino, Soveria Mannelli
61. Forgione F (2009) Mafia export: come 'Ndrangheta, Cosa Nostra e Camorra hanno colonizzato il mondo, 2nd edn. Baldini Castoldi Dalai, Milano
62. Campana P (2013) Understanding then responding to Italian organized crime operations across territories. Policing. doi:10.1093/police/pat012
63. Varese F (2006) How Mafias migrate: the case of the 'Ndrangheta in Northern Italy. Law Soc Rev 40:411–444
64. Varese F (2011) Mafias on the move: how organized crime conquers new territories. Princeton University Press, New Jersey
65. Arlacchi P (1992) Gli uomini del disonore: La mafia siciliana nella vita del grande pentito Antonino Calderone. Mondadori, Milano
66. Paoli L (2007) Mafia and organised crime in Italy: the unacknowledged successes of law enforcement. West Eur Polit 30:854
67. Savona EU (2012) Italian Mafias' asymmetries. Transnatl. Organ. Crime Mod. World. Springer, New York, pp 3–25
68. Malafarina L (1978) Il Codice Della 'Ndrangheta. Parallelo 38, Reggio Calabria
69. Transcrime (2013) Progetto PON Sicurezza 2007-2013: Gli investimenti delle mafie. Rapporto Linea 1. Ministero dell'Interno, Milano
70. Tribunale di Milano (2011) Ordinanza di applicazione di misura coercitiva con mandato di cattura—art. 292 c.p.p. (Operazione Infinito)
71. Natarajan M, Belanger M (1998) Varieties of drug trafficking organizations: a typology of cases prosecuted in New York City. J Drug Issues 28:1005–1026
72. Heber A (2009) The networks of drug offenders. Trends Organ Crime 12:1–20
73. Natarajan M, Zanella M, Yu C (2010) How organized is dug trafficking?
74. Borgatti SP, Everett MG, Freeman LC (2002) Ucinet 6 for windows: software for social network analysis. Analytic Technologies, Harvard
75. Freeman LC (1979) Centrality in social networks conceptual clarification. Soc Netw 1:215–239
76. Baker WE, Faulkner RR (1993) The social organization of conspiracy: illegal networks in the heavy electrical equipment industry. Am Sociol Rev 58:837–860. doi:10.2307/2095954
77. Bonacich P (1972) Factoring and weighting approaches to status scores and clique identification. J Math Sociol 2:113–120
78. Watts DJ (1992) Small worlds. Princeton University Press, Princeton
79. Calderoni F (2011) Strategic positioning in mafia networks. Third Annu, Illicit Networks Work
80. Xu J, Chen H (2008) The topology of dark networks. Commun ACM 51:58–65
81. Burt RS (1992) Structural Holes: the social structure of competition. Harvard University Press, Cambridge
82. Morselli C (2005) Contacts, opportunities, and criminal enterprise. University of Toronto Press, Toronto
83. Block AA (1983) East side, west side: organizing crime in New York 1930–1950. Transaction Publishers, New Brunswick

Macrosocial Network Analysis: The Case of Transnational Drug Trafficking

Rémi Boivin

Abstract The social network perspective is a fruitful way to understand criminal organizations. Most analyses are based on networks of relations between individuals or small groups of individuals. Yet contextual (or macrosocial) factors may be crucial for a complete understanding of criminal activity. This chapter aims to put criminal activity in the larger context in which it takes place. More precisely, it is argued that country-level features may complicate or facilitate legitimate business as well as criminal activities. A brief review of the literature on macrosocial network analysis and a framework for the study of transnational criminal activities are provided. Various sources of information are described and commented.

Keywords Macrosocial • Network analysis • Drug trafficking

1 Introduction

A growing number of studies recognize the embeddedness of human relations: individuals function within social groups [22]. This idea is central to Social Network Analysis (SNA) and important as well for other research methods. While most research has focused on relational networks based on individuals or small groups of individuals, contextual (or macrosocial) factors are often crucial for a complete understanding of social phenomenon. This chapter discusses criminal activity in terms of the larger context within which it takes place.

Macrosociology is the analysis of social systems on a large scale or at high level of abstraction [9]. In this chapter, it will be understood as the study of global

R. Boivin (✉)
International Centre for Comparative Criminology, University of Montreal, Montreal, Canada
e-mail: remi.boivin@umontreal.ca

A. J. Masys (ed.), *Networks and Network Analysis for Defence and Security*, Lecture Notes in Social Networks, DOI: 10.1007/978-3-319-04147-6_3,

processes, particularly country-level analyses. Countries are usually thought of as the widest unit of analysis in (quantitative) sociology, although various authors have commented on the evolution of larger regions of the world. This chapter provides a brief review of the literature on macrosocial network analysis and a framework for the study of transnational criminal activities. The main argument is that country-level features can complicate or facilitate both legitimate and criminal activities. The focus on individual networks, while providing important information, neglects the idea of social embeddedness and fails to take into account the ways in which social relations can influence individual behavior. For example, criminals operating in high-risk settings do not behave in the same way they would in less dangerous settings, no matter what their personal characteristics and networks. An empirical study of drug trafficking [5] is used as an illustration of the theory of macrosocial analysis presented here, but the framework is meant to be applicable to all sorts of "dark networks."

1.1 Macrosocial Network Analysis

Macrosocial network analysis has been used for years in the study of infectious diseases. Steven Soderberg's movie *Contagion* illustrates the process involved: an American businesswomen is infected by a disease in Hong Kong, brings it back home and infects people in Chicago and Minneapolis, who in turn infect people in other parts of the United States and London. An infection that started in China quickly extends throughout the United States and the United Kingdom. The spread of diseases such as H1N1 and HIV is known to be related to patterns of mobility and migration between countries [17, 32] and network analysis has been used to trace the origins of outbreaks of these diseases [19]. In such cases, the interest is both on who individuals have contacted (for sexually transmitted diseases, for example) and where they have been (for airborne diseases). Studies of infectious diseases that use network analysis are thus interested in both micro-networks and macro-networks. This interest involves more than the mere aggregation of individual patterns as mobility flows are also studied in order to understand disease propagation and develop probability models [4].

Network analysis has also been used by researchers interested in the evolution of social phenomenon. The use of micro-level network analysis in studies of defence and security is discussed in length in other chapters of this book and several detailed reviews of its use in criminology are available elsewhere (e.g. [10, 28, 29]). Macro-level network analysis, although used in some fields, is seldom used in criminology.

1.2 Globalization

In recent years, the research community has become increasingly interested in the phenomenon of globalization, a loose label used to refer to an increasing level of interdependence between national systems in terms of trade, military alliance and domination, and culture [35]. In other words, people are increasingly connected to each other, and borders are increasingly blurred. The idea of globalization has appeared in many fields of research. For example, co-authorship in academic research has been shown to have grown exponentially in recent decades, suggesting the rise of new "research networks" [1]. Adams argues that co-authorship is a sign of increasing collaboration between countries and also provides evidence that collaboration among authors in countries from the Asia-Pacific and the Middle-East regions is growing in parallel with cooperation between European and US authors. He shows as well that the world is globalizing, but some parts of it are developing more dense relations than others.

Proponents of the world-trade system perspective argue that the globalization process started decades, if not centuries, ago and has accelerated in the second half of the 20th century, a proposition supported by empirical evidence [27, 36]. A key aspect of globalization is the growth of international trade. If the globalization hypothesis is true, trade should not only be occurring increasingly outside country boundaries—a proposition now widely accepted—but should also be occurring with a larger number of countries. In other words, the world trade network should show a growing density of relations between countries. De Benedictis and Tajoli [14] tested this idea with import data reported by the International Monetary Fund and found that the world trade network actually became considerably denser between 1950 and 2000, from 0.067 to 0.388.

A related issue is the evolution of the networks within the global trade network. Many authors argue that countries tend to occupy similar positions over time and that radical changes are anecdotal [27], and it is true that a small number of core countries are the source of most exchanges of legal commodities. The global network is structured in a way that favors stability, even if countries tend to limit their imports and to develop foreign markets in order to improve their position in the system [34]. Empirical evidence supports the idea of global stability but also highlights the fact that the overall number of trade partners has increased because of a drastic change in the distribution of trade partners per country: in 2000, there was still a small group of countries importing from a very large number of partners (almost the same distribution as in 1950) but there were an increasing number of countries importing from an "average" number of countries [14]. World trade has changed primarily because these intermediate countries are more numerous and are reducing the distance between themselves and the "top" countries. The global network evolves slowly, and some countries in it are evolving more quickly than others.

2 A Framework for the Study of Transnational Crime

One aspect of the global trade is business in illegal commodities, such as recreational drugs, arms, wildlife, etc. Such commodities are traded through "dark networks," groups that do their best to conceal themselves and their activities from the authorities [15]. On this definition, human trafficking, insofar as it involves individuals who are exploited over a period of time—i.e., treated as a commodity—is also such a business [3]. Transnational trafficking is of great national concern for reasons ranging from violence to border protection but macro-analyses of transnational criminal activities are often basic or anecdotal. Network analysis tools have the potential to help develop general and systematic models of criminal activities in their wider context.

The following sections report on our work in constructing and analyzing country-level trade networks for cocaine, heroin, and marijuana over a ten-year period [5]. This work is then used to illustrate the proposed framework for macro-social network analysis of transnational illegal activities. Drug trafficking is a convenient illustration for several reasons. First, cocaine and heroin have been smuggled through borders for over a century as the cocaine and heroin used in most parts of the world still comes from only a small number of countries. Cocaine, derived from a plant (coca) grown exclusively in South America [24], is used largely in North America and Europe [33]. Heroin is derived from large crops of opium poppies grown in South and Southeast Asia, and increasingly in Central and South America, but is used across the world. Second, these drugs have been a public concern since the time their addictive properties were discovered and their use was prohibited. Numerous organizations are dedicated to monitoring the "drug problem", which means that a considerable amount of data is available. Many aspects of the issue, including production, trafficking, and consumption, are fairly well-documented and information is available on prices, consumption rates, trafficking routes, drug seizures, enforcement efforts, etc. Third, several authors have tried to construct global drug trade networks [2, 6, 11, 12, 16, 20, 31].[1] The existence of structured drug flows between countries is widely accepted and the concept is currently used in the influential World Drug Report published by the United Nations Office on Drugs and Crime (UNODC).

Four general steps in the methodology are described: definition of the unit of analysis, definition of relations between nodes, data collection, and analysis. Macro-analysis is distinguished from the analysis of individual networks at each of these steps.

[1] Most of these works did not use network analysis techniques although they employ network-related vocabulary.

2.1 Definition of the Unit of Analysis

The most distinguishing feature of macro-social network analysis is that, in contrast to micro-social analysis, it usually begins with a complete list of nodes. Most individual-level network analyses use a "snowball" process: data collection starts with a single individual or a list of persons of interest (level 0; for example, a list of gang members) and determines the relations between them. Contacts that these individuals have with others who are not already in the network are then identified (level 1), linked with the original nodes, and relations between them are identified. Level 2 identifies individuals who are connected to Level 1 nodes but not to Level 0 nodes, and so on, in an almost perpetually expanding network. Networks built this way grow exponentially and often include irrelevant nodes, creating the "boundary specification problem" discussed at length in the literature. In theory, the maximal number of nodes of any individual network is equivalent to the number of living humans on Earth (approximately seven billion as of 2013), but, despite claims that there are only "six degrees of separation," the number of levels of connections necessary to map a global network including all individuals remains unknown. The important point here is that the number of nodes (and consequently of ties) in an individual network is often unknown at the beginning of the data collection and researchers often have to define the boundaries of the network.

The situation is reversed for macro-social network analysis as the precise number of nodes in a macro-network is usually known at the beginning of the analysis. In this case it is necessary to clarify the criterion of inclusion (or exclusion). For example, one could start with the members of the United Nations, 193 sovereign states. Nodes could then be excluded on the basis of their location (e.g., isolated Pacific Islands) or other characteristics (e.g., nonexistent drug use). In the case of drug trafficking, it is possible to picture the complete set of nodes simply by looking at a world map. The most important part of network construction is then to determine the actual ties between existing nodes.

2.2 Definition of Relations Between Nodes

Once the nodes are identified, it is necessary to determine the actual relations between them—in other words, to connect the dots. Dichotomization is often used to build individual networks as it is generally easy to determine if relations are present or absent. Dichotomization is commonly used in network analysis and is even necessary for some sorts of calculations. Still, one needs to define what kind of relation is of interest. Imagine a famous crime figure named John; researchers studying his genealogy may well be interested in knowing who his relatives are but criminologists are much more likely to be interested in who his co-offenders are.

Defining the kind of relations to be studied is an important part of network construction.

Keep in mind that country-level networks are aggregates at a very large scale. A potential problem of macro-networks is therefore their great density as at some point all countries are connected to each other. Trade networks in which every node trades with every other node require limited explanations. Incomplete networks, in which all nodes are not equivalent, often require more explanation. In this case, definition of thresholds is common; for example, Kim and Shin [26] consider that there is a trade relation between two countries if the total amount of trade between them was (1) over $1 million or (2) over $10 million. Fixing a threshold has the advantage of simplicity because the value involved remains dichotomous (presence or absence of a relation). For example, it could be decided that drug trafficking occurs between two countries if at least one seizure of drugs coming from one country and entering the other is reported. There is anecdotal evidence of drug seizures occurring between most countries, but it would be risky to conclude that one seizure is indicative of a trend. A more conservative threshold would be that trafficking exists between two countries if there have been at least two (or more) seizures of significant quantity.

Dichotomization, while useful, provides only a rough picture of any social phenomenon: it indicates that there is a relation between two nodes, but it considers every relation to be equivalent. Drug trafficking has been observed between Canada and Australia, but trafficking between Mexico and the United States is considerably more important; despite this, both relations would be coded in the same way. Giving a value to relations is an alternate way to define relations between nodes. In the preceding example, a more precise measure would give a higher value for Mexico-US than for Canada-Australia. Most network software allows the analysis of valued relations, but it is not common for at least two reasons. First, many tools have not been designed to work with valued relations. For example, the most familiar statistical measure in network analysis, degree centrality, cannot be calculated without dichotomization. Analysis of valued networks creates a methodological challenge that has not yet been seriously addressed in the social sciences. Second, it is sometimes difficult to give a value to relations, especially in dark networks. Networks are built on the basis of available information and for criminal activities this data often reflects, at least to some extent, law enforcement efforts. Drug seizures are one preferred indicator of trafficking because information related to them is systematically collected and widely available. However, as the widespread use of drugs indicates, only a relatively small percentage of drugs is seized. While it is not possible to precisely measure interception rates—because the actual quantity being shipped is unknown—it is often assumed that interception rates are considerably higher for cocaine and heroin than for marijuana and synthetic drugs, that border crossing is the most risky stage of drug trafficking, and that the extent of law enforcement efforts is closely related to the number of seizures [37]. In addition, the size of the global drug market has been the subject of debates for decades; precise measures of drug flows between countries are, predictably enough, unavailable. The proportion of drugs seized appears to vary significantly over time, across space, and among types of drugs. No one has yet found any satisfactory measures of undetected criminal activities, meaning that valued networks of drug trafficking (or any other criminal activities) are at best very rough estimates.

2.3 Data Collection

Network analysis requires relational rather than attributive data and studies of global phenomenon often rely on official data compiled by various international organizations. Trade networks, for instance, are usually based on import and export data from the International Monetary Fund or the United Nations Comtrade. These data are relational because they indicate how much trade occurred between countries A and B. A related challenge is to find fairly complete datasets so that relations between nodes are established on a comparable basis. Missing data can be an important issue in network analysis.

There are three categories of relational data—inputs, outputs and "undefined"—all of which have pros and cons in terms of their use in networks analysis. Inputs are reported by receiving countries and are usually considered to be the most reliable because authorities are primarily concerned with what (or who) comes into their territory. Imports are an example of inputs: in network vocabulary, input corresponds to indegree, i.e., the number of ties directed to a node. An obvious difficulty occurs when some countries or nodes do not report their inputs, as is the case for trade. Outputs—outdegree—can be used as a corrective and used to validate or to complement input data. However output nodes report where commodities are sent with limited knowledge of the actual final destination. As well, output data is less systematically compiled and thus less available for analysis. Finally, it is possible to build networks with relational data from unknown sources. International organizations typically report relations without mentioning where the information comes from. Such statements are ultimately based on inputs or outputs, or both. For example, marijuana trafficking was observed between Mexico and the United States. This statement could be based on reports by the United States (input), by Mexico (output), or by an international organization (unknown).

Data collection is of particular concern when it involves dark networks. Few organizations compile large datasets on illegal trafficking and the datasets that exist are often not designed in a way that is conducive to the study of relations. For example, the UNODC, Interpol, the European Monitoring Center for Drugs and Drug Addiction, and the American Bureau of International Narcotics and Law Enforcement Affairs, among others, publish various indicators of trafficking at the country-level, but most of the data concerns attributes: the number of seizures and quantities seized, wholesale and retail prices, etc., while network analysts are more interested in trafficking routes and price markups.

Researchers have therefore had to rely on imperfect data or strong assumptions to build global networks of exchange. For example, Paoli et al. [31] developed a global model to "keep track of opium flows." Their model is based on attributive seizure data and can be summarized as the result of a series of simple subtractions and additions. The general idea is that drugs that are produced must at some point either be consumed or seized. Paoli et al. propose a "distribution model": because the potential production of cocaine and heroin is fairly well known (see the UNODC's World Drug Report), one simply has to follow drugs down commodity chains to establish the amount available in various countries. Imagine that a total of 500 tons is produced in Latin America. Consumption surveys and law enforcement reports indicate that 50 tons are seized or consumed in Latin America, meaning that 450 tons are sent abroad. One hundred tons are sent towards West Africa, where 25 tons are seized or consumed; the remaining 75 tons are therefore sent towards Europe, where they are seized or consumed. Similar calculations can be done for other routes. This model offers a crude estimation of drug flows between countries or regions. It is based on the strong assumption that national surveys on drug use are comparable and appropriate measures of the amount of drugs consumed (see [25]). Also, the model offers little insight on the structure of drug trafficking because trafficking routes are assumed rather than actually determined.

Chandra et al. [12] offer an alternative model in an attempt to empirically define trafficking routes. Their title highlights the fact that their model is also based on strong assumptions: *Inferring Cocaine Flows in Europe* (see Chandra and Barkell [11] for a similar approach to heroin flows). Their first assumption is that "if the price of cocaine for two countries is highly correlated, we infer that the cocaine markets in those two countries are integrated" (p. 6). In other words, they assume that correlated prices mean that there is a relation between two countries. The second assumption is that cocaine flows from countries where prices are lower to countries where prices are higher—the cocaine trade is directed towards countries in which wholesale prices are higher. While both assumptions are reasonable, the network is restricted in that it cannot be expanded to include countries for which wholesale prices are not available. The use of a single measure to account for an entire country is also debatable.

In the absence of complete datasets on drug flows, researchers have found ways to use attributive data to determine trafficking routes [11, 12] or to quantify the flow of drugs between countries [31]. Their models are supported by anecdotal evidence. In my recent attempt to model three global drug markets [6], two different modes of collection of relational data were used. The first—seizures of "significant" quantities of drugs that occurred between 1998 and 2007 as reported by the UNODC[2]—is a mix of inputs and outputs that is rarely analyzed by scholars

[2] The data is compiled through the "Significant drug seizure report" and available on the UNODC website. The UNODC defines thresholds of significant quantities used as 1 kg or more for marijuana and 100 g or more for cocaine and heroin.

because it suffers from a considerable missing data problem. This dataset provides detailed information on a large number of cases, including origin and/or destination countries (n = 20,527 dyads). When Spanish authorities seize drugs coming from Venezuela and headed to France, they collect information on a network of 3 nodes (Spain, Venezuela, and France) and 2 relations (Venezuela-Spain, Spain-France). The accumulation of such information allows the construction of a network that covers the world. However, seizures are reported to the UNODC on a voluntary basis. As a consequence, key players in the drug trade are not included in the UNODC dataset.

The second dataset is based on reports published by various international organizations involved in the monitoring of drug trafficking.[3] In some cases, organizations reported their source (e.g., "there was a large seizure between X and Y"), but most sources were undefined. The combination of those two datasets made it possible to build separate trade networks covering most countries of the world (n = 173). Despite various attempts, I was not able to provide satisfactory estimates of drug flows but it was possible to construct dichotomous directed networks. Dyads were collected as a list of relations, a format that is now supported by various software, such as UCINET and Pajek.

2.4 Analysis

Network analysis is the study of structures built on relations. In that sense, it adds to traditional attributive analysis in two ways. First, it allows comparison of the structural features of whole networks. For example, the cocaine market can be compared to the heroin trade network on various global measures. Dark networks can be compared to equivalent legal networks. Second, network analysis adds a relational dimension to the study of nodes, allowing patterns of relations to be compared and grouped.

Macro-networks are small-scale in the sense that the amount of information on which they are based is not massive: they include a relatively small number of nodes—in this case, 173—and a small number of relations—here, a maximum of 29,756 ties (173 * (173−1)) for the complete network. Available and user-friendly network software packages can usually handle such networks (see Huisman and van Duijn [23] for a review). Visualization is also relatively easy, especially if the focus is on a specific part of the network (Fig. 1).

The first thing that is apparent about illegal drug markets is that they have a very loose structure as compared to legal commodity trades. The global trade network has been found to be quite dense (38.8 %; [14, 27]) and networks of

[3] Based on a systematic review of information contained in 48 annual reports and country overviews published by the UNODC, the Bureau of International Narcotics and Law Enforcement Affairs (BINLEA), the International Narcotics Control Board (INCB), and the European Monitoring Center for Drugs and Drug Addiction (EMCDDA).

Fig. 1 Visualization of the marijuana trade, 1998–2007 (*source* Boivin [5])

addictive plant-based commodities—coffee and chocolate—which are often used in comparison with drug markets, were also found to have a high density (21.8 and 14.7 %, respectively; [7]). Trade networks for cocaine (3.2 %), heroin (2.1 %), and marijuana (0.5 %) are much less dense. These results are not completely surprising but a potential implication is that dark networks should probably be analyzed in terms of effectiveness rather than number. Drug trafficking is a risky business: traffickers can be arrested, loads can be seized, violence can be used to settle disagreements, etc. Only one efficient route is necessary to secure a drug supply, even if a variety of routes might make drug markets more resilient [8]. Network density is the expression of the efficiency/security trade-off in criminal networks [30].

Another interesting feature of illicit drug networks is that the direction of trade appears to be inverted: core countries of the legal world-economy are less dominant and peripheral countries are more involved [7]. Countries that are peripheral in most economic activities are the source of most exchanges of illegal drugs, and a significant share of exchanges is directed towards core countries. In sharp contrast with legal trades, core countries are dependent on others for their supply of illegal drugs—or at least for their supply of cocaine and heroin. For various reasons, starting with more active local drug law enforcement, core countries are not able to meet the national demand for drugs. Moreover, a country's position and role in global drug markets is also closely associated with the value of drugs. Consistent with the "risks and prices" model, prices increase more sharply when drugs are headed to countries where law enforcement imposes higher costs on traffickers [6]. Price markups are lower if the destination country is, for example, a transit point in the trade to large potential markets.

Network analysis is also useful in finding equivalences between nodes. Block modeling involves a set of statistical tools that can be used to detect groups of nodes on the basis of their relational similarities. It is a form of aggregation into larger units—from micro to macro. Relations between clusters of nodes are then analyzed to detect general patterns. For example, block modeling has been used in tests of the world-system perspective to examine the existence of a core and a periphery composed of countries with similar patterns of trade but which are not necessarily close in terms of geographic distance [27]. Similarly, block modeling highlights the existence of different roles at the country-level in drug trafficking networks. Transit countries can be separated into three categories: regional, gatekeepers, and exit points [21]. These roles are based on the location of a country within a continent: drugs usually pass through regional transits by land and are more prevalent in the heroin trade network (e.g., the Balkan route) than in the cocaine trade network. Gatekeepers typically refers to countries that import from source countries to export to neighboring countries. Spain is the classic example of a gatekeeper transit point for cocaine between South America and Europe. Exit points are the exact opposite: drugs imported from neighboring countries are destined for foreign regions. Venezuela, for instance, was been an exit point to foreign markets for cocaine in the early 2000s [18].

3 Conclusion

Macro-social features are largely neglected in most network analyses, with more emphasis put on individual characteristics that may have a direct impact on the structure of social relations. However, macro-network analysis is a relevant tool for understanding global criminal markets. Transnational crime is supported by a set of relations between individuals and groups that operate in larger networks of countries.

The framework presented in this chapter highlights a number of interesting features. Macro-networks are often based on a complete list of nodes, i.e., the number of units in the final network is known by the beginning of the analysis. This factor eliminates the boundary specification problem, reduces concerns about missing data, and generally keeps the network construction "under control." The resulting networks can usually be analyzed with common and easy to handle software packages. On the substantive side, macro-network analysis adds a relational layer to traditional studies of social phenomenon. Global structures can be compared, resulting in more complete explanations. Nodes can be examined on the basis of their relations, providing additional evidence that geographical proximity is not the only source of connection.

The most challenging part of macro-network analysis is finding suitable data. Country-level data are widely available but rarely provide information on relations between units. The UNODC gathers a great deal of data from various sources, including some relational information on trafficking routes, but it is far from exhaustive.

An interesting issue that hasn't been addressed in the current chapter is the evolution of macro-level dark networks over time. Time is an important aspect of many studies of globalization, and particularly of the global trade [27]. Resilience and stability are often implicitly assumed for illegal markets; for example, because historical factors contributed greatly to the establishment of opium poppy cultivation in Afghanistan [13], it is expected that this cultivation will be maintained unless major changes occur. Pessimistic—or perhaps realistic—observers also acknowledge that the so-called "War on Drugs" was doomed from the start because the demand for recreational drugs will continue despite efforts to control supplies. The containment of one supply route simply forces traffickers to use another existing route more frequently or to create a new one. However, traditional plant-based drugs (cocaine, heroin, marijuana) are increasingly less used or transported in favor of relatively new synthetic drugs (methamphetamine, ecstasy) which are more easily produced domestically in many countries. The rise of domestic production contributes to changing the face of global drug markets: drug trafficking is no longer necessarily transnational. Network analysis can not only help provide important insights on developing drug markets but can also be used to study declining ones.

Acknowledgements The author would like to thank Pierre Tremblay and Carlo Morselli for their comments on previous versions of this paper.

References

1. Adams J (2012) The rise of research networks. Nature 490:335–336
2. Anthony R, Fries A (2004) Empirical modelling of narcotics trafficking from farm gate to street. B Narcotics LVI(1–2):1–55
3. Aronowitz AA (2001) Smuggling and trafficking in human beings: the phenomenon, the markets that drive it and the organizations that promote it. Eur J Crim Policy Res 9(1):163–195
4. Balcan D, Gonçalves B, Hu H et al (2010) Modeling the spatial spread of infectious diseases: the global epidemic and mobility computational model. J Comput Sci 1:132–145
5. Boivin R (2011) The flip side of legitimate markets: an empirical analysis of transnational drug trafficking. Paper presented at the 11th annual conference of the European society of criminology, Vilnius, Sept 2011
6. Boivin R (2013) Risks, prices, and positions: a social network analysis of illegal drug trafficking in the world-economy. Int J Drug Policy
7. Boivin R (2014) Drug trafficking networks in the world-economy. In: Morselli C (ed) Crime and networks. Routledge, London
8. Bouchard M (2007) On the resilience of illegal drug markets. Glob Crime 8(4):325–344
9. Calhoun CJ (2002) Dictionary of the social sciences. Oxford University Press, Oxford
10. Carrington PJ (2011) Crime and social network analysis. In: Scott J, Carrington PJ (eds) The SAGE handbook of Social network analysis. Sage Publications, London, pp 236–255
11. Chandra S, Barkell M (2013) What the price data tell us about heroin flows across Europe. Int J Comp Appl Crim Just 37(1):1–13
12. Chandra S, Barkell M, Steffen K (2011) Inferring cocaine flows across Europe: evidence from price data. J Drug Policy Anal 4(1):1–18
13. Chouvy PA (2010) Opium: uncovering the politics of the poppy. Harvard University Press, Cambridge

14. de Benedictis L, Tajoli L (2009) The world trade network. Working paper
15. Everton S (2012) Disrupting dark networks. Cambridge University Press, Cambridge
16. Farrell G, Mansur K, Tullis M (1996) Cocaine and heroin in Europe 1983–1993: a cross-national comparison of trafficking and prices. Brit J Criminol 36(2):255–281
17. Field V, Gautret P, Schlagenhauf P et al (2010) Travel and migration associated infectious diseases morbidity in Europe, 2008. BMC Infect Dis 10:330
18. Figueira D (2006) Cocaine and heroin trafficking in the Caribbean, vol 2. The case of Haiti, the Dominican Republic and Venezuela. iUniverse, Lincoln
19. Goh K, Cusick ME, Valle D et al (2007) The human disease network. Proc Natl Acad Sci U.S.A 104(21):8685–8690
20. Gootenberg P (2006) Cocaine in chains: the rise and demise of a global commodity, 1860–1950. In: Topik S, Marichal C, Frank Z (eds) From silver to cocaine: Latin American commodity chains and the building of the world economy, 1500–2000. Duke University Press, Durham and London, pp 321–351
21. Gould RV, Fernandez RM (1989) Structures of mediation: a formal approach to brokerage in transaction networks. Sociol Methodol 19:89–126
22. Granovetter M (1985) Economic action and social structure: the problem of embeddedness. Am J Sociol 91(3):481–510
23. Huisman M, van Duijn MAJ (2011) A reader's guide to SNA software. In: Scott J, Carrington PJ (eds) The SAGE handbook of Social network analysis. Sage Publications, London, pp 578–600
24. Karch SB (2006) A brief history of cocaine, 2nd edn. Taylor and Francis, Boca Raton
25. Kilmer B, Pacula RL (2009) Estimating the size of the global drug market: a demand-side approach: report 2. Rand Corporation, Santa Monica
26. Kim S, Shin EH (2002) A longitudinal analysis of globalization and regionalization in international trade: a social network approach. Soc Forces 81(2):445–471
27. Mahutga MC (2006) The persistence of structural inequality? A network analysis of international trade, 1965–2000. Soc Forces 84(4):1863–1889
28. McGloin JM, Kirk DS (2010) Social network analysis. In: Piquero AR, Weisburd D (eds) Handbook of quantitative criminology. Springer, New York, pp 209–224
29. Morselli C (2009) Inside criminal networks. Springer, New York
30. Morselli C, Giguere C, Petit K (2007) The efficiency/security trade-off in criminal networks. Soc Netw 29(1):143–153
31. Paoli L, Greenfield VA, Reuter P (2009) The world heroin market: can supply be cut?. Oxford University Press, New York
32. Quinn T (1994) Population migration and the spread of types 1 and 2 human immunodeficiency viruses. Proc Natl Acad Sci 91:2407–2414
33. United Nations Office on Drugs and Crime (2013) World drug report. http://www.unodc.org/wdr/. Accessed 15 Jul 2013
34. Van Rossem R (1996) The world-system paradigm as general theory of development: a cross-national test. Am Sociol Rev 61(3):508–527
35. Wallerstein I (1974) The modern world-system I: capitalist agriculture and the origins of the European world-economy in the sixteenth century. Academic Press, New York
36. Wallerstein I (1979) The capitalist world-economy. Cambridge University Press, Cambridge
37. Wilson L, Stevens A (2008) Understanding drug markets and how to influence them. Beckley Foundation Drug Policy Programme, Oxford

Policing the Hackers by Hacking Them: Studying Online Deviants in IRC Chat Rooms

David Décary-Hétu, Benoit Dupont and Francis Fortin

Abstract The Internet is now an irreplaceable source of news and communication. While traditional data-gathering techniques such as surveys and interviews have proven useful in this environment, we believe that the monitoring of online communities will provide new and innovative datasets that will greatly enhance our comprehension of the criminal phenomenon in a virtual setting. This chapter describes how social researchers can tap into the Internet Chat Relay (IRC) to collect information on cyber deviants and build activity logs, social graphs and do content analysis. The research also provides an analysis of a dataset that was generated through this methodology and concludes on the limits and ethical problems that such a technique poses.

Keywords Hackers · Data-gathering · IRC · Social networks

1 Introduction

Criminologists have faced very unique problems in their search of research subjects in the past. Criminals are by nature distrustful of others and unwilling to share their experiences. Fortuitous meetings such as Sutherland's professional thief [1], Hobb's street crooks [2] or Venkatesh's [3] gang leader have produced impressive results in the field of criminology but such encounters are unfortunately far and few in between.

Criminologists have adopted many strategies to increase the odds of such meetings happening. They have gone in prisons [4] to interview convicted

D. Décary-Hétu (✉)
University of Lausanne, Lausanne, Switzerland
e-mail: david.decary-hetu@unil.ch

B. Dupont · F. Fortin
University of Montreal, Montreal, Canada

A. J. Masys (ed.), *Networks and Network Analysis for Defence and Security*,
Lecture Notes in Social Networks, DOI: 10.1007/978-3-319-04147-6_4,

criminals and measure criminal performance. They also used participant observation to explain the role of women in outlaw motorcycle gangs [5]. The strategy behind these research projects is simple: go and find criminals wherever they are.

As more and more criminals have moved to the Internet, criminologists have needed to adapt their methodology as this migration has provided them with new opportunities. Researchers can now interact with delinquents from the safety of their home or their office which allows them to reach out to criminals that may have been difficult to reach in the past. Furthermore, delinquents tend to leave more traces on the Internet where their messages, IP addresses and behavior is recorded on a multitude of servers across the globe.

Past research [6] has shown that the Internet Relay Chat (IRC), a worldwide instant messaging system, has been and is still one of the favorite meeting grounds for cybercriminals. Unfortunately, criminologists have yet to fully take advantage of this network of chat rooms where millions of individuals interact every day. This chapter's aim is to develop a framework that allows criminologists to easily tap into this pristine source of data on cybercriminals.

This framework, while useful in the context of academic research, will also be very valuable to law-enforcement agencies. Already, police organizations are using more and more the Internet as a policing tool. News articles are reporting that social networks like Facebook are used by police organizations to "post mug shots and let the public know what their department is up to" [7]. Interpol successfully used social networks in 2010 to arrest over 100 criminals who had been on the run for some time [8]. Internet services like social networks have allowed police officers to reach large segments of the population and to broadcast informations widely and efficiently. Such services have also provided easy to use tools which even computer neophytes can master rapidly. This framework will enable law-enforcement agencies to go beyond these social platforms and to track delinquents in underground chat rooms where they hid in security through obscurity. If law-enforcement agencies are to gather the best intelligence, they will therefore need to monitor IRC networks just as some of them have been doing with social networks like Facebook and Twitter.

This chapter begins with a review of how criminologists have addressed the question of cybercrime so far. We present the inherent limits to the current tools they use and demonstrate that gathering data on hackers with IRC has many advantages including access to unbiased sources of information and increased efficiency. This chapter then showcases the versatility of IRC logs in criminological research and details both the solutions and the new challenges that such a methodology brings forward. Following is a discussion of the ethical and legal issues raised by IRC research. We conclude this chapter with a real-life example of a how IRC data can be used to study hackers.

2 Gathering Data on Hackers: The Old Fashioned Way

There is no denying that a growing body of literature is focusing on the subject of hackers and hacking. A simple search on Google Scholar shows that nearly 113,000 papers and articles mention the word *hacking*. To date, these papers have used four main sources of information: interviews, surveys, ethnographies and case studies. Each of them were adapted to the context of the Internet but not all of them succeeded at the same level.

Interviews are one of the oldest tools used in the field of social research. Many papers have been written on the advantages of using interviews and on how such a tool should be used by social science researchers [9–11]. The technique has been used in studies on hackers with varying success. Some of the best uses of this data-gathering technique [12, 13] showed the quality of the in-depth data that it can provide. Jordan and Taylor [13], for example, highlighted the community aspects of the hacking world and the existence of a hacking culture. The authors, through their interaction with hackers, came to understand the hacker ethos and concluded that hackers "have become the nightmare of information societies despite very few documented cases of upheaval caused by hackers" [13].

Surveys have also been used in studying hackers, although with less success. Researchers [14, 15] have had a hard time finding "black hat" (malicious) hackers willing to answer their questionnaires. Hackers' fear of being identified and the difficulty in distinguishing security experts from criminals seem to be the other problems encountered by researchers. Both security experts and criminals often behave in the same way but in the former case, individuals have the authorization to test their target's security. The results of these studies (and of others based on surveys) must thus be taken with a grain of salt since researchers have no way to validate the identity of the person answering their online questionnaire. Researchers are thus faced with a catch twenty-two: in order to provide respondents with the anonymity they insist on, they must trust that they are indeed black hat hackers. If they use tracking tools or questions that are too inquisitive, hackers will walk away from their survey. When surveys are administered face-to-face at hacking conventions (such as Defcon, Blackhat, Hope, or Schmoocon), most respondents are security professionals who are not representative of black hat hackers. Although a valuable tool in more traditional social research, surveys have yet to show their usefulness in the study of cybercriminals.

As the more successful studies show, it is possible to adapt traditional data-gathering techniques to the study of hackers. Many researchers have used ethnographies [16–19] as a way to become a part of and to understand the hacking culture and its members. Ethnographies are comprehensive works that detail the behavior of participants, interactions between them, and the beliefs and moral justifications held by the community. As shown in these studies, hacking communities can be surprisingly open and welcoming when it comes to researchers and new members who are interested in joining hacking groups. Rehn [16] in particular has demonstrated how the community that illegally distributes intellectual property (known as the "warez scene") has evolved into a gift community where hacking groups participate in a continual tournament to gain recognition from their peers.

Finally, a review of the current state of research on hackers would not be complete without mentioning case studies. Dreyfus [20], Kleen [21], and [22] studied different types of hackers, ranging from industrial spies to nation-state hackers. These case studies draw from different sources of data, such as media reports, biographies or legal documents and generally provide detailed and exhaustive details on the hacker or the hacking groups under scrutiny. Although rarely used at the moment, police and court records should be a growing source of information for researchers, given that the number of arrests and trials is bound to explode in the next few years. Recent studies [23, 24] clearly demonstrate the usefulness of such data for case studies dealing with a variety of criminal offences such as warez trading, botnets, and carding.

3 Learning from the Past

Our review of the data-gathering techniques used in the cybercrime field describes how researchers have found ways to adapt their methodology to the problem of hackers. Because of the nature of the Internet, however, there are limits to what can be achieved through interviews, surveys, ethnographies and case studies.

The first problem deals with research subjects. Finding cybercriminals willing to participate in academic research has been difficult at best so far. Hackers are often suspicious by nature and unwilling to take the risk of talking to an outsider who could be trying to find their real identity. Researchers who target this type of subjects will inevitably end up with a very low and unrepresentative number of participants in their studies. Furthermore, there is no reliable method to differentiate nefarious hackers from "script kiddies" (unskilled beginners) and "white hat" hackers (benevolent actors) given the limited amount of information available on participants and the limited amount of interaction between researchers and hackers. These categories are often confused in the current literature [14] but need to be segregated to really understand the specific motivations, skills and customs of each group.

The second problem is one of self-selection. Using the same tools over and over leads researchers to make use of the data sources that work best for these tools, thus running the risk of constantly repeating themselves. This can be seen in the studies on the warez scene for example, where tens of studies have looked at the motivations of students who pirate software and music [25, 26]. Researchers are also missing out on important data sources such as online forums and websites where massive amounts of documents can be harvested. In one paper [27], researchers did access these resources but manually copied hundreds of web pages in Word documents. Such a tedious method only gives access to a fraction of the data available and highlights the need for new and innovative tools.

Third and last comes the question of resources, time and energy. Ethnographies, surveys or case studies all need extensive preparation in order to succeed: contacts have to be established in the targeted community, law-enforcement agencies or defense attorneys must be convinced to share investigation reports and questionnaires need to be tested and validated beforehand. All these imply time and resources which are diverted from the analysis of data towards the gathering of raw material. While Rehn [16] and others' ethnographers have greatly improved our knowledge of delinquent behavior online, not all researchers have the personal characteristics needed to immerse themselves in a criminal community. Combined with the high level of uncertainty associated with the access to research subjects, ethnographies display a high risk factor which limits their use in the context of social science research.

Social research isn't the only field that has invested time and energy in the study of hackers. So-called security experts have also looked at hackers from a more practical and technical point of view, using their programming and computer skills to understand the behavior of both malware applications and their creators. Antivirus software companies like Panda Security[1] and Symantec[2] now have labs dedicated to the study of hackers. These outlets regularly produce reports on the state of the hacking world such as the *Data Breach Investigations Report* that is produced annually by Verizon [28] in collaboration with the United States Secret Service. Based on the past year's investigations, this report aims to present the state of computer insecurity and hacking particularly in the business world. It would be easy for academics to discard such papers as mere tools that take advantage of scared customers and businesses. We believe however that what these studies lack in theoretical frameworks and scientific rigor is compensated for by their computer skills and knowledge.

Criminologists who have kept up-to-date on such research are well aware of one particular trend in computer research that seems particularly promising: Internet Relay Chat monitoring. Invented in the late 1980s, IRC or *Internet Relay Chat* is a powerful communication tool that allows group communication, private messaging, and file exchanges [29]. It is divided into networks that in turn host thousands of chat rooms. The IRC protocol allows for both public and private messages. All communications are synchronous, meaning that users have to be online to read and receive messages—there is no central repository of messages that can be searched and archived. Each chat room can designate administrators who manage both the users and the discussions. These administrators, also known as operators, have the power of "life and death" over IRC users and can censure and even ban users for disorderly behavior. IRC users can be identified using their

[1] Some examples of white papers from Panda Labs can be found at http://press.pandasecurity.com/press-room/reports/.

[2] Some examples of white papers from Symantec can be found at: http://www.symantec.com/business/security_response/whitepapers.jsp.

IP address, their nickname (or handle) as well as the username they use to register their IRC client software.

The IRC protocol allows also for the creation of programs commonly called bots that simulate the behaviour of real users [6]. These bots are controlled via messages and serve four primary purposes: (1) make conversation, (2) provide files on demand, (3) provide information, and (4) monitor public messages. Bots are usually built to deal with redundant and annoying tasks, such as monitoring conversations for curse words and banning users who are not respectful of others.

Monitoring discussions on IRC chat rooms (or channels) has been described and used in many technical papers [6, 30, 31]. Mutton [32] even developed an automated tool called PieSpy which is an IRC bot that allows researchers to monitor multiple chat rooms simultaneously and to periodically create a social graph of all active participants. While PieSpy is unable to track private messages, it can detect (with a varying degree of certainty) the conversation patterns of IRC users. To do so, PieSpy uses three heuristics: (1) direct addressing which happens when a user mentions another user's online handle (IRC nickname) in a message; (2) temporal proximity or the time lapse between messages and; (3) temporal density which is a series of messages posted by two users in a very short period of time. This relational data is used by the software bot to build images of networks.

The automation that PieSpy provides is definitively very helpful, especially to less computer-savvy researchers. It unfortunately also comes with a hefty price: the unavailability of the raw data and the poor tracking of targets. The tool does not provide the raw metrics on which the social graphs are built. Since the only outputs of the bot are graph images, it is difficult to compare or even to analyze these graphs which limits the usefulness of such a tool in a research project. Furthermore, PieSpy only identifies IRC users through their IRC nickname, their IP address and the username associated with their IRC software. All of these identifiers tend to change regularly and PieSpy incorrectly assumes that an individual who has the same nickname and username is a different person if he logs from a new IP address.

This chapter will draw on Mutton's [32] methodology of IRC monitoring and social network analysis with the aim of improving his research framework. The methodology presented in this chapter will provide a new approach to target monitoring on IRC and will lower the relative anonymity of IRC users. It will also provide researchers with the raw data collected in the IRC chat rooms which will enable researchers to parse, analyze and read all the public messages posted online and to generate their own social networking metrics. Finally, this research will go over and beyond Mutton's [32] chapter and provide a detailed description of both how to set up an IRC monitoring system as well as how to analyze the data collected through this new methodology.

4 Gathering Data: The IRC Way

Researchers who are interested in the data that can be harvested through IRC channels need to follow a three step process that involves the search of relevant channels, the installation of an IRC bot and the trimming of duplicate profiles.

Step 1: Identifying the relevant channels

The first step—and possibly the hardest of the three in this framework—is to identify the networks and channels that warrant monitoring from a criminological point of view. Hackers rely on security through obscurity since there are a countless number of chat rooms active at any given time on IRC. To circumvent this problem, a multilevel approach has proven very effective in the past. First, IRC search engines such as Netsplit (http://irc.netsplit.de) can perform keyword searches of IRC networks and channels. Such search engines have robots that crawl through chat room descriptions and return a list of all channels that fit a search query. Netsplit also tracks the number of users in each of the channels providing a measure of each chat room's popularity.

A little more work is needed to find more underground and exclusive chat rooms. Hackers often spend time in specialized hacking chat rooms but also visit more general channels such as #linux or #perl (the pound (#) sign indicates the beginning of a channel's name). Befriending users in these chat rooms can sometimes provide inside information on what channels the hackers are currently using. Hacking forums on the Internet can also be of some help to researchers. Many of them have IRC channels for live discussion as well as links pointing to interesting IRC chat rooms. A snowball-type technique can also be applied to this process once a few hacking chat rooms have been identified. The IRC protocol allows users to query each other in order to list all the channels they are connected to. This information can be obtained for all of the hackers connected to hacking chat rooms, providing researchers with a fresh list of channels to monitor.

Step 2: Setting up an IRC bot

While building a list of hacking chat rooms does not require any technical skills, the same cannot be said of the second step of this data-gathering process, the development on an IRC bot. As mentioned before, the Internet Relay Chat protocol is synchronous, meaning that messages sent are not stored on central servers and cannot be read at a later time. Rather than keeping an open IRC window on their desktop, researchers can take advantage of the latest developments of IRC bots to monitor and store events (people joining or leaving channels, messages posted, nickname changes). These bots are specialized software that logs into chat rooms and listens on all public conversations. They can be instructed to store all communications, commands and events of a channel in a database. Bots can also store each user's username used to register an IRC client, their IRC handle and their IP address.

Many bots are freely available on the Internet and they can be divided into two categories: those programmed in JAVA and those programmed in PHP or Perl. Both classes can be run on Windows, Mac OS X and Linux and share the same functionalities. JAVA, however, usually offers a graphical user interface that is more user-friendly. PHP and Perl bots often require some knowledge of programming languages and must be run from a command-line interface. They are also more customizable and easier to modify. Less computer-savvy users should experiment with JAVA-based IRC bots first before they move on to PHP- and Perl-based bots.

Configuration of IRC bots is usually done through text files which come with the software. Users must enter their IRC nickname, username, server as well as the channels they wish to join. Some bots can monitor multiple chat rooms simultaneously but in most cases, an instance of the bot must be run for each room that warrants monitoring. JAVA, PHP and Perl bots usually have text configuration files but some PHP- and Perl-based bots store their configurations settings inside their source code. It is however very easy even for neophytes to open the source code in a text editor and to configure the software as most programmers embed many comments in their code.

IRC bots store the data they have collected in databases of text files. Databases are much more efficient at managing vast amounts of data and allow for easy visual inspection of the data. They however require that researchers have a basic knowledge of databases such as MySQL, the common free open-source tool used by IRC bots. Data stored in text files uses the comma separated values syntax which means that each entry is saved on a row and specific information about that entry is separated by a comma. Learning how to set up, manage and use MySQL databases can be done in a matter of hours and vastly reduces the time spent on analyzing the data.

Step 3: Preparing the data for analysis

The third and last step of this three-step process focuses on the data- processing of the harvested data. Monitoring IRC channels will generate a wealth of conversations as well as activity profiles such as who joins and quits a channel. No matter how rich this data is, it is utterly useless unless it can be connected to individual hackers. At first glance, such a procedure seems trivial: researchers only need to match the username, nickname, and IP of a message to those of a hacker. Unfortunately, IRC users tend to use many usernames, nicknames, and IP addresses. Creating a single profile for each hacker is thus very tricky since one hacker can easily use tens if not hundreds of different nicknames and IP addresses over the course of a year. Tools like the previously mentioned PieSPy do not address this problem even though it directly affects the validity of the data as an IRC user could have many distinct profiles in a database built by an IRC bot.

To offset this identification problem, we developed a four-step decision tree. We implemented it in a PHP script that parses the data collected on IRC in a chronological order. The order in which the data is processed is important as new analysis must take advantage of prior knowledge whenever possible. Figure 1 displays the decision tree.

For each entry in a database of IRC events/messages, we first check if we have already seen the same username/nickname/IP combination in the past. If so, we

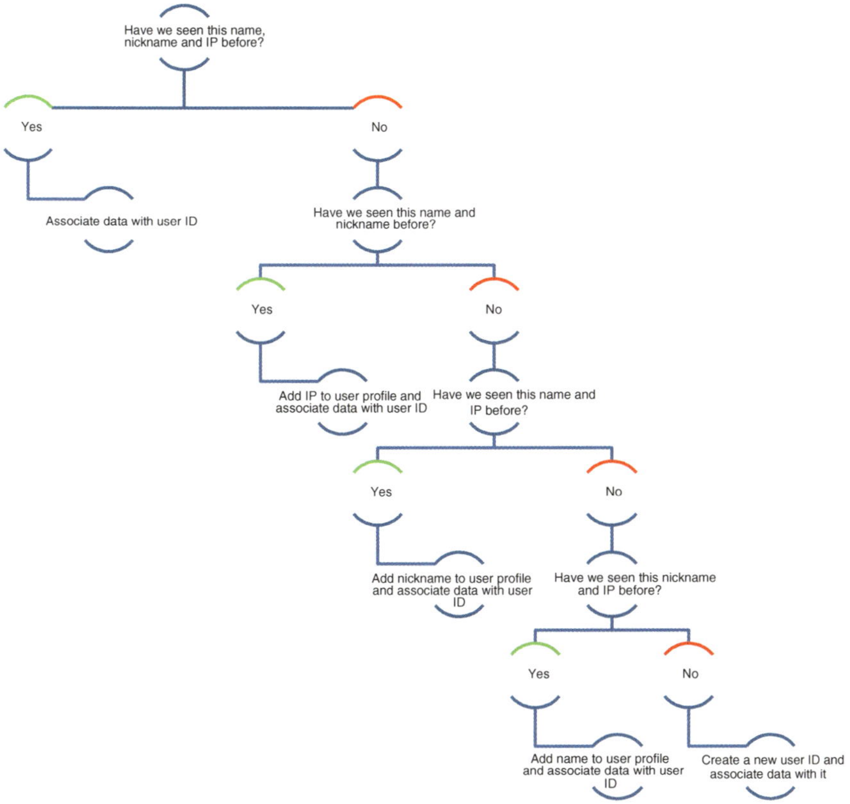

Fig. 1 Four-step decision tree

can be 100 % certain that this is the same IRC user and that this new event/message can be associated with his profile. If we haven't seen this combination in the past, we then check if we have seen the same username/nickname duo in the past. If so, we assume that the user is merely connecting from a new IP address; we add this new IP address to his profile and associate the rest of the data with his profile. If this username/nickname combination is unknown to us, we successively test if we know the username/IP address and nickname/IP address duos and add the appropriate information to their profile. If all of this combinations are still unknown, we then assume that this is a user we have never seen in the past and we create a new user ID and associate the information with this new profile.

While not perfect, this four-step decision tree highlights the similarities between users and works to reduce the number of duplicate accounts. These steps were designed to take into account the probability that a username, nickname or IP address has changed over time. Very few users will change the username they have registered their IRC client software with but will often connect from various IP addresses. This is why the first parameter that is validated is the IP address and the last one is the IRC username.

Using this framework, researchers should be able to create a database of events and messages as well as some basic information on individuals (ex: date of first message, number of messages posted). These metrics can easily be transferred to qualitative and quantitative software for analysis. The following section will describe how this data can be analyzed.

5 IRC Logs' Versatility

There are countless ways to take advantage of data collected by IRC bots. In this section, we will focus on three specific types of analysis where this data's potential is extremely promising.

5.1 Activity Profile

IRC bots keep track of all the individuals who join and quit a channel, allowing researchers to track the number of active users in a chat room. The popularity of IRC chat rooms can be used to decide which channel to monitor and serves as an early warning sign that hackers are moving on to new chat rooms. Furthermore, knowing when and where hackers are online allows researchers to better understand the importance of social interactions in the hacking world as well as the patterns of activity of hackers. It is possible, for example, to guess the time zone a hacker lives in simply by looking at his peak hours of activity.

5.2 Social Graphs

IRC chat rooms are first and foremost communication channels between individuals. In the case of hackers, they are the preferred place to hang out, meet new people, boast, and even look for help. Since most hackers never meet their hacking friends face to face [33], many of their dearest relationships are based on IRC discussions. This gives researchers an incredible insight at the social structure of the hacking community through unbiased data. Using IRC logs, researchers can build a representation of each hacker's social graph by looking at the people he or she is interacting with. IRC chat rooms usually revolve around public discussions, meaning that many people tend to share their thoughts and feelings simultaneously on a specific subject. The above mentioned methodology used in PieSpy [32] can be used to identify the interlocutors of each conversation.

The analysis of social networks is becoming more and more prominent and has been integrated in many police-oriented software suite such as i2's Analyst

Notebook and Palantir. Its objective is to determine the structure of a network as well as the positions of its actors [34]. Metrics such as centrality, betweenness and power which are derived from the individuals' ties have been used to model criminals' social recognition [23] and their performance [35]. IRC promises to give even more momentum to this research field.

5.3 Content Analysis

The content of the messages themselves also holds troves of information on hackers. Researchers have the opportunity to learn about hackers' lives as well as their hacking needs, techniques, and behavior. This untainted source of information is unique in that it can offer a window into the hackers' reality over an extended period of time. The time and effort needed to perform such content analysis can be greatly reduced by performing keyword searches or automated lexical analysis, although the idiosyncrasies of hackers' talk and ways of writing (Leet) can be challenging for automated software (more on this later).

6 Solutions and New Challenges

Gathering data on cybercriminals the old-fashioned way raised, as we mentioned before, concerns on three levels: the subjects, the tools and the resources. As we will demonstrate, the framework presented in this chapter solves if not eliminates most of these issues.

The first challenge was finding subjects willing to participate in a research on cybercrime. Using IRC logs, researchers bypass completely this issue by gaining access to an endless supply of hackers who share their thoughts and behavior in online chat rooms. To find new participants, researchers merely have to add new channels to their monitoring systems or to wait a few weeks for the IRC users to engage in more conversations. Prior research also failed to differentiate between black hat and white hat hackers. IRC logs solve this problem by giving researchers more insights into each individual's mind frame. Messages can be analyzed to determine whether their author is indeed involved in illegal activities. More underground channels may also be less popular with people from the computer security industry and provide better access to cybercriminals. IRC data simply provides more information on each user allowing researchers to better understand who their subjects are.

The tools used in prior research were also problematic as they forced repetitiveness and limited the online harvesting of data. We have demonstrated in the previous section just three examples of research areas where IRC logs can be used. Willing researchers could push further ahead and measure criminal performance by monitoring the sale of illegal goods and services in IRC rooms. Information sharing in the computer underground could also be studied to determine how innovation and technical skills are transferred from an individual to another. The automation of data gathering through IRC bots also vastly reduces the energy and resources needed to start projects. These robots can monitor multiple channels simultaneously producing a wealth of information on any given hacking community. This automation of research is, we believe, one of the main features of this proposed methodology. It is true that setting up IRC bots and an implementation of our framework will need some investment of time and energy on the researcher's part. However, this setup only needs to be created once and given the static nature of IRC, there will never be any need to update the bots once they are operational. Furthermore, researchers could and should always use one of the many freely available bots on the Internet to speed up the initial phase of research on hackers who communicate via IRC channels.

Given the specificity of each research project, it is impossible for us to provide exact recommendations on how long each channel should be monitored or what number of events are adequate to reliably represent a hacking community. Each chat room has a level of activity that varies over time and so most researchers will want to monitor each channel for a period of weeks if not months. They should also try not to focus on only one or two channels at a time to increase their odds of gathering useful data. Since monitoring more channels only requires that the researcher adds new lines in the configuration file, as many channels as possible should be included in any research project.

As helpful as it is, the use of IRC data still poses two technical problems. First, IRC users tend to use online jargon (ex: newb, p0wn) and to misspell works when talking to each other. Furthermore, some keywords such as *virus* can also have many meanings depending on the context. This creates a problem for automatic content analysis through the use of keywords as the risk is missing valuable data or to find false positives are both increased. To address this issue, researchers should include many variations of the same word when doing keyword matching. For example, instead of just using *serial number* as a keyword, they should use *serial number* and *serialz* and *serials*. Keyword lists should also contain expressions which give more context. Again, using *windows virus* will give better results than just using the word *virus*. Finally, software packages that specialize in qualitative data analysis such as NVivo, Leximancer and Atlas.ti offer various tools to identify the topics of discussion and to measure the prevalence of words or expressions in IRC logs. These software have built-in error correction that ease the difficulties of working with raw conversations.

The second technical problem is the presence of spammers and other bots that saturate public channels with the same useless messages (flooding the in IRC jargon). Best practices would suggest that researchers measure the ratio of unique messages to the number of messages sent by each user to identify possible spammers and bots. These suspicious users could then be automatically erased from the IRC logs or stored in a quarantine file which could be used to detect new bots or spammers. This method has proven very effective in our own research and should be used before analyzing any datasets from IRC.

7 Ethical and Legal Issues

Conversations over IRC are inherently public and anyone can join and listen on discussions at any given time. Nevertheless, many issues have been raised regarding the ethics of using IRC logs for academic research. Ethics guide highlight the need to maintain "confidentiality and participant anonymity, as well as respecting the right to privacy, to dignity and whether or not to participate in research" [36].

Maintaining the anonymity of participants can be easily guaranteed through the use of pseudonyms or IDs (N1, N2, N3, N4) to represent each individual. Special care must be taken to further protect the identity of IRC users by replacing any and all references to participants' names in the citations used in chapters with a pseudonym or an ID and by avoiding to release the name of the networks and the channels where the data was collected [36]. These steps, although necessary, do limit the external validation of findings given that other researchers will not be able to reproduce the studies. In this case however, this limit is outweighed by the importance of protecting the participants' identity.

Regarding the subject's right to participate or not in research, Lawson [36] suggests that participants should be allowed to choose between five levels of involvement which range from giving access to their messages anonymously to the full disclosure of their name and messages. In the context of IRC, obtaining such consent may pose certain logistical problems. First, with the high turnover of users in channels, contacting each person who logs into a chat room to ask for their permission to use their messages may be impractical and even be considered as harassment by channel administrators. Second, many hackers may be reluctant to agree to share their messages with an unknown researcher who can hardly prove his true identity and motivation. Obtaining participants' consent may be possible in the case of a smaller community being monitored over an extended period of time for the purpose of ethnography. In other contexts, obtaining the explicit permission to use all users' messages may very well be impossible. This highlights the need for updated ethical guides which could apply to online research. Such guides have been developed in the past by universities [37, 38] and by associations of researchers [39]. Some guides, such as Standford's, are quite comprehensive and require a very high level of commitment from researchers who wish to engage in online social research. They mention federal regulations and state strict guidelines that must be followed at all times. Others, such as York University's, are much less formal and only require oversight when the data comes from protected or private forums or chat rooms.

This should not, however, stop researchers from monitoring IRC channels as the rights of individuals are protected by maintaining the strict anonymity of participants and ensuring that no information leaks about the true identity of the subjects. Moreover, by keeping secret the name of the channels where the data is collected, identifying the real author of a citation may be next to impossible.

The current literature has not studied so far the legal implications of the monitoring IRC channels by academic researchers. Cate [40] describes the increasing use of online data mining by governments and highlights the need for a new legal framework that could frame this type of information gathering but does not mention how the current legal framework could apply to the Internet Chat Relay and researchers. Anyone can create a new chat room on IRC but the creator loses all his rights over the channel as soon as he exists the room. It is possible to assign administrative rights but these are temporary and can be changed by anyone who becomes an administrator. In this context, no one can be considered as the owner of a channel. Given the lack of ownership of chat rooms, the content on open and public channels should be considered as being a part of the public domain, a position that has been adopted by others beforehand [36, 41] as no law could forbid the copying of something that does not belong to anyone. Once again, this license to use IRC data does come with some constraints. In order to foster good relationships between the IRC community and the academic world, special care should be taken to anonymize the citations and the data collected in IRC chat rooms. This simple solution should be the main concern of researchers willing to follow ethical and legal guidelines as closely as possible.

8 Real-Life Example of IRC Data

In order to illustrate how IRC logs can be used in the context of criminological research, we contacted a North-American law-enforcement agency who has been monitoring hacking channels on IRC for the past few years and who agreed to share with us some of the IRC logs they had collected. We were given 74 days' worth of logs from 17 hacking channels that were active between February 20th 2009 and November 26th 2010. This sample included 25,940 events which can either be messages posted online, users joining or leaving chat rooms or users changing their nicknames.

To evaluate the effectiveness of our decision-tree to identify the unique profiles of hackers, we created a list of all the unique combinations of IP addresses, usernames and nicknames. Our data contained 3,542 of these unique combinations. Using the four-step process described earlier in this chapter, we managed to reduce the number of users by 67 % to 1,165. This indicates that at one point or another, all of these individuals changed one of their online identifiers which artificially increased the number of users in these chat rooms.

To better understand who these individuals are, we present in Table 1 the descriptive statistics for some of the characteristics of these users.

The first three rows refine the above mentioned findings on usernames, nicknames and IPs. Users who visited hacking chat rooms tended to change their nicknames more often (mean = 2.16) than their IPs (mean = 1.86) and their usernames (mean = 1.07). Since nicknames are the most visible identification of IRC users, modifying it is obviously the easiest way to hide one's true identity. Users were not active for extended periods of time in chat rooms as the lifespan's (number of days of activity) mean is limited to 4.45. The number of channels is

Table 1 Descriptive statistics of IRC users

	Minimum	Maximum	Mean	Median
Nb of usernames	1	8	1.07	1.00
Nb of nicknames	1	80	2.16	2.00
Nb of IPs	1	47	1.86	1.00
Lifespan (days)	1	46	4.45	1.00
Nb of channels	1	11	1.30	1.00
Nb of messages	0	465	3.89	0.00
Nb of known IPs	0	42	1.24	1.00
Nb of missing IPs	0	29	0.62	0.00
Ratio of missing to known IPs	0.00 %	100.00 %	37.59 %	0.00 %
America	39.20 %			
South America	3.70 %			
Europe	19.20 %			
Africa	2.50 %			
Russia	0.50 %			
Asia	3.70 %			
Australia	0.70 %			
Unknown	30.50 %			

also very limited with a median of 1 and a mean of 1.30. Although some users sent hundreds of messages (maximum = 465), over half of them did not send a single public message (median = 0) and the average message count was lower than four (mean = 3.89). Users did use more open IPs than hidden IPs (routed through anonymizing proxies) and their ratio of hiding to open IPs is relatively low at 37.59 %; most users thus did not try very hard to hide their true origin and identity. We found that out of those who had used a proxy services to hide their real IP addresses, 16.92 % had also connected to IRC without using an anonymizing service which allowed us to trace them through their open IP address. These individuals would be of higher interests in future research as their decision to proactively hide their identity suggests they may have something to hide. In this case study, our framework thus managed to reduce the signal to noise ratio in the number of users and to enhance our ability to track down users.

We also used IPInfoDB's service to geolocate the different IPs used by individuals who visited the monitored channels. 39.20 % of our sample originated from North America. Europe followed with 19.20 %. All other origins (South America, Africa, Russia, Asia and Australia) accounted for less than 4 % each. Almost a third of users used IP addresses which could not be traced.

This general portrait of IRC users who visited the monitored hacking chat rooms proves that most individuals were casual visitors who were not involved in the community and who mostly listened to what others were saying. Those who changed their identifiers often and who used proxies to hide their true origin should warrant careful attention from researchers however.

IRC data can also be used to determine the social graph of actors. We adopted Mutton's [32] approach for the creation of ties between users of IRC. We created a

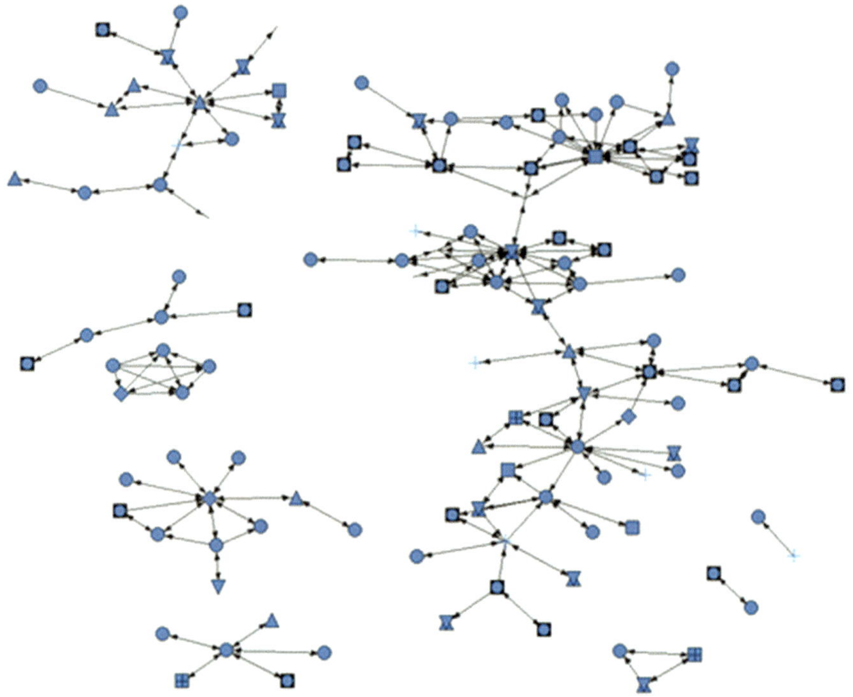

Fig. 2 Network of IRC users

relational matrix where we stored links between individuals. We considered that two individuals were tied if they posted messages less than two minutes apart from each other in the same channel. While more robust methodologies could and should be developed to model tie creation in the context of IRC, we opted for a simplified version given the constraints of this limited case study.

Figure 2 represents the network of conversations between IRC users. Arrows indicate the direction of ties between dyads of actors and each distinct shape represents the country of origin of the IP address of an actor. Figure 2 demonstrates that individuals visiting hacking channels on IRC are not limiting their contacts to individuals from the same country as them. In fact, further analysis shows that individuals are slightly more likely to interact with a foreigner than with someone from their own country (average homophily of contacts > 0), therefore highlighting the international nature of the hacking community. Figure 2 also showcases a network with many fragments of different sizes and shapes which take advantage of brokers or bridges, individuals who sit between a great number of actors—a position associated with power in past literature [42].

To complete this study of a real-life IRC dataset, we provide in Table 2 the correlation matrix between the level of activity of IRC users and their identifiers, social capital, origin and personal characteristics. The level of activity is measured

Table 2 Correlation model of the level of activity in IRC channels

Identifiers	
Nb of usernames	0.231[b]
Nb of nicknames	0.255[b]
Nb of IPs	0.562[b]
Social capital	
Betweenness	0.468[b]
OutDegree	0.635[b]
InDegree	0.603[b]
Homophily	−0.105
Origin	
North America	0.094[b]
South America	−0.029
Europe	0.022
Africa	−0.044
Russia	−0.012
Asia	−0.026
Australia	−0.031
Personal characteristics	
Lifespan	0.780[b]
Nb of channels	0.320[b]
Nb of messages	0.581[b]
Nb of IPs from known location	0.481[b]
Nb of IPs from unknown location	0.445[b]
Ration unknown/known IP location	0.058[a]

[a] $p < 0.05$
[b] $p < 0.01$

by the number of events registered (messages posted, joining/leaving a chat room, changing a nickname).

IRC users who often changed their identifiers (usernames, nicknames and IPs) tend to have a higher level of activity on IRC. This could be a result of them using multiple online personas or of their need to hide they true identity as they would be more involved in the world of hacking.

The social capital of IRC users is measured using four different social networking metrics. Betweenness measures the extent to which an individual sits between two other people, a metric associated with the notion of brokers and bridges. Outdegree measures the numbers of outgoing ties and the Indegree measures the number of incoming ties for each individual. Finally homophily analyzes whether the ties of an individual are homogeneous or heterogeneous based on the country of origin of their IP address. Betweenness, Outdegree and Indegree are all positively and significantly correlated to the level of activity meaning that those with better networking are the most active in this network of IRC users.

Only one origin is correlated with the level of activity and that was North America (Mexico, United States and Canada). The fact that the discussions in chat rooms are all in English and French may have contributed to this finding.

Finally, all personal characteristics are significantly and positively correlated with a higher level of activity on IRC. The lifespan or number of days online is the most correlated variable of the set. Users that tend to use proxies instead of direct connections to log on to IRC are also more active, meaning that the most active hackers in these channels are also the most careful to hide their tracks.

Our model presented in this case study targeted mostly North American hackers who have been active between 2009 and 2010 in hacking channels. We have found that the level of activity is correlated with many variables which include the identifiers, the social capital and personal characteristics. These findings already draw a somewhat tentative profile of the individuals most active on IRC channels that will need to be validated through other samples. Still, these preliminary results are consistent with what would be expected of hackers connecting to each other in public chat rooms.

9 Conclusion

Research on hackers has grown by leaps and bounds over the last two decades and researchers have adapted many of the techniques used in traditional social research with varying success. We have shown in this chapter that while some traditional tools have produced chapter that have shed much-needed light on the hacking community and its members, a technique used by computer security experts, IRC monitoring, can be adapted by social researchers to provide new datasets. We have also noted some of the strengths and weaknesses of this approach. Although we did not address these possibilities in our case study, qualitative studies could also benefit greatly from this data-gathering method. Access to the raw communication logs would allow anthropologists to conduct specific studies on hacker behavior and social organization. This data would provide them with an overview of the hacking community, permitting qualitative studies that would greatly enhance our knowledge of the computer underground.

On a more practical note, law-enforcement agencies may also increase their effectiveness at identifying and investigating cybercriminals by integrating IRC monitoring onto their day-to-day workflow. Simple and automated tools are freely available on the Internet and allow police officers to follow specific individuals or to monitor whole chat rooms. This first-hand and unbiased information will be valuable to police investigations targeting such dark networks. Law-enforcement agencies would benefit from the vast amount of data available but would also need to pay attention to the management such information. Hundreds of messages are posted each minute in the most active IRC chat rooms and indexing and storing this data requires some serious planning. This is still a very small price to pay for gaining an inside look at criminal networks and being able to understand their structure. And with the framework we developed above in this research, police officers will be one step closer to getting a more complete profile of all the cybercriminals under investigation.

While the focus of this chapter is on IRC monitoring, researchers should definitively look beyond this technique to consider other tools and methods used by security experts. Forthcoming work from the authors will present the web-scraping technique that makes it possible to download a copy of online forums and websites in order to analyze interactions between actors. P2P networks and online forums are also excellent sources of data and by focusing on them researchers can develop a new kind of social research on hackers. The Internet has brought in its wake a whole new range of crime and criminals and for the first time criminals are using communication tools that are widely and easily accessible to others. Researchers should tap into these communications to understand how this emergent breed of criminals is organizing and conspiring against citizens, corporations, and governments.

References

1. Sutherland EH (1947) The professional thief. University Of Chicago Press, Chicago
2. Hobbs D (1995) Bad business: professional crime in modern Britain. Oxford University Press, Boston
3. Venkatesh SA (2002) American project: the rise and fall of a modern ghetto. Harvard University Press, Boston
4. Morselli C, Tremblay P (2004) Criminal achievement, offender networks and the benefits of low self-control. Criminology 42(3):773–804
5. Hopper CB, Moore J (1990) Women in outlaw motorcycle gangs. J Contemp Ethnography 18(4):363–387.
6. Brumley D (1999) Tracking hackers on IRC. Retrieved May 19th 2011 on: https://db.usenix.org/publications/login/1999-11/features/hackers.html
7. Hugues L (2012) Law enforcement getting on the social media Bandwagon. Retrieved 22nd Oct 2012 on: http://www.ktre.com/story/19884594/local-police-are-using-facebook-to-keep-community-informed
8. Sayer P (2010) Interpol uses facebook to hunt for most-wanted suspects. Retrieved 22nd Oct 2012 on: http://news.techworld.com/personal-tech/3229834/interpol-uses-facebook-to-hunt-for-most-wanted-suspects/
9. Thompson J, Demerath NJ (1952) Some experiences with the group interview. Soc Forces 31:148–154
10. Foddy W (1994) Constructing questions for interviews and questionnaires: theory and practice in social research. Cambridge University Press, Cambridge
11. Newman J, Des Jerlais DC, Charles T, Jay G (2002) The differential effects of face-to-face and computer interview modes. Am J Publ Health 92:294–297
12. Denning P (1990) Computers under attack: intruders, worms and viruses. Addison-Wesley, New York
13. Jordan T, Taylor P (1998) A sociology of hackers. Sociol Rev 46(4):757–780
14. Schell B, Dodge J (2002) The hacking of America: who's doing it, why, and how. Quorum Books, Westport
15. Holt TJ (2010) Exploring strategies for qualitative criminological and criminal justice inquiry using on-line data. J Crim Justice Educ 21(4):466–487
16. Rehn A (2003) The politics of contraband: the honor economies of the Warez scene. J Socio-Econ 33:359–374
17. Coleman S, Normann E (2000) New media and social inclusion. Hansard Society, London

18. Poier S (2008) Fighting on words, fighting on worlds: brief ethnography of hackmetting. In: Verga M (ed) Ais—Sezione di Sociologia del diritto. Università Degli Studi Di Messina, Messina
19. Blankwater E (2011) Hacking the field: an ethnographic and historical study of the Dutch hacker field. Master's thesis, Department of Sociology, Universiteit van Amsterdam
20. Dreyfus S (1997) Underground hacking, madness, and obsession on the electronic frontier. Random House, Sydney
21. Kleen L (2001) Malicious hackers: a framework for analysis and case study. Master's Thesis, Department of Operations Research, Air Force Institute of Technology
22. Winkler IS (1996) The non-technical threat to computing systems. Comput Syst 9(1):3–14.
23. Décary-Hétu D, Morselli C, Leman-Langlois S (2011) Welcome to the scene: a study of social organization and recognition among Warez hackers. J Res Crime Delinquency 49(3):359–382
24. Basamanowicz J, Bouchard M (2011) Overcoming the Warez paradox: online piracy groups and situational crime prevention. Policy Internet 3(2):1–25
25. Hinduja S (2001) Correlates of internet software piracy. J Contemp Crim Justice 17(4):369–382
26. Funkhouser N (2006) Software piracy among students in Taiwan: the ethical decision-making process and attitudes toward the use of pirated software. IMBA Master's Thesis, National Cheng Kung University
27. Holt TJ, Lampke E (2010) Exploring stolen data markets online: products and market forces. Crim Justice Stud: A Crit J Crime, Law Soc 23(1):33–50
28. Verizon (2012) 2012 data breach investigations report. Retrieved 2nd May 2012 on: http://www.verizonbusiness.com/resources/reports/rp_data-breach-investigations-report-2012_en_xg.pdf
29. Simpson C (2000) Internet relay chat. Educ Media Technol Yearb 25:62–65
30. Spitzner L (2003) Honeytokens: the other honeypot. Security Focus
31. Poulsen K (2005) Hacker penetrates t-mobile systems. Retrieved on 19th May 2011 on: http://www.securityfocus.com/news/10271
32. Mutton P (2004) Inferring and visualizing social networks on internet relay chat. In: Proceedings of the international conference on information visualization. Canterbury, UK
33. Calce M, Silverman C (2008) Mafiaboy: how i cracked the internet and why it's still broken. The Viking Canada, Toronto
34. Sparrow M (1991) The application of network analysis to criminal intelligence: an assessment of the prospects. Soc Netw 13:251–274
35. Morselli C, Giguere C (2006) Legitimate strengths in criminal networks. Crime, Law Soc Chang 45(3):185–200
36. Lawson D (2003) Blurring the boundaries: ethical considerations for online research using synchronous CMC forums. In: Buchanan EA (ed) Readings in virtual research ethics: issues and controversies. Information Science Publishing, Hersey
37. York University (2012) Surveys and research in an online environment. Retrieved 2nd May 2012 on: http://www.yorku.ca/research/support/ethics/
38. Standford (2012) Internet research ethics. Retrieved 22nd Oct 2012 on: http://plato.stanford.edu/entries/ethics-internet-research/
39. Ess C, AoIR Ethics Working Committee (2002) Ethical decision-making and internet research. Retrieved 19th May 2011 on: http://aoir.org/reports/ethics.pdf
40. Cate FH (2008) Government data mining: the need for a legal framework. Harvard Civil Rights-Civil Liberties Law Rev 43(2)
41. Rafaeli S, Sudweeks F, Konstan J, Mabry E (1998) ProjectH: a collaborative quantitative study of computer-mediated communication. In: Sudweeks F, McLaughlin M, Rafaeli S (eds) Network and netplay: virtual groups on the internet. MIT Press, Cambridge, pp 265–281
42. Morselli C (2009) Inside criminal networks. Springer, New York

Why Terror Networks are Dissimilar: How Structure Relates to Function

Christian Leuprecht and Kenneth Hall

Abstract Theories on international terrorist networks are wrought with contradiction. On the one hand, networks that support or facilitate politically motivated violent extremism are thought to pose a threat because they are centralized and hierarchical. On the other hand, the same networks are thought to pose a threat because they are decentralized and operate autonomously. Social networks analysis (SNA) makes it possible to resolve this apparent contradiction by controlling across countries for characteristics and structure of networks linked to the same terrorist organization relative to different functions that such networks perform. One terrorist organization for which sufficient open-source data exist to mount a systematic comparison is Al-Shabaab (AS). Comparing traits such as brokers, centrality characteristics of nodes, international linkages, and use of funds, the chapter compares AS networks as they relate to recruitment, fundraising and attacks across the United States and Australia with corroborating evidence from Canada, the United Kingdom, the Netherlands and Denmark. Although networks differ markedly across these attributes, unrelated networks performing similar functions are consistent in their nature and structure. These findings suggest that networks are functionally differentiated insofar as they serve as strategic repertoires. This is a significant finding. Knowing how a network's function is related strategically to its structure means being able to infer a network's function if only its structure is known and, conversely, being able to infer a network's structure if only its function is known. Not only does SNA thereby facilitate detection and dismantling of networks, it also suggests that recruitment, fundraising and attack networks require differentiated approaches by defence and security agencies insofar as SNA shows them to be distinct phenomena.

C. Leuprecht (✉) · K. Hall
Royal Military College of Canada and Queen's University, Kingston, Kingston, Canada

A. J. Masys (ed.), *Networks and Network Analysis for Defence and Security*,
Lecture Notes in Social Networks, DOI: 10.1007/978-3-319-04147-6_5,

1 Introduction

How do individuals within networks connect, remain connected, bring others into the fold, connect to other networks, and execute their ultimate objectives while striving to remain undetected? Answers to these questions are pertinent to contain and disrupt both the genesis and diffusion of networks that facilitate terrorist ends. Accordingly, this chapter gauges the extent to which the nature and characteristics of terror networks are a function of their purpose.

This claim is difficult to investigate. To generate hypotheses about terror networks, one needs to be able to control for time and space. Terror networks, however, are notoriously impervious to examination. Moreover, much of their activity transpires in hostile environments that are inhospitable to scholarly fieldwork. Nonetheless, that need not deter us from exploiting available information to generate knowledge about terrorist networks. To this end, this chapter applies Social Network Analysis (SNA) to compare variation across various al-Shabaab (AS) networks. AS is a particularly opportune subject as it is the one terrorist organization that operates across different countries while maintaining a degree of common cause and connection. Although the number of known AS cases is limited, the quality of the available open-source data is sufficient to allow for comparative analysis.

Terror networks such as AS are commonly compared to multinational corporations: hierarchical with centralized command and control structures. Yet, a growing body of research is skeptical about these assumptions. This chapter scrutinizes these assumptions empirically in an effort to stimulate a more nuanced approach to terror organizations and their networks. To this end, it examines variation in AS activity in the United States, Australia and Canada, and compares associated networks and their purposes. Recruitment and financing networks turn out to operate surprisingly independently and different types of activities in the same locale spawn quite different networks. Although these networks sometimes overlap in time and space, they are not connected.

Concern among governments and security intelligence services across democratic countries about "foreign fighters"—residents and citizens who go off to fight in causes abroad—and the consequent risk of homegrown foreign-trained militants returning from abroad and committing violent acts of terrorism is widespread and growing [1, 116]. However acute the problem, exceedingly little is known about local recruitment networks in Western countries. Since open-source intelligence is hard to come by, empirical SNA research in this area is accordingly scant. In this light, AS recruiting and fundraising from the Somali diaspora is a critical case study: the *n* of cases may be limited and the networks relatively small, but the available data are comparatively good, robust and comprehensive so as to allow for methodologically rigorous comparative empirical research.

An estimated 1200 foreigners or ethnic Somalis with a foreign passport are thought to be fighting with AS, including upwards of 40 American, 100 British and 20 Canadian expatriates since 2007 [2]. When the U.S. House Homeland Security

Committee released a report detailing AS' threat to America, Committee Chairman Peter T. King declared: "Right now Al-Shabaab has to be our main concern because of the fact of such easy travel back and forth, because there is a large number of (U.S. recruits), and the fact that there is such open recruitment" [3]. Concerned that the organization will soon be able to muster an attack on American soil, U.S. federal prosecutor W. Anders Folk as well as the U.S. Department of State warned the U.S. government to "take Al-Shabaab seriously" [4, 5].

Expatriates from Maryland to California have reportedly travelled to Somalia to join AS, but most hail from the 'Twin Cities' region of Minnesota, home to America's largest Somali diaspora, whose concentration of 25,000 members represent about one-third of the total Somali diaspora in the United States [6–8].[1] The evidence suggests that a group of men concentrated in Minneapolis and St. Paul formed a network, the activities of which ultimately saw 18 of them leave for Somalia, and 11 of them die there [9–12]. However, reliable data on these individuals are sparse. There are three American networks affiliated with AS for which open-source data are available, the largest of which has 23 verifiable nodes.

Australia also has a sizable Somali diaspora; approximately 11,000 Somali–Australians live in and around the city of Melbourne, where the network under investigation was concentrated [13]. This network was responsible for sending two men to Somalia to train with AS, and was plotting to attack the Holsworthy Barracks near Sydney at the time of their arrest.

This chapter begins with an overview of current scholarship on how terrorist networks are structured and how their members interact, as well as an operationalization of key concepts, such as Morselli's concept of 'brokers' within networks primary. These concepts are central to formulating the subsequent hypotheses that were generated by an earlier version of this study that drew on fewer cases and less complete data. A discussion of the importance and applicability of SNA and small-*n* research to this topic, as well as the methodology used to select, analyze and present these cases follows. After the cases have been presented and analyzed, the hypotheses will be scrutinized using the expanded data, and avenues for future research discussed.

2 Terror Networks: Overview of the State of Knowledge

Insofar as they link actors who are working towards common goals, networks are important means to terrorist ends [14, 15]. They make it possible for terrorists to overcome collective-action problems arising out of complexity and the uneven distribution of assets that they need to carry out attacks. In the 'global Salafi jihad' "the distribution of assets seriously affects its mission against the United States"

---

[1] Accurate figures for the total US Somali population are hard to come by and range from 35,000 to 150,000.

[16, p. 145]. Networks are used by terrorist groups to recruit, train and prepare for an attack to compensate for inadequate resources, identity, culture, emotions, elite access, ideological support, and recruits [17–21] and to "provide flexibility, adaptability, deniability, multidimensionality, and the capacity to do things at a distance, often through surrogates [22].

This chapter collates two strands of the literature, namely the inter- and intra-organizational connections between networks. The first investigates how larger, discrete, yet loosely affiliated networks interact with each other, such as the connections between al-Qaeda (AQ) and its primary counterpart in South-East Asia, Jemaah Islamiya [23, 24]. The second investigates how a covert or peripheral network operating apart from a central network with which it identifies interacts with this central network and potentially with other networks under the same umbrella organization. Krebs' investigation into the network of the 19 individuals who carried out the 9/11 attacks is an example [25].

Although the very notion of a network contrasts with the notion of top-down decision-making and arbitration and although "networks are never managed by a single (central) authority", the notion of hierarchy pervades research on terrorist networks [17, p. 65]. Different forms of hierarchical control posited in the literature have different implications for the structure and dynamics of the organization and its networks: (1) top-down decisions made about to the goals, objectives and/or function of individual networks (and the dissemination of the information required to carry these out), (2) ideas pertaining to the general goals of the entire organization and the reasoning behind these goals (this could be considered the ideology of the group), and (3) the distribution of funds and resources to networks to advance either the particular or general goals of the group. This chapter addresses only assumptions about the first and third forms of hierarchy. It is less concerned with the more "macro" or "abstract" [26, p. 36] level of terrorism, than with what cells do to further the ideologically motivated goals of a terrorist entity, and how these activities are funded.

It is sometimes assumed that a core network directs, at least to an extent, the operation of peripheral networks. This could include instructions on means and targeting, when to engage or desist from activities such as recruitment or fund-raising, quality and quantity of the membership of the network, and members' characteristics. The 9/11 network was already quite well formed prior to its arrival in America [25, p. 49], and it appears that the plot's targets and even the structure of the network were determined well in advance by AQ authorities outside of the immediate attack network, namely Khalid Sheikh Mohammed, who allegedly proposed the idea for the attacks to Osama Bin Laden as early as 1996 [27]. While 9/11 was notorious for its complexity and the duration of its planning phase, other studies of terrorism also presume a substantial degree of external control over the activities of a particular network. A quote from Ilachinski captures several assumptions about the operation of illicit networks: "the manpower mission requirement is an explicit goal that must be accomplished by the leader of the cell to which a given target is assigned" [28, p. 51]. Ilachinski not only assumes that a network has a *de facto* or *de jure* leader, and that s/he sets, controls and accomplishes "manpower mission requirements" (which in turn assumes that these networks coalesce in accomplishing a series of discrete tasks), but that the mission towards which this manpower is directed is assigned exogenously.

Similar assumptions are ubiquitous: "a Global Terrorist Organization (GTO) determines the nature and level of terrorist attacks in each country indirectly through its choice of representatives associated with the local terrorist group" [29, p. 238]. Leistedt also assumes that "in most cases, there is one person, typically the founder or cofounder, at the top of the terrorist organisation and structure, and s/he centralises decisions," and proceeds to speculate about their psychological state [30, p. 24]. Similarly, Corman's study of a hypothetical model of AQ depicts the specialized function of some networks as operating at the behest of a central organization, to the point where these specializations correspond with individual top-level council members [26, p. 40], [31, p. 66]. In the same vein, successful networks are thought to require strong individual leaders who can devise strategies in response to rapidly changing conditions and impose their will on sub-ordinates to operationalize their decisions [32, p. 106]. This is reflected in some suggestions for counter-terrorist policy that understands networks to be organized into hierarchical hubs where the removal of the central node or leader will cause disarray and the dissolution of the network. Conversely, the removal of "grassroots," lower-level actors supposedly leaves the leader stranded and powerless [33, p. 70, 34, pp. 1016–1017].

The sort of specialization Corman ascribes to networks is an aberration in the literature on terrorism. Instead, networks are assumed to be multitasking: the same network recruits, fundraises and attacks. The division of labour proceeds by actors, not networks. According to this conception, individuals hold rigid positions within a network with responsibilities for specific tasks, and people are recruited into networks (by a recruitment specialist, of course) to fill particular roles [35, p. 63]. Yang and Sageman [36, p. 301], for instance, assumes that different individuals play highly specified and agreed-upon tasks within a single network. For example, "some key members may act as leaders that control the activities of the whole group while others may serve as gatekeepers to ensure the communication and coordination between different groups of a larger network". This degree of specialization between individual nodes is premised on central decision-making and delegation to define and assign roles.

This underlying assumption is that networks are preoccupied with planning and executing attacks coincident with their location. Tupman, for example, observes: "Western Europe still remains a recruitment target and perhaps a target for a spectacular atrocity, as does the USA," due to the presence of terrorist cells covertly embedded into the fabric of Western society [17, 18]. Another variant of this line of argument acknowledges the post-9/11 flattening of AQ, which has led to an increased focus on autonomous, self-funded groups (discussed below in greater detail). Nonetheless, it assumes that networks, formally disconnected from an umbrella organization but informally linked through ideological solidarity and self-branding, are ultimately focused on perpetrating attacks in Western democracies.

Even when a decentralized network structure is assumed, the “parent organization” is thought to infuse “start-up capital” into fledgling peripheral cells [37, p. 47]. This scenario has a post-9/11 AQ supplying “money to underwrite conflicts in many parts of the world,” including those involving separate but affiliated terrorist entities, and perpetually masterminding or executing attacks [37, p. 27]. Shapiro and Siegel’s application of rational choice theory to terrorist funding, for instance, is predicated on this very assumption even though they themselves think it is no longer applicable given the structural changes to AQ post-9/11 [38, p. 426].

Hierarchy, however, runs counter to the very characteristics of networks, which are heralded as “temporary, dynamic, emergent, adaptive, entrepreneurial and flexible structures”, a “cutting-edge design” [26, p. 35] [31, p. 66]. Similarly, the rigid depiction of networks contrasts starkly with networks as an organizational structure that consists of “operatives [who] are highly adaptive, compartmentalised [and] mobile” [31, p. 45]. These observations contradict. On the one hand, terrorist networks pose a threat to Western interests because of their resilient connections to declared enemies abroad, such as AQ. On the other hand, such networks are a threat domestically due to their very ability to operate autonomously by being able to complete the variegated tasks of planning, funding, and ultimately executing attacks with little more than ideological support passively offered by a central organization. That is, the emphasis on offensive networks risks obscuring how peripheral networks support a central organization in varied ways, with each manner of support entailing a different relation to the central organization.

3 Variables: Network Structure and Centrality

This section defines and explains terms that are crucial to structuring and analyzing these groups, and thus vital in formulating and expressing our hypotheses. First, network structure matters because it affects the flow of information and resources.

Networks can take different forms—chain, hub, multi-player, all-channel—but only two concern this chapter’s findings. A “hub” network features a single node or very small cluster of nodes at the center of three or more other nodes which have few—if any—links. Nodes on the hub’s periphery are likely only connected to each other through the center of the hub, which, as a consequence, has a disproportionately large influence on the flow of information and resources through the network. By contrast, an “all-channel” network exhibits a much more horizontal formation which decentralizes the flow of information and resources. Only a few if any individuals in an all-channel network are seen as substantially more influential and well-connected than the rest, and nodes are generally connected to three or more other nodes in the network.[2]

Three inter-related concepts are useful in describing and analyzing how nodes influence the movement of information and resources within and between networks: brokerage, degree centrality and betweenness centrality. Brokers are conferred positional advantage in a network insofar as they bridge structural holes—two unconnected groups of actors—by virtue of having greater access to information, opportunities and skills [39–41]. Morselli's study of members of the Hell's Angel's motorcycle gang in Quebec found that elite members of the group were directly connected to only a few other members of the network (i.e. low degree centrality) while at the same time many "efficient paths pass[ed] through [the] given node" (i.e. high betweenness centrality) [42, p. 187] [43, pp. 385–386]. These are precisely the traits of a broker: a node with few but influential connections. Ergo, an 'ideal broker' is an autonomous link between a single node in each of two networks where such a link constitutes the only connection between them [43, p. 386].

Brokers are advantageous because they can manage the flow of information and resources between two groups to their benefit and that of the networks they link [44, 72–73]. Especially in illicit situations, members in each network can minimize detection by minimizing connections to illicit individuals [43, p. 384], while maximizing opportunities to further their objectives through potential access to the resources of the other group via the broker [41, p. 347, 353]. In turn, the autonomous ideal broker—the most knowledgeable about both of the networks of all nodes involved—can act opportunistically, in this case by connecting transnational legal and illicit markets. As a result, brokers tend to maximize monetary returns from illicit activity [43, p. 385].

However, such an actor is described as 'ideal' for good reason; more often, one will observe "one or two participants who are high in both degree and betweenness centrality," especially in smaller networks [43, p. 388]. While AS data does not allow for the quantitative precision of Morselli's study, Morselli's 2 $\times$ 2 matrix of the two varieties of centrality outlines four types of actors and their relation to the nodes of these networks offers a useful typology [43, p. 388] (Table 1):

This matrix is useful because the evidence presented below shows that nodes within recruitment and fundraising networks differ substantially in degree and betweenness centrality. They also differ in the special functions that some nodes serve and that are crucial to the overall function and dynamics of those networks. Differences in network structure and the centrality characteristics of nodes aside, their international linkages and the way they use funds are also distinct.

[2] Like hub networks, defining all-channel networks as generally having one or two connections per node avoids conflation with "chain" networks.

Table 1 Centrality matrix

High in degree centrality and low in betweenness centrality	High in both betweenness and degree centralities
Low in both centralities	High in betweenness centrality but low in degree centrality

4 Hypotheses

4.1 Functional Differentiation

The illicit networks studied here reveal patterns that challenge assumptions of current scholarship and lend themselves to generating hypotheses about the relationship between the ways in which terrorist networks in the West are structured and how they function. Corman applies the concept of Activity Focus Networks (AFNs) to terrorist networks, whereby networks are organized around 'activity foci,' defined as categories of activities to which resources and manpower are assigned strategically to achieve the organization's goals [26, p. 38]. He (correctly) acknowledges that networks operating outside of the central network can have specialized functions [26, p. 39], but falls short of inferring that such specialization can have consequences for the very structure of these networks. An initial goal of this study was to identify how existing literature predicts funding and recruiting-oriented networks to be structured, and subsequently to compare these expectations against the empirical evidence presented here. However, the apparent absence of such prediction in the literature suggests that the very notion of a relationship between network structure on the one hand and network function on the other hand is underexplored (if not unexplored) in the scholarship of terrorist networks.

The most striking observation to be gleaned from these findings is the correlation between network structure and its functional objectives. The two recruiting networks described below follow an all-channel structure. Removal of nodes from this network as individuals left for Somalia (and upon the subsequent death of some of the actors there), even in significant numbers over a short period of time, neither dismantled these networks nor compromised their function of recruiting men and funding their travel to Somalia until the last wave of recruits departed in October 2009.

No individual or small group of individuals remained behind to act as a 'conveyor belt,' helping groups of men to radicalize and leave for Somalia under their supervision. While the departure of nodes did not impair the ability of remaining recruits to network and eventually leave for Somalia, the lack of new recruits and apprehension of some nodes by law enforcement dismantled the network. This indicates that there were no nodes, detected or not, that were primarily and/or specifically tasked with recruitment; recruiters that were above the fray of traveling to Somalia to join the insurgency do not appear to exist within these networks.

Contrary to much of the literature on terrorism, the network's function was highly specialized, while individual functions were undifferentiated. Two hypotheses follow from these observations:

H1 Terrorist networks are functionally differentiated

H2 The structure and characteristics of terrorist networks is a function of their purpose

4.2 *Modes of control*

Stohl and Stohl assert that terrorist groups such as AQ "[do] not maintain control over who is or is not admitted into the organization" [15, pp. 105, 115].[3] In effect, control over who was allowed into the network appears to have been informal and decentralized. Not only did no one person or group within the network regulate who was to be recruited and groomed for travel to Somalia, the central organization set no goals or quotas as to the quantity and quality of Somalia-bound individuals. Just as no one was controlling who was a part of the network, no particular person seemed to have domain over any specific task.

The recruiting network was also geographically concentrated. All actors lived in or around the Twin Cities area, mostly in Minneapolis, St. Paul and in adjacent suburbs such as New Brighton [45]. This dense distribution of the nodes of the network enabled in-person meetings between a variety of nodes in an assortment of venues including mosques, restaurants, private residences and a university campus. Court documents do not indicate any interactions over telephone or the Internet between actors in Minneapolis and note only one conversation between an actor in Minneapolis and another in Somalia [10]. This is especially noteworthy given the scholarly interest in the role of the Internet in radicalization and recruitment [46, p. 18, 47, p. 222, 48, 49, pp. 205–208, 50, p. 41]. While it is entirely possible that members of the network were exposed to radical videos circulating on the Internet, including those of Al-Awlaki, it seems that these face-to-face meetings were crucial to establishing trustworthiness, essential in convincing these individuals to travel overseas, and devising the plans and procuring the funds for them to do so [32, p. 41]. This is not to say that the role of the Internet is unimportant; Internet propaganda positions AS and its adherents as ideological authorities who ought to be esteemed by aspiring members of this network [51].

---

[3] Furthermore, this point does not countenance the hierarchical assumptions raised in the introduction, perhaps most succinctly put in Matthew and Shambaugh's dictum that "Networks are easy to create, but hard to control" (2005, pp. 621–622).

However, this ideological hierarchy did not entail top-down controls over the specialized function of the network, namely the recruitment of combatants for AS.

The two fundraising networks both exhibit a similar structure, and one that is distinct from the delinquency homophily—the tendency of individuals to associate with others of the same kind—exhibited by the recruitment network. Not only are the nodes situated in America strewn across at least two states in both cases, but each network also contains individuals located in Somalia who may have never been to America. The connections between these foreign nodes and others in Somalia are shadowy in both cases, but their individual contact with American nodes is well-documented and integral to the successful function of the network, which in both of these cases was to channel money to Somalia.[4] Unlike the recruitment network, some donors were solicited with jihadi rhetoric, and some funds were donated with full knowledge of their illicit purpose [52, pp. 6–7]. Similar rhetoric may have been used in the recruitment network, but there is no discernible effort within either of the fundraising groups to encourage or facilitate travel to Somalia.

Both fundraising networks exhibit a 'hub' network pattern: in each case, a single node (the broker) was primarily responsible for communicating with an AS leader in Somalia and relaying pertinent information to the rest of the American nodes, although the identity of this node can change over time, as we will see. In one case, the group of contacts in Somalia appeared to form a hub-like structure of their own, while in both cases U.S.-based nodes are arranged in a single hub pattern or multiple hubs that are linked to each other by brokers. The broker between the American and Somali nodes need not be the same individual who brokers between U.S.-based hubs. These 'hub' network structures "introduce an element of hierarchy" to the network [17, p. 12], with those positioned at the center having access to information and control over the flow of information and resources that make them de facto leaders of sorts and analytically special within the network in ways discussed below.

Three more hypotheses follow from these observations:

H3 Recruitment-oriented networks rely on domestic all-channel networks that are geographically concentrated (that is, for the purpose of recruitment, proximity matters)

H4 Fund-raising networks rely on transcontinental hub networks (that is, proximity does not seem an impediment)

H5 Control over access to recruitment networks is informal and decentralized

[4] This is, of course, relative. Plane tickets for the Minneapolis recruits reportedly to cost between $1,500 and $2,000 each; but the Minneapolis and St.Louis/San Diego fundraising group each channeled approximately $8,500 and $30,000 to AS operatives in Somalia.

4.3 Presence of brokers

The recruiting networks were comprised of many individuals who are high in both degree and betweenness centrality, as well as a few peripheral nodes that are low in both. In this case, no one node or small groups of nodes appear to be brokers, especially if all or even some of the implied but unspecified linkages described above obtain. That is, this network is notable for a lack of actors who are able to control the flow of information and resources to the extent that they can substantially determine or even influence the activities of the group.

Conversely, the fundraising networks include many individuals with minimal degree centrality (i.e. one link), which logically results in a betweenness centrality of zero. However, within each group a pair or pairs of interacting nodes exhibit high degrees of both betweenness and degree centrality. The link between these pairs constitute the crux of the fundraising operations; without these links, the funds would have to find an alternate sender or receiver: they comprise the main conduit of information and resources for this network.[5] Information (e.g. account numbers) travelled exclusively in one direction (from Somalia to America), while funds travelled exclusively in the other.

The hypotheses that follow are:

H6 Fundraising networks rely heavily on the actions of 'brokers';

H7 Recruitment networks do not rely on brokers.

4.4 Financial dimensions

Most scholarship on hawala centres focuses on how government policy and media attention towards Islamic remittance practices is misplaced and/or futile to stop the minority of hawala transactions that are criminal in nature (see [53, 54, pp. 514–515, 55, pp. 166–167]. The specialized structure of some fundraising networks in the U.S. as elucidated here can give rise to equally specialized strategies for network identification and dismantlement, hence drawing attention away from the vast hawala system and towards targeting individual networks. As Tupman notes, "it is difficult to typologise by financing, as groups resort to a variety of financing activities over time," [56, p. 198]. This potentially holds for AS, as they may utilize other means of accruing funds within Somalia.

Technically, the Minneapolis 'recruitment' network explicitly engaged in fundraising as well as radicalization/recruitment, but the pattern of fundraising and the use of these funds differ qualitatively from the two fundraising networks.

[5] Given, as is the case, that the receiving node controls the information (i.e. the account numbers that correspond to his sub-ordinates) that allows the sending broker to successfully transfer funds to these nodes.

Within the recruitment network, all known fundraising activity was in the form of door-to-door solicitation under false pretense, including sending one of the subjects to Saudi Arabia to study the Koran and supporting orphanages of Somalia. Almost all of these funds were spent in the Minneapolis area to purchase airline tickets for the men travelling to Somalia. Even in the two cases where this rule does not hold, funds were intended for exclusive use by members of the recruitment network while in Somalia.

Fundraising activities coordinated and carried out by this group were necessary to the recruitment function, and in this case the effective use of such recruits required an expensive travel itinerary. What few other funds were raised or distributed by this group were used towards arming specific recruits after their arrival in Somalia. This funding and spending model is not mentioned in the literature: an autonomously funded group disinterested in domestic attacks, sending manpower rather than money to support the central organization. The opposite is true of the other two networks. Funds raised were not consumed by the Western portion of the network; instead, they were transferred directly overseas to members of AS' administrative network in Somalia for general disbursement. When funds were destined for specific purchases, their quantity and use were determined by the central group. Overall, the recruitment network did not raise funds for use beyond the network, while the American nodes of the fundraising networks did the exact opposite. The fundraising networks represent a nuanced form of hierarchy between the centre and the periphery, where the ideological authority of the centre compelled actors in the West to mobilize on behalf of the centre, which in turn relied in part on funds raised by the periphery to achieve objectives in Somalia. This interdependence hinges on ideological authority, or in Bakker et al.'s terms, external legitimacy, which a grievance-driven group, such as AS, needs to maintain to convince people to risk legal prosecution by offering financial support [57, p. 54].

There is no evidence that any of these networks, irrespective of their structure or function, received funds or material resources from AS affiliates outside of the network, nor did they expect such assistance.[6] This financial isolation did not result in a turn to criminal activity to procure funds [7, 58].[7] This is especially interesting in the case of the recruitment network, which contained several individuals with prior criminal records, including at least one individual indicted on charges of theft [59, 60]. The use of hawala services for remitting funds to Somalia and the manner in which these funds were obtained represents an aberration that is under-represented in the literature. This is especially pertinent to the case of AS

[6] While it may have been more expeditious for the individuals in the recruitment network to receive money from AS for airfare, no actors in this network seem to have stayed behind for lack of funds. For example, Abdiweli Isse planned to depart in December 2007, but was held back because his identification needed to travel internationally was not up-to-date. He departed successfully in October 2009.

[7] Door-to-door solicitation (under false pretense or not) for an organization recognized as a foreign terrorist entity is a criminal activity; but door-to-door solicitation separate from the terrorist element is not, unlike activities such as theft or trafficking narcotics.

and Somalia because the failed state does not support Western banking/wiring services such as Western Union; so, this group cannot effectively make use of the Western banking system, as terrorist groups such as Al-Qaida have in the past [54, p. 517, 61, p. 289]. A final hypothesis follows:

H8 Transfer of funds from the central network to peripheral networks is not necessarily indicative of the pursuit of terrorist ends

Table 2 summarizes key differences and similarities between recruitment and fundraising networks observed in this section:

Table 2 Summary of fundraising and recruitment characteristics by network

Network Type	Fundraising	Recruitment
Network Structure	Hub	All-channel
Select nodes function as brokers	Yes	No
Centrality characteristics of nodes	Brokers: High betweenness centrality, low degree centrality All other nodes: Low betweenness centrality, low degree centrality	High betweenness centrality, high degree centrality
International linkages	Yes	No
Intent to commit domestic attacks	No	No
Use of funds	Remittances: American donors to AS contacts in Somalia	Internal domestic activities: mostly to purchase airfare

5 Method/Social Network Analysis

The nature of the evidence perforce imposes limits on Social Network Analysis (SNA) that has implications for its application to terrorist networks: the number of data points is insufficient to employ conventional network metrics. As a result, the application of SNA in this chapter relies more on concepts and visualisation than on quantitative measures. However unconventional, the results warrant the application of SNA. This represents the ideal opportunity for a small-*n* study; cases are few, information is scant but has had its validity tested by courts in multiple jurisdictions, and a study can be carried out with the available information, rather than waiting for more detailed information that might never materialize [62, p. 348].

Owing to their smaller scope, small-*n* studies can be undertaken in a timely manner using government documents [62, p. 347]. While the generation of these documents usually requires those involved to be convicted, this approach can be operationalized without interviews and other information gleaned directly from terror suspects, a daunting task whether they are imprisoned or not. The government documents to which this study has access focus on recorded interactions between suspects as evidence showing act and/or intent. This kind of information is essential to SNA, which is primarily concerned with determining, mapping and analyzing the structures created by patterns of interaction between individuals [63, p. 2].

Many insights about terrorist networks, including their genesis, purpose, and the way they work, stem from interviews with incarcerated and former terrorists or their associates [64–72]. Yet, such subjective evidence needs to be corroborated.[8] Furthermore, victimization surveys and crime-reporting data were never designed to capture terrorism-related offences; offender reports are difficult to access, and data held by intelligence services are hard to come by [74]. In light of these methodological challenges, this study relies instead on readily replicable data.

The illicit nature of these networks as well as the necessarily covert nature of investigative methods to detect and dismember them encourages secrecy by network members and government officials alike. Identification of the temporal and geospatial patterns of the stochastic networks in this study relied exclusively on open sources, including newspapers, academic research, court records, think tanks, governments and NGOs, and the Internet (caveat emptor). The nature of these sources imposes certain limitations. For example, the network diagrams below represent only links between nodes that are confirmed by these sources. Additional links between an individual and other nodes of a network are known to exist, but court documents are too vague to draw these links reliably.[9]

5.1 Case and Node Selection

This analysis expands an earlier pilot study that investigated the structures and functions of three AS networks operating inside the U.S. between 2007 and 2010

[8] First, interviewees are subject to sampling bias and, consequently, information gleaned from interviews subject to omitted variables and less-robust results, since convicted terrorists are difficult to access and most refuse to be interviewed [73]. Second, we cannot just take the claims of interview subjects at face value without corroborating evidence. Third, *ex post facto* interviews are prone to the psychological phenomenon of hindsight bias: an interviewee's memory is susceptible to distortion when asked to recollect and reconstruct content. They may also intend to deceive. Fourth, interviews may suffer from the Hawthorne effect: people change their answers by virtue of the fact that they know that they are being studied. Fifth, interview results are subject to coding bias. Sixth, information gained through interviews is subject to a priming effect that is inherent in the way questions are posed and the order in which they are posed.

[9] The networks mapped herein include individuals who have interacted and/or coordinated with each other for purposes related to AS. They exclude friends and family members who may have interacted with these individuals during their recruitment yet remained oblivious to their involvement with AS until after their departure and/or death. As a result, not all known connections between radicalized individuals and members of the Somali diaspora are mapped and analyzed; we required a strong indication that advancing of AS-related objectives was part of the relationship between two individuals.

[75].[10]. Not only has more information since come to light on these three networks, but sufficient information has surfaced on an AS network operating in Australia to warrant expanding the scope. International comparison now makes it possible to control for factors endogenous to the United States. Addition of the Australian case also increases the reliability of the findings by virtue of a comparison case for the Minneapolis-based recruiting network.

These are not the only known AS networks, but information about the others is too vague to map the network and perform SNA, especially with a reasonable degree of certainty. For other cases, the absence of court documents significantly hinders the confirmation of biographical detail and interactions between network nodes on which SNA is premised [116].

Toronto, for example, home to a majority of Canada's Somalia diaspora estimated at 150,000, has been linked to an apparent AS network [79]. At least five men left for Somalia to fight with AS where at least two of them died. All originated in the Rexdale/Dixon Park area, worshipped at the same mosque and departed Canada for Somalia within only a few days of each other in October 2009 [80]. However, only one arrest has been made in the case, and the trial is under a publication ban.[11] Our discussion of this group is thus based primarily on media reports and will consequently not be accompanied by a network map and play only a supplementary role.

Similarly, AS networks have been reported in the U.K. and the Netherlands. British authorities reportedly dismantled a network that smuggled Khat from the U.K. to the U.S. and Canada, where it is a controlled substance used almost exclusively by Somali males [82, 83]. This is of relevance to this study because it appears to be a fundraising network that, at first glance, deviates from the fundraising networks based in the United States. Besides brief mentions of a Netherlands-based AS network in the media, a group of Somali men that included the father of AS' then leader Farhan Mohamed Kahiye were arrested in late 2010, but ultimately turned out to centre on a local case of extortion with only loose links to AS, if any [84, 85, 117].

---

[10] By definition, a network contains >2 nodes. For this reason, "lone wolves" as well as those who worked in pairs are excluded from this study. This is worth noting because as a consequence several prominent AS operatives will be excluded from this study. This includes about a dozen American cases, most notably Omar Hammami, who grew up in Atlanta with Syrian and American parents to become AS' top English-language propagandist before he and AS became disillusioned with each other in early 2013. This also excludes Danish cartoonist Kurt Westergaard's attacker, who acted alone in his attempt to murder Westergaard for his depiction of the prophet Mohammed, as well as two Danish-Somali brothers arrested in 2012 [76–78]

[11] The sole arrested in connection to this group, Mohammed Hersi, has been set to stand trial since December 2011, but a publication ban is in place, and no new information has come to light [81].

6 Evidence

6.1 Minneapolis Recruitment Network

This network consists of 23 individuals, all of whom were American citizens or legal permanent residents of Somali descent, save one American-born Caucasian [86]. The purposes of this network were to (1) radicalize, (2) encourage its members to travel to Somalia to wage *jihad* with AS, and (3) raise funds to defray the costs of travel. All 23 men left for Somalia during two waves about a year apart, with others leaving in-between and later. The first of these waves occurred in early December 2007, when six men aged 22 to 26 boarded planes en route to Somalia in the space of eight days [87, 88] [89, pp. 4–5]. For three months prior, they had met in mosques, restaurants, and private residences to discuss their plans and to co-ordinate fundraising activities to pay for airfare. All meetings involved Mahamud Said Omar (MSO) and Omer Abdi Mohamed (OAM) who encouraged the others to travel to Somalia; MSO also helped raise funds for airline tickets and at least partially funded the tickets for these men [89] (Government's Total Briefing, 2012; pp. 12–16). Of the six, two pairs of men shared the same travel itinerary to Somalia [89, pp. 4–6]. Others left sporadically over the following months—MSO himself in January 2008, Zakaria Maruf (ZM) in February 2008, and Mustafa Ali Salat (MAS) and Mohammed Abdullahi Hassan (MAH) in August 2008, on the same flight [89, pp. 4–6, 90].

The second wave occurred in 2008 when six men aged 17 to 26 vanished from Minneapolis, including at least three in the period of 1–4 November [91–93]. At least four of these men had met regularly in the Carlson building on the University of Minnesota campus in Minneapolis where two of them worked as security guards [91]. The last known departures before federal authorities closed in on the conspiracy occurred in October 2009 when five men left Minneapolis destined for the U.S.-Mexico border at San Ysidro, which three of them ended up crossing en route to Somalia [9, 94, 95, p. 6]. Two other young men have reportedly left Minneapolis since for AS-related reasons, although no connections to this or any other network can be established [87]. MSO and OAM played an important role in the initial wave of departures without departing themselves at that time; in most of these later departures they had less of a role, despite the high interconnectedness of all those who left. More broadly, while OAM, MSO, ZM and CAF helped to radicalize recruits and co-ordinate fundraising roles, they all travelled to Somalia at some point (CAF and ZM died there). This suggests that no node of the network consistently acted as a 'conveyor belt' to recruit individuals and possibly assist in funding their voyage to Somalia.

At first glance, this pattern of departures resembles a series of unrelated conspiracies. SNA, however, reveals a high degree of interconnectivity pointing to an all-channel network[12] among the co-conspirators: as Fig. 1 shows, of 23 individuals implicated[13] in the plot, only three do not have a confirmed or highly suspected link to another member of the plot.

Additional links, which could not be mapped due to less reliable information that proved more difficult to replicate, corroborate the all-channel nature of this network:

Cabdulaahi Ahmed Faarax (CAF) travelled to Somalia in early 2007, apparently under his own volition, and court documents claim that between September and December 2007 he told some involved in sending the six men to Somalia in December 2007 that he experienced "true brotherhood" there [9, p. 9]. Court documents do not stipulate with which of these eight men he met, but the evidence suggests that CAF is probably connected to more of these nodes than is apparent from Fig. 1.[14]

Although the eight men involved in the December 2007 departures are confirmed to have met multiple times, it is unclear whether all men were at all meetings; therefore, links in Fig. 1 among the six men who left reflect only confirmed interactions and common travel itineraries. Yet, these six men were probably more interconnected than is apparent, lest part of Fig. 1 be (mis)interpreted to depict a 'hub' network pattern surrounding MSO and OAM.

Court records indicate that MSO had contact with at least some of the men who left in November 2008, although the vagueness of these documents makes it difficult to establish definitive links between nodes [89]. Court documents also indicate that he travelled to Somalia between January and April 2008, supplying funds to unspecified members of the network so that they could procure firearms [96].

Three nodes lack any verifiable link to any other node in the network. However, the degree of confidence that these nodes were indeed part of this network is high because:

Court records for Adarus Abdulle Ali (AAA2) have him meeting with a group of men to discuss plans to travel to Somalia for the purposes of assisting AS and accompanying one of these men to the airport [97, p. 1]. However, the court report does not indicate the group of men with which he attended that meeting or which individual he accompanied.

Links between Adbikadir Ali Abdi (AAA) and the other members of the plot cannot be confirmed, although the fact that he was indicted in the same document

[12] All-channel networks tend to be horizontal with a high interconnectivity of nodes. If any nodes appear to be especially central and interconnected, they are few and only marginally more influential than the rest.

[13] 'Implicated' means that the individual was either indicted by an American court on terrorist charges related to these activities, and/or the individual has personally admitted involvement or is widely acknowledged by AS, but was killed before she could be indicted. Burhan Hassan, who departed in November 2008, exemplifies the latter.

[14] This refers to the six men that departed in December 2007, as well as Mahamud Said Omar and Omer Abdi Mohamud.

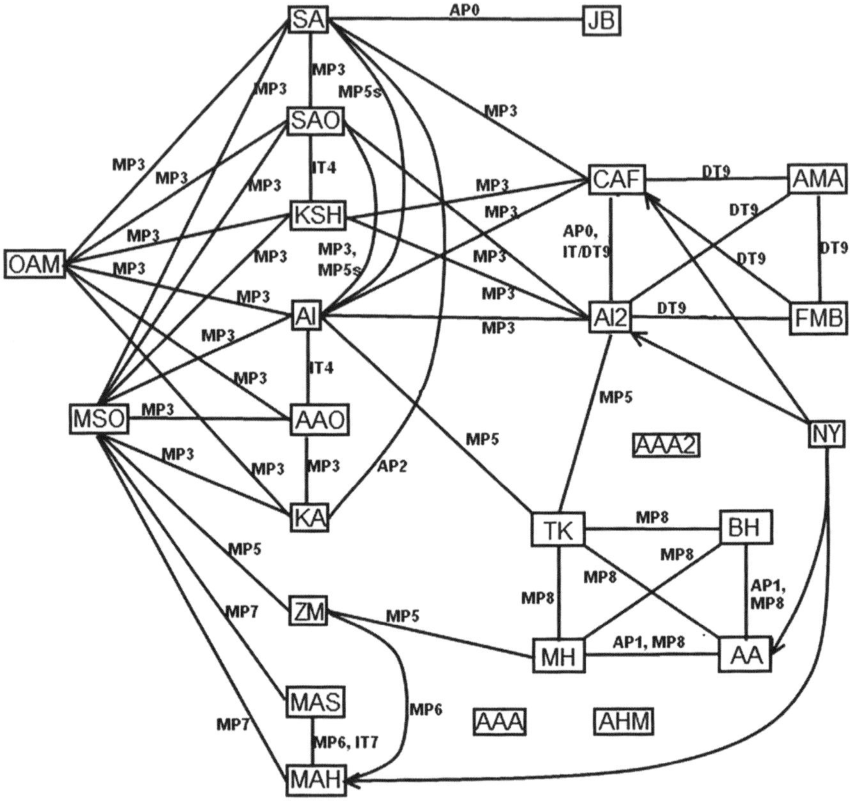

Figure 1 Legend

Node Identity			Link Quality		Link Duration	
AA	Abdisalan Ali		IT	international travel for the furtheranceof illicit objectives	1	2005 - 2008
AAA	Abdikadir Ali Abdi					
AAA2	Adarus Abdulle Ali		DT	domestic travel for the furtherance of illicit objectives	2	March 2007
AAO	Ahmed Ali Omar					
AHM	Ahmed Hussein Mahamud		AP	associates prior to involvement in illicit network	3	September – December 2007

Fig. 1 Minneapolis recruitment network (It should be noted that the links here represent the quality and duration of links *between* people; therefore, international or domestic travel independent of other network nodes is not indicated on this sociogram, but is indicated in the text.)

as ten of the other individuals here as well as his departure during November 2008 intimate links to at least some of the nodes of the network [12].

AHM is confirmed to have raised funds for several of the men who departed between August and November 2008. On three separate occasions between April 2009 and April 2010, once they had travelled to Somalia, he also provided two of the men with money at their behest [98, pp. 2–3]. However, court documents do not indicate to whom AHM was linked.

AI	Abdifatah Isse
AI2	Abdiweli Isse
AMA	Abdow Munye Abdow
BH	Burhan Hassan
CAF	Cabdulaahi Ahmed Faarax
FMB	Farah Mohamad Beledi
JB	Jamal Banna
KA	Khalid Abshir
KSH	Kamal Said Hassan
MAH	Mohamed Abdullahi Hassan
MAS	Mustafa Ali Salat
MH	Mohamud Hassan
MSO	Mahamud Said Omar
OAM	Omer Abdi Mohamed
SA	Shirwa Ahmed
SAO	Salah Ahmed Osman
TK	Troy Kastigar
ZM	Zakaria Maruf

→	(placed at receiving end of link) transfer of funds for furtherance of illicit objectives	4	December 2007
i	placed by arrow, indicates **i**nternational transfer	5	Early 2008
d	placed by arrow, indicates **d**omestic transfer	6	February – August 2008
'n'	Number placed by arrow indicates the number of transfers (if n>1)	7	August 2008
MP	≥1 **m**eeting in **p**erson for the furtherance of illicit objectives	8	Fall 2008
MT	≥1 **m**eeting by **t**elephone for the furtherance of illicit objectives	9	October 2009
FP	furtherance of network objectives under **f**alse **p**retenses	0	Unknown/unverifiable
UK	unknown/unverifiable		

Fig. 1 (continued)

Stohl and Stohl hypothesize that ties of friendship and acquaintance can form the bedrock of a resilient network, and are often strong ties that appear to be weak when scrutinized [31, pp. 101–102, 25, p. 49]. There are a few confirmed links between individual nodes that pre-dated any known radicalization or illicit activity, but such antecedent ties probably existed, especially considering the geographic concentration of the nodes as well as the substantial number of subjects who attended the same educational institutions as well as the Abubakar As-Sadique mosque in Minneapolis. One confirmed example are Shirwa Ahmed and Jamal Bana. The latter has no other verifiable links to the network, despite his departure concurrent with five others in November 2008.

Within the recruitment network, all known fundraising activity was in the form of door-to-door solicitation under false pretense, including sending one of the subjects to Saudi Arabia to study the Koran and supporting orphanages of Somalia. Almost all of these funds were spent in the Minneapolis area to purchase airline tickets for the men travelling to Somalia. Even in the two cases where this rule does not hold, the funds were intended for use by individual members of the network while in Somalia. This includes MSO's trip to Somalia between January and April 2008, where he supplied money to help unspecified members of the MRN to purchase weapons, and AHM's transfer of small sums of money (three transactions totaling $200) to help unspecified members of this network to purchase firearms.

In addition to raising money for themselves and others within the MRN to travel to Somalia, one node within the network funded several nodes' trips to Somalia without the intention of leaving for Somalia herself. Nima Yusuf (NY) funded individual combatants from the MRN while they were in Somalia, but unlike MSO, this node is a woman who never intended to travel to Somalia. She is confirmed to have sent money to four nodes within the MRN (AA, AI2, CAF, and MAH) [99, pp. 2–4]. These connections reinforce the all-channel structure of the network; NY did not fund just a single sub-group of the larger network (e.g. the group of AI2, AMA, CAF, and FMB that met together repeatedly at the University of Minnesota), but two individuals within this group as well as two others with no documented contact with A12, CAF or each other. While this funding is distinct from the larger sums directed towards the general cause that fundraising networks generate, this seems to be relatively common behaviour in recruiting networks, undertaken by MSO, NY, as well as SEA in the Australia Recruiting Network.

In sum, the MRN appears to be the largest publicly known AS network in the West. Although almost two years separate the first known departure from the last, a majority of nodes are highly interconnected and exhibit no discernible hub or chain patterns. In terms of their activities and objectives (1) there is no indication that they were in any way plotting an attack on American soil, (2) funds raised by the network were solely to facilitate members' travel to Somalia and (3) many nodes raised at least a portion of their own travel funds [12, 100, pp. 12–16].

6.2 Australia Recruiting Network

At first glance this group of individuals, who were indicted on charges of conspiring to attack the Holsworthy Barracks near Sydney, Australia in 2009, appears to run counter to trends established by the MRN. In stark contrast to the MRN, members of this network attempted to hatch a domestic plot; as opposed to funding the travel of recruits to Somalia to be under the command of AS leaders there, this network sought to plan its own offensive and execute it domestically [101]. However, an investigation of the group reveals that its initial intentions were in fact very similar to those of the MRN. In late 2008, Walid Osman Mohamed (WOM) left Australia for Somalia to fight for AS (Fig. 2).

Wissam Mahmoud Fattal (WMF) accompanied WOM with the same intent, but lacked a proper visa for the first leg of the voyage to Kenya [102]. The plan for the domestic attack was concocted only once WMF had determined that securing the appropriate visa would take too long [102]. Furthermore, Yacqub Khayre (YK) travelled to Somalia to train with AS from April until shortly before his arrest in August 2009 [103]. Why he returned is not known, but it does not appear that this training was for use back in Australia. The network includes conspirators who never attempted to travel to Somalia, but it is unclear whether these individuals were oriented towards a domestic attack prior to WMF's inability to travel to

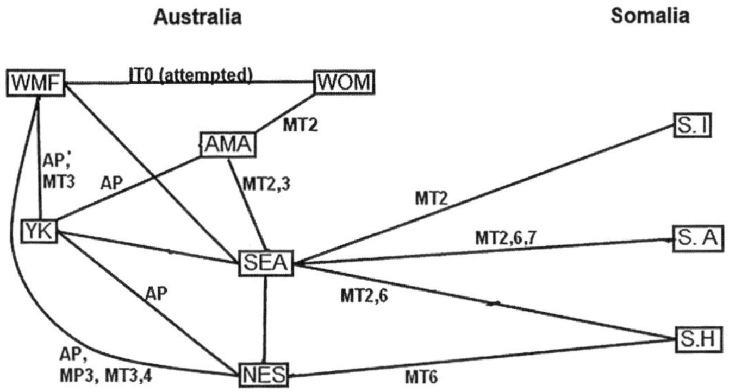

Figure 2 Legend

Node Identity		Link Quality		Link Duration	
AMA	Abdirahman Mohamud Ahmed	**IT**	**in**ternational **t**ravel for the furtherance of illicit objectives	**0**	December 2008
NES	Nayef El Sayed	**DT**	**d**omestic **t**ravel for the furtherance of illicit objectives	**1**	January 2009
SA	Sheikh Abdirahman	**AP**	**a**ssociates **p**rior to involvement in illicit network	**2**	February 2009
SH	Sheikh Hayakallah	→	(placed at receiving end of link) transfer of funds for furtherance of illicit objectives	**3**	March 2009
SI	Sheikh Ikrama	**i**	placed by arrow, indicates **i**nternational transfer	**4**	April 2009
SEA	Saney Edow Aweys	**d**	placed by arrow, indicates **d**omestic transfer	**5**	May 2009
WMF	Wissam Mahmoud Fattal	**'n'**	Number placed by arrow indicates the number of transfers (if n>1)	**6**	June 2009
WOM	Walid Osman Mohamed	**MP**	$\geq$1 **m**eeting in **p**erson for the furtherance of illicit objectives	**7**	July 2009
YK	Yacqub Khayre	**MT**	$\geq$1 **m**eeting by **t**elephone for the furtherance of illicit objectives		
*identified as such in court documents		**FP**	Furtherance of network objectives under **f**alse **p**retenses		
		UK	unknown/unverifiable		

Fig. 2 Australia recruitment network

Somalia. Saney Edow Aweys (SEA) supported WOM before and after his travel to Somalia, and he was also the main conduit among a trio of AS-affiliated sheikhs (SA, SH, and SI) situated in Somalia [102]. However, it was unclear how he came into contact with the sheikhs, and their first documented interaction took place in February 2009, two months after WMFs aborted departure from Australia [102].

Interaction with AS authorities in Somalia is documented in the two fundraising networks, but not in the MRN. Interaction between the Somali and Western networks is mediated by a broker, a single node situated in the Western network which receives and discriminately discloses information from Somalia to other

Western nodes. While Naser el-Sayed (NES) at one time joined SEA in a conference call to SH in June 2009, NES never spoke to the sheikhs alone; SEA was responsible for all other interaction with the sheikhs and thus is clearly a broker for the ARN receiving instructions and advice from authorities within the central AS network [102]. Had the ARN followed the advice of the sheikhs, the function of the group would have remained limited to sending able men to Somalia to fight. After months of deliberation and hesitation, SA, SH, and SI decided to deny the ARN the fatwa they had repeatedly requested for the attack on the Holsworthy Barracks [102]. The plot targeting the Barracks was not far along when Australian authorities intervened; the most substantial action had been WMF'S trip from Melbourne to Sydney to perform some naively superficial surveillance on the Barrack's exterior walls [102]. This attack was actively discouraged by the central AS network as an alternative to travelling to Somalia, which was deemed too costly. Notwithstanding the network's domestic plot, it is comparable to the MRN.

The overall failure of the ARN to facilitate travel of potential AS combatants to Somalia aside, the network is similar to the MRN in key respects. While the MRN has many more nodes than the ARN, the Western portions of both are unequivocally all-channel networks, in contrast to the fundraising networks' distinct hub patterns. All six of the Australian nodes have documented connections to at least two other domestic nodes, and four of the six have connections to at least three. While some connections between nodes are stronger than others; for example, it was noted in court that WMF and SEA had less contact with each other than their co-accused, NES, had with either of them [102]. It also seems that the all-channel structure of this network was reinforced by documented regular interaction between nodes outside of their illicit activities in a more substantial manner than observed in the MRN. Not only did media reports claim that several of the six men attended Preston Mosque in Melbourne together, but NES, WMF and YK worked together as bricklayers beginning early in 2009 until their arrest in August of that year [102, 104].[15]

In addition, funds were sent to Somalia (by SEA to WOM) to help fund combatants who had travelled there after joining the domestic network [102]. This differs from the two fundraising networks insofar as money was raised and sent to Somalia to aid AS' general cause as determined by authorities there, and not by nodes within the Western network. Furthermore, the MRN and the domestic nodes of the ARN are geographically concentrated, unlike those of the fundraising networks. Just as the MRN is concentrated in Minneapolis and St. Paul, Minnesota, all of the ARN's domestic nodes lived in Melbourne and surrounding suburbs.

[15] While accompanying one another to the Preston Mosque is worth noting, reports of this behaviour were not specific enough to confirm interactions between specific nodes in this regard.

6.3 Toronto Recruitment Network

A group consisting mostly of young men[16] in Canada also have links to AS, and while information about them is too scant to render a network map, what little data is available suggests an all-channel network focused on sending young men to fight in Somalia. Like the MRN and ARN, the individuals who make up this group are quite geographically concentrated; besides all residing in Toronto, the majority lived in the Rexdale/Dixon Park area of Toronto, a region known for its high concentration of Somali-Canadians [105, 107, 108]. Among these are six young men who left for Somalia in October 2009. At least two have since been killed while fighting for AS, one as recently as April 2013 [109, 110]. For reasons aforementioned, information on these individuals and their interactions are scarce. The only locus at which all of these nodes intersect is the Abu Huraira Centre in Rexdale, which three of the six had begun attending before their departure, and where they were described as "occasional" congregants [107].

While there is no indication that authorities at the mosque facilitated or condoned these activities, the mosque presents a social atmosphere where interactions away from religious leaders are common and vital to the community. Similarly, all six Australian members of the ARN attended Preston Mosque in Melbourne, and many MRN members attended Abubakar As-Saddique in Minneapolis. In all three cases, no mosque leaders have been connected to these networks, and information about whatever interactions may have occurred is too vague to draw precise, node-to-node links. Mohamed Hersi, who also worshipped at Abu Huraira, was arrested as he attempted to leave Canada for Somalia in March 2011, about a year and a half after the initial group of six [108]. It is (currently) unknown if he is connected to this group, but he is also the only person to have been arrested in Canada in connection to AS. Ergo, information from his trial will likely be crucial to understanding more about any networking activity amongst the Toronto recruits, and whether he is actually linked to the others.

6.4 Minneapolis Fundraising Network

At about the same time as the MRN, three individuals in Minneapolis and one in Columbus, Ohio conspired to provide financial support to AS. However, they did

---

[16] Most in this group are young men, with the notable outlier Abdulli Ali Afrah a.k.a. "Aspro", who rose through AS' ranks during its first few years to be killed in a mortar attack in early 2008 at age 56. Furthermore, two young Somali-Canadian women from Toronto have also mysteriously left for Somalia and are rumoured to be there aiding AS. One of these women is Asli Nur, the 19-year-old niece of Mohamed Abdullahi Mohamed, former Prime Minister of Somalia. None of these individuals have been connected to the Abu Huraira Islamic Center or to the other recruits in any other known way, and so these cases may be lone wolves unconnected to a potential network likely based around Abu Huraira [105, 106].

not also conspire to leave the U.S. for Somalia. Their activities focused solely on fundraising and transferring funds to established contacts in Somalia for the purpose of furthering unspecified AS objectives. This network appears to consist of two hub networks, situated in Somalia and the U.S., respectively.

Beginning in September 2008, Amina Farah Ali (AFA) of Minneapolis was confirmed to have been in contact with an AS militant in Somalia, described in court documents as "UC1", a financial representative for the organization who was promoted to an administrative governor of several AS-controlled regions in February 2009 [52, pp. 2–3]. Court documents identify four other contacts in Somalia who were subordinates of UC1 and who do not appear to have interacted with one another, three of whom oversaw accounts to which AFA transferred funds [52, pp. 2–3]. The account numbers corresponding to these individuals were supplied to AFA by UC1 with whom AFA was in contact repeatedly between September 2008 and July 2009 [52, pp. 6–8]. Court documents have AFA corresponding directly with two of these subordinates, interacting with one only once in May 2009, and contacting the other in October 2008 to arrange for him to be a guest speaker at a fundraising teleconference that same month [52, pp. 6–8] (Fig. 3).

In America, AFA was in contact with three individuals, one of whom assisted with bookkeeping and recording pledges, while the other two collected funds from donors and directed them to AFA for transfer to Somalia. One of these actors was explicitly instructed by AFA to collect funds under false pretense, while she tasked the other with collecting pledges made during one of the teleconferences [52, pp. 2–3].[17] The available information suggests that these three nodes never interacted. Unlike the MRN, there are no detectable ambiguities to suggest further links. Notwithstanding the geographic overlap between the MRN and some of the nodes of the MFN, an exhaustive search revealed no apparent connections between these networks. The conclusion to be drawn is that the MFN appears to be composed of two hub networks with AFA as the broker between the Minneapolitan and Somali hubs. While the individuals in Somalia may have other unknown functions in the larger AS network there, the conspirators in America appear to be concerned exclusively with supplying funds to be used at the discretion of AS operatives in Somalia.

[17] Of this network, only Ali and the book-keeper (Hawo Mohamed Hassan) were indicted on charges by the United States government. Information about un-indicted co-conspirators was crucial to justifying these indictments and is important here (and in the third network, to be discussed shortly) in accurately portraying the nature of this network's activities and the structure of the network necessary for these activities. Information on non-indicted individuals in the Minneapolis recruitment network is not included because no such individuals can be credibly implicated in any of the network's illicit activities, despite numerous calls from the community alleging complicity of the religious leadership of the Abu-Bakar As-Saddique mosque in Minneapolis.

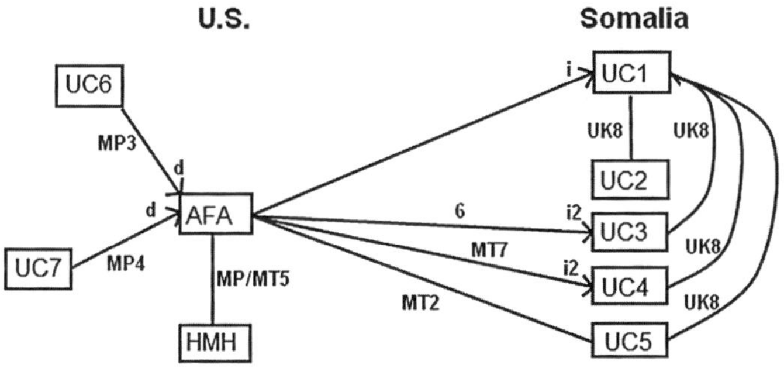

Figure 3 Legend

Node Identity		Link Quality		Link Duration	
AFA	Amina Farah Ali	**IT**	**i**nternational **t**ravel for the furtherance of illicit objectives	1	September 2008 – July 2009
HMH	Hawo Mohamed Hassan	**DT**	**d**omestic **t**ravel for the furtherance of illicit objectives	2	October 2008
UC1	Uninidicted Conspirator 1*	**AP**	**a**ssociates **p**rior to involvement in illicit network	3	November 2008
UC2	Uninidicted Conspirator 2*	→	(placed at receiving end of link) transfer of funds for furtherance of illicit objectives	4	January 2009
UC3	Uninidicted Conspirator 3*	**i**	placed by arrow, indicates **i**nternational transfer	5	February 2009
UC4	Uninidicted Conspirator 4*	**d**	placed by arrow, indicates **d**omestic transfer	6	April 2009
UC5	Uninidicted Conspirator 5*	**'n'**	Number placed by arrow indicates the number of transfers (if n>1)	7	May 2009
UC6	Uninidicted Conspirator 6*	**MP**	≥1 **m**eeting in **p**erson for the furtherance of illicit objectives	8	Unknown/ unverifiable
UC7	Uninidicted Conspirator 7*	**MT**	≥1 **m**eeting by **t**elephone for the furtherance of illicit objectives		
*identified as such in court documents		**FP**	Furtherance of network objectives under **f**alse **p**retenses		
		UK	unknown/unverifiable		

Fig. 3 Minneapolis fundraising network

6.4.1 St. Louis/San Diego Fundraising Network

Akin to the MFN, the SL/SD FN, active between January 2008 and March 2009, was concerned exclusively with raising funds and transferring them to contacts in Somalia for use there, including the purchase of a vehicle to transport AS militants [111, pp. 10–12]. Akin to the MFN, the function of the SL/SD FN relied on repeated contact with AS operatives in Somalia, which were more demanding than those in the MFN: they requested specific amounts of money for particular purposes. While court documents are unclear on how these funds were raised, the manner in which these funds were transferred to Somalia is indicative of a network similar in structure to the MFN. The main actor in this network was Mohamud Abdi Yusuf (MAY) of St. Louis, MO. From January to July 2008, he was repeatedly in contact with Basaaly Saeed Moalin (BM), based in San Diego, California. In February 2008 BM communicated with Aryow (A), then the leader of AS, and BM received funds with instructions from MAY to transfer them to "Omar Mataan" (OM) based in Somalia [111, pp. 4, 9, 12–13]. Yusuf himself sent five installments to Duane Mohamed Diriye (DMD) in Somalia, and interacted with "Sheeik Saaid" (SS),[18] a contact in Somalia introduced to him by DMD [111, pp. 8–10]. MAY also had multiple conversations with an unknown acquaintance in Somalia with whom he discussed skirmishes between AS and Ethiopian forces [98, pp. 12–13].

Court documents unsealed in early 2013 show that the development of the SD/SL FN is more complex than information available at the time of the pilot study indicated. It now seems that BM acted as a broker among three men collecting funds for al-Shabaab in and around San Diego and Ayrow, who was purported to be a top leader within AS until he was assassinated on May 1, 2008 [112, p. 14]. Independent of MAY or the other nodes in St. Louis, BM co-ordinated a number of hawala transfers totalling several thousand dollars in the first several months of 2009 [112, pp. 8–14]. While BM was crucially aided by one of the network nodes (ID), who worked at a San Diego hawala service and assisted both BM and MM in transferring funds to A in Somalia, all information concerning these transfers (e.g. the aliases the funds were transferred under, when they were sent and how much they total) passed through BM [112, p. 13].

Not only did BM broker information in a manner observed in the MFN and the network BM and his co-conspirators would link up with after A's death, but this network also exhibits a hub pattern, with little interaction among the three fundraisers that depended on BM for information. Only a single phone call between ID and MM is documented, and it concerns a hawala transfer requested by A and co-ordinated by BM [112, p. 12]. Thus, this network exhibits the structure and dynamic posited earlier as characteristic of fundraising networks, qualities these nodes maintain after Ayrow's death.

Despite BM's apparent autonomy in communicating with Ayrow and co-ordinating transfers of funds to Somalia, he had been in contact with MAY in St. Louis about sending funds to AS prior to Ayrow's death, as early as January 2009 [111, p. 9]. After A's death, BM was given money by MAY to transfer to one of MAY's contacts in Somalia [111, pp. 12–13]. This activity relied on MAY's brokerage of information, whereas BM's role was simply to act on MAY's instructions. MAY became the broker of the network because of his resilient contacts in Somalia, where BM relinquished his status as broker due to A's death. Interestingly, BM remained a broker between his associates in San Diego and the

---

[18] This name as well as "Omar Mataan" are always in quotations in court documents, leading to the suspicion that they are known to be code names by American authorities. This further obscures the nature of relations between the nodes situated in Somalia, which are unknown except for the link between Diriye and "Sheeik Saaid".

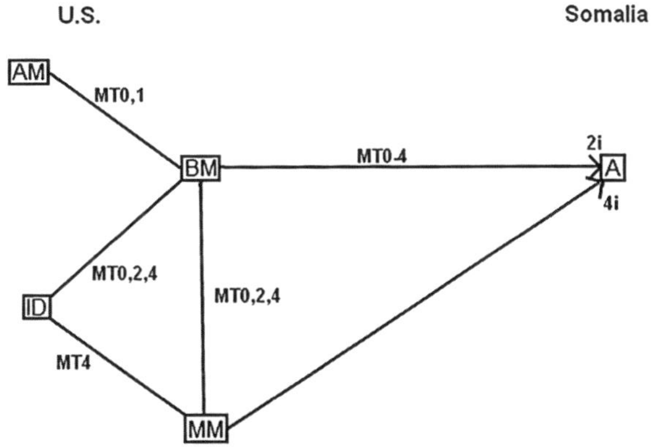

Fig. 4 St. Louis/San Diego fundraising network—before Ayrow's death

hub centered around MAY in San Diego. MAY only spoke with nodes in San Diego other than BM on one occasion, a conference call between MAY, BM, ID and MM in July 2008 [112, p. 16]. Otherwise, MAY and BM were the centers of their respective hubs, each being the primary contact for the other. Who served as a broker between fundraising efforts in the West and AS leaders in Somalia appears to be largely determined by the strength of pre-existing ties to these authorities. However, there is no information on how these initial contacts were formed, or how BM in San Diego came to associate with MAY in St. Louis [112, p. 14] (Figs. 4 and 5).[19]

From May 2008 until March 2009, Yusuf was also linked to Abdi Mahdi Hussein (AMH), an employee of Qaran Financial Express, LLC, a *hawala* remittance firm with a branch in Minneapolis. AMH agreed to make 14 remittances to an unknown contact in Somalia in a manner that obscured the identities of the sender and the receiver, and in amounts small enough to avoid requiring the sender to provide and verify their identity. However, court documents and the criminal charge against AMH indicate that he was unaware of any connection to AS [111, pp. 20–23]. MAY interacted with four other individuals about various aspects involved in sending money oversees to support AS, and AMH met one of them once in May 2008 [111, p. 21]. This is the only direct link between individuals in America with whom Yusuf discussed his illicit activities.

The activities of the SL/SD FN were coordinated by nodes broken down into hubs. MAY is the most active node on the American side, connected to five other nodes of which only two had contact on a single occasion. One of these contacts, BM, was also at the center of a hub with six individuals aside from MAY, also with no documented connections except through BM. As in the other fundraising network, no node expressed any desire to travel to Somalia. Connections between America and Somalia other than through MAY are inconsequential; BM had a

[19] While a contact of BM's identified as "Kay" did successfully refer BM to Mahad Karate, an AS member in Somalia, no money was ever remitted to him from the Western nodes.

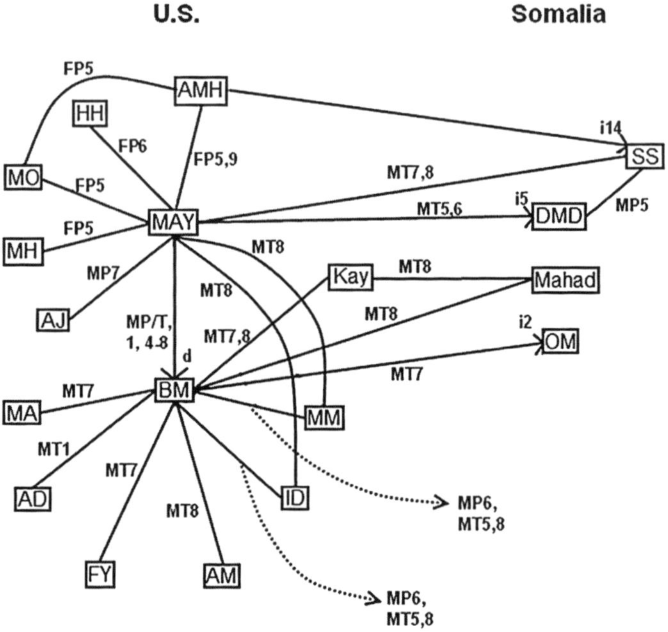

Figure 4-5 Legend

Node Identity		Link Quality		Link Duration	
A	Aryow	**IT**	**i**nternational **t**ravel for the furtherance of illicit objectives	**0**	December 2007
AD	*	**DT**	**d**omestic **t**ravel for the furtherance of illicit objectives	**1**	January 2008
AJ	*	**AP**	**a**ssociates **p**rior to involvement in illicit network	**2**	February 2008
AMH	Abdi Mahdi Hussein	→	(placed at receiving end of link) transfer of funds for furtherance of illicit objectives	**3**	March 2008
BM	Basaaly Saeed Moalin	**i**	placed by arrow, indicates **i**nternational transfer	**4**	April 2008
DD	Duane Mohamed Diriye	**d**	placed by arrow, indicates **d**omestic transfer	**5**	May 2008
FY	*				
HH	*	**'n'**	Number placed by arrow indicates the number of transfers (if n>1)	**6**	June 2008
ID	Isse Doreh				
Kay	*				
Mahad	*	**MP**	≥1 **m**eeting in **p**erson for the furtherance of illicit objectives	**7**	July 2008
MA	*				
MAY	Mohamed Abdi Yusuf	**MT**	≥1 **m**eeting by **t**elephone for the furtherance of illicit objectives	**8**	August 2008
MH	*				

Fig. 5 San Diego/St. Louis fundraising network—After Ayrow's Death

MM	Mohamed Mohamed Mohamud	FP	furtherance of network objectives under false pretenses	9	March 2009
MO	*	UK	unknown/unverifiable		
NM	Ahmed Nasir Taalil Mohamud				
OM	Omar Mataan*				
SS	Shiekh Saaid*				

Fig. 5 (continued)

single conversation with a high-ranking AS official and sent funds to Somalia, but the quantity and destination of these funds were determined by MAY with some instruction from his two contacts in Somalia. Of the nodes in Somalia, DMD was the most important; while SS seems to exert influence over the quantity, timing, and use of the Somalia-bound funds, MAY was introduced to SS through DMD.

The SL/SD FN matches Morselli's predictions for small networks. The individuals connected to MAY and DMD who were situated in America exhibit a low degree centrality and almost non-existent betweenness centrality. Here, brokers play a slightly less crucial role than observed in the MFN because there are multiple links across which funds and information flowed between America and Somalia within this network. However, one pair of brokers, MAY in America and DMD in Somalia, appears to be pivotal to the network's function. While both Hussein and BM sent money to Somalia, and BM had a relatively high betweenness centrality due to his links to MAY as well as A and in Somalia, MAY received and controlled the flow of information from the AS operative in Somalia and controlled the timing and quantity of funds remitted after A's death. Like the Minneapolis network, the removal of the link between MAY and DMD and/or SS (a link forged due to MAY's prior link with DMD) compromised the functionality of the network.

7 Findings and Discussion

The addition of the ARN and the TRN to the initial pilot application of SNA largely confirm the initial hypotheses. The two most striking differences between the MRN and the ARN are the presence of brokers as well as intent to commit a domestic attack. These will be addressed in turn.

In both cases, fundraising networks required a pair of brokers—one in Somalia and one in the U.S. respectively—to facilitate a flow of information that in turn allowed for the remittance of funds to Somalia. As the brokers are essential to the function of these networks (as evidenced by the changes to the SD/SLFN following the death of Ayrow, who was the Somali component of the pair of brokers at the center of that network), fundraising networks are thought to depend on the actions of brokers. The ARN did not rely on brokers in this sense, as they did not rely on (or even follow) information and advice given to them by SA, SH, and SI. Had this advice been heeded, this network's actors would have oriented their efforts towards recruitment. However, individuals joined the group and departed for Somalia before any documented contact with SA, SH, and SI; so, they did not rely on brokers for this function either. The ARN (and perhaps recruiting networks in general) may contain nodes that act as brokers, but nowhere are they seen to rely on these brokers to maintain their network or achieve their objectives.

The presence of a domestic plot may strike more deeply at the initial hypotheses, because it seems to indicate that networks need not be specialized at all; this network exhibits the two distinct functions of recruitment (for foreign training and combat) and planning a domestic attack. However, this discrepancy is mitigated by other available information. As discussed above, the function of recruitment began chronologically prior to a domestic plot being hatched, and the failure to travel abroad is confirmed to have encouraged the idea of an attack within Australia. This presents a counter-intuitive case insofar as security measures (in this case, the particular agreement between Australia and Kenya as to what documentation is required to travel between the two states) actually encouraged an attack on Western soil rather than preventing it. However, it also serves to demonstrate that the recruiting function was the primary aim of this group, and that it fits the scope conditions of a recruitment network. It is possible that terrorist networks aiming to perpetrate a domestic terrorist attack will also organize into an all-channel network, but as of yet there is no known network that fits the parameters of this study; so, further analysis is not yet possible.

Table 3 distinguishes attributes of fundraising and recruitment networks and compares these traits to the ARN:

The ARN deviates somewhat from the structure and dynamic of a recruitment network as posited by the pilot study. While these deviations, for reasons discussed above, do not fundamentally challenge the hypotheses, they introduce certain nuances into the discussion.

H1 Terrorist networks are functionally differentiated

H2 The structure and characteristics of terrorist networks is a function of their purpose

These hypotheses are confirmed by the ARN, and potentially hold true for the TRN. These networks were focused on recruiting and sending individuals to fight in Somalia until the travel plans of a high inter-connected node were frustrated. The ARN exhibited an all-channel structure reminiscent of the MRN, and distinct from the two fundraising networks. The TRN appears to have a similar structure.

H3 Recruitment-oriented networks rely on domestic all-channel networks that are geographically concentrated (that is, for the purpose of recruitment, proximity matters)

Table 3 Asssessment of fundraising and recruitment characteristics by type of network, with comparison to ARN

Network Type	Fundraising	Recruitment	Australian Recruitment Network[a]
Network Structure	Hub	All-channel	Hub
Select nodes function as brokers	Yes	No	Yes
Centrality characteristics of nodes	Brokers: High betweenness centrality, low degree centrality All other nodes: Low betweenness centrality, low degree centrality	High betweenness centrality, high degree centrality	High betweenness centrality, high degree centrality
International linkages	Yes	No	Yes
Intent to commit domestic attacks	No	No	Yes
Use of funds	Remittances: American donors to AS contacts in Somalia	Internal domestic activities: mostly to purchase airfare	Internal domestic activities: mostly to purchase airfare

[a] Due to that aforementioned limits on what can be known about the TRN, only the ARN is compared in this table. The TRN will be discussed along with the ARN in the ensuing comments

H4 Fund-raising networks rely on transcontinental hub networks (that is, proximity does not seem an impediment)

H5 Control over access to recruitment networks is informal and decentralized

The TRN appears to be confined to a single city, much like the MRN. Much of the ARN was concentrated in the city of Melbourne, but also contained three nodes who resided in Somalia during the ARN's operation. Although the ARN contained international linkages, it did not rely on these linkages for recruitment activities or the preparation for a domestic attack that followed. The contention that "for the purpose of recruitment, proximity matters" still obtains, as neither the presence nor the activities of SA, SH, and SI in Somalia affected recruitment activities. The importance of geographical proximity to different types of terrorist activity makes this hypothesis particularly useful to the study of illicit networks, and the interaction between SA, SH, and SI with the Australian nodes of the ARN ought not detract from this finding.

The additional information on BM and the fundraising network he brokered prior to his involvement with MAY shows that fundraising networks adopt a hub pattern even on very small scales. Furthermore, there is no evidence to suggest that anyone from any of the networks under investigation controlled or even made an attempt to control the membership of the ARN; admittance to both fundraising and recruitment networks appears informal and decentralized across all cases.

H6 Fundraising networks rely heavily on the actions of 'brokers';
H7 Recruitment networks do not rely on brokers

Like the MRN, the TRN currently appears devoid of brokers. The claim that recruitment networks do not contain brokers whatsoever cannot be made in light of the ARN, but useful claims can still be consistently made as to the function of brokers within different types of networks. The ARN did not rely on brokers, but instead contained a node (SEA) who had substantially more contact than any other node with SA, SH, and SI. Ergo, SEA may have been in a position to broker the flow of information between the network's Somali and Australian nodes, but this information appears to have been inconsequential to the ARN's tactics, especially since they disregarded the sheikhs' advice regarding the proposed domestic attack. As before, funding networks rely heavily on brokers to coordinate international transfers of funds. Juxtaposing the activities of brokers in the fundraising networks with those in the ARN shows that the latter's brokers are irrelevant.

H8 Transfer of funds from the central network to peripheral networks is not necessarily indicative of the pursuit of terrorist ends

This final hypothesis is strengthened by the ARN. Not only did the central network (SA, SH, SI) not send money or other resources to the peripheral network, but they discouraged the domestic attack and advised the group to specialize solely in recruitment. At least in the case of AS, the dynamics of support between the centre and the periphery run counter to prevailing assumptions about hierarchical, top-down flows of funds and personnel. Conversely, the central AS network relies on networks situated within Western states for infusions of cash and manpower. While the successful transfer of funds requires coordination between the peripheral and central networks, recruitment networks appear to function rather independently, recruiting individuals and funding their travel without advice or expectations from the central network.[20]

8 Conclusion

In a large-scale study of terrorist groups, Piazza divides them into two broads categories based upon the nature of their ultimate objectives and the means used to achieve them. These types are "universalist/abstract" and "strategic". The former is characterized by esoteric goals that are often broad in scale, and by contrast the latter by goals which can be empirically measured and are often more directly

[20] This should not indicate that members of the recruitment networks were not exposed to propaganda available on the Internet that has been produced by AS operatives in Somalia such as Omar Hammami. However, this is obviously distinct from communicating with the central network to co-ordinate the activities of the peripheral network.

achievable because of these concrete goals and a typically smaller scope [113, p. 65]. In turn, the former attempts to achieve these goals mostly through acts designed to draw broad attention and unite some under an ideology while demonizing others, while the latter achieves their objectives through more localized, tactical activities [113, p. 65].

AS is a critical case precisely because it straddles these two categories: not only is AS intent on establishing an Islamic caliphate in Somalia by raising an international militia and combating states' armies, but it also engages in high-profile kidnappings of Westerners and sporadically refers to global jihad [114, 115, 2, pp. 203–204]. It is difficult to think of another terrorist group with the combination of substantial activity among multiple separate networks and reliable documentation of these interactions and activities with the presence of both a central network and subsidiary networks in the West that is hypothesized and fantasized about by much of the current literature on terrorism and many Western politicians.

On the one hand, the sample in this study may be small, thus limiting the inferences that can be drawn from the findings and their generalizability. On the other hand, the evidence on which the study draws is relatively robust. To compare networks across the same time and space makes it possible to control for similarities and differences in ways that would otherwise be more difficult methodologically if context and conditions were held less constant. The initial hypotheses need further empirical scrutiny and validation, both through comparison to other AS networks and through comparison to other terror networks about which reliable information is available, so that brokers can be identified where they exist, linkages confirmed, and an accurate model of the entire network and its relations to a central organization constructed. The fact that the great majority of the nodes in these networks are Somalis living in Western diasporas raises the importance of diasporas and ethnic capital as means of decreasing marginal and transaction costs as an issue that also warrants further study. Ethnic identity compounded by radical Islamist/jihadist ideology certainly had a hand in congealing these networks.

The findings of this study of all known Western networks connected to AS challenges some of the conventional wisdom surrounding the structure and function of terrorist networks, especially those in the West. Most importantly, this study finds that such networks have specialized functions, and that the structure of such networks seems to correlate with these functions. These different functions determine the nature of their relationship with the central organization. This has implications for law enforcement and counter-terrorism.

First, information about the function of a network, even when many of its nodes and linkages remain obscure, can be indicative of its structure and, therefore, how best to intercept it. For example, knowledge that the network is oriented towards raising and remitting funds would warrant the search for a 'broker' node whose disruption would debilitate the function of the network, at least temporarily. By contrast, networks specializing in recruitment appear to be more robust and resilient to the removal of even multiple nodes. As Bakker et al. confirm, much work remains to be done on how networks replace nodes, re-establish links or re-route flows of information and/or resources through other nodes; so, it is difficult to predict how effective the removal of nodes would be over time [57, pp. 56–57]. However, the possibility that a network's function and structure are related is a promising step towards a more nuanced strategy to contain and deter such networks: not all terror networks are alike. This is a significant empirical finding for counter-terrorism. Knowing the function of a network makes it possible to counter it by detecting and debilitating its nodes. Conversely, knowing the structure of a network makes it possible to surmise its purpose.

References

1. Zelin AY (2013) European foreign fighters in Syria. International Centre for the Study of Radicalization, http://icsr.info/2013/04/icsr-insight-european-foreign-fighters-in-syria-2/. Accessed 23 June 2013
2. Shinn D (2011), Al-Shabaab's foreign threat to Somalia. Orbis 203–215
3. Robertson N and Cruikshank P (2012), Somali AQ's western reach, CNN, 24 Feb 2012. http://homeland.house.gov/news/cnn-somali-AQs-western-reach. Accessed 21 Sept 2012
4. IPT News (2011) Prosecutor warns not to ignore al Shabaab, investigative project on terrorism (Dec 6). http://www.investigativeproject.org/3323/prosecutor-warns-not-to-ignore-al-shabaab-threat. Accessed 21 Sept 2012
5. Labott E, State dept: as core weakens, al Qaeda affiliates are top terror threat CNN Security Clearance, http://security.blogs.cnn.com/2012/07/31/state-dept-as-core-weakens-AQ-affilia tes-are-top-terror-threat/. Accessed 21 Sept 2012
6. Dunbar E (2010) Survey: nearly 1 in 3 US Somalis live in Minnesota. Minnesota Public Radio News, USA
7. BBC (2012) Former U.S. soldier Craig Baxam 'helped al-Shabab' (Jan 9)
8. Stickney R, Kreuger P (2010) Accused terrorist was "kind, peaceful man": friends. NBC, San Diego, 6 Aug 2010
9. District of Minnesota, state of Minnesota (2009) Affidavit of Michael N. Cannizarro Jr., United States of America vs. Abdiweli Yassin Isse and Cabdulaahi Ahmed Faarax (Oct 8)
10. New York Times (2011) Joining the fight in Somalia (Oct 30)
11. United States District Court, District of Minnesota (2009) Affidavit in Support of Request for Extradition, United States of America v. Mahamud Said Omar (20 Nov 2009)
12. United Sates District Court, District of Minnesota (2009) Third superseding indictment, United Sates of America v. Ahmed Ali Omar et al (23 Nov 2009)
13. Zwartz B, Cooke D (2009) A battered faith. The Age (6 Aug 2009)
14. Powell WW (1990) Neither market nor hierarchy: network forms of organization. Res Organ Behav 12(1):295–336
15. Stohl C, Stohl M (2007) Networks of terror: theoretical assumptions and pragmatic consequences. Commun Theor 17:93–124
16. Sageman M (2004) Understanding terror networks. University of Pennsylvania Press, Philadelphia
17. Eilstrup-Sangiovanni M, Jones C (2008) Assessing the dangers of illicit networks: why al-Qaida May be less dangerous than many think. Int Secur 33:7–44
18. Giraldo JK, Trinkunas HA (2007) Introduction. In: Giraldo JK, Trinkunas HA (eds) Terrorism financing and state responses: a comparative perspective. Stanford University Press, Stanford, pp 7–20
19. Gunning J (2009) Social movement theory and the study of terrorism. In: Jackson R, Smyth MB, Gunning J (eds) Critical terrorism studies: a new research agenda. Routledge, London

20. Jereon Gunning J (2008) Terrorism, charities, and diasporas. In: Biersteker TJ, Eckert SE (eds) Countering the financing of terrorism. Routledge, London, pp 93–125
21. Matthew R, Shambaugh G (2005) The limits of terrorism: a network perspective. Int Stud Rev 7:617–627
22. Sheffer G (2005) Diasporas, terrorism and WMD. In: Blum A, Asal V, Wilkenfeld J (eds) Nonstate actors, terrorism and weapons of mass destruction. Int Stud Rev 7:133–70
23. Abuza Z (2003) Terrorist funding in Southeast Asia: the financial network of Al Qaeda and Jemaah Islamiya. Contemp SE Asia 25:169–199
24. Magouirk J, Atran S, Sageman M (2008) Connecting terrorist networks. Stud Confl Terrorism 31:1–16
25. Krebs VE (2002) Mapping networks of terrorist cells. Connections 24:43–52
26. Corman SR (2006) Using activity focus networks to pressure terrorist organizations. Comput Math Organ Theor 12:35–49
27. BBC (2003) Suspect 'reveals terrorist planning' (Sept 22)
28. Ilachinski A (2012) Modelling insurgent and terrorist networks as self-organised complex adaptive systems. Int J Parallel Emergent Distrib Syst 27:45–77
29. Siqueira K, Sandler T (2010) Terrorist networks, support, anddelegation. Public Choice 142:237–253
30. Leistedt SJ (2013) Behavioural aspects of terrorism. Forensic Sci Int 228:21–27
31. Stohl M (2008) Networks, terrorists, and criminals: the implications for community policing. Crime, Law Soc Change 50
32. Vertigans S (2011) The sociology of terrorism: people, places and processes. Taylor and Francis, New York
33. Asal V, Rethemeyer RK (2006) Researching terrorist networks. J Secur Educ 4
34. Keller JP, Desousa KC, Lin Y (2010) Dismantling terrorist networks: evaluating strategic options using agent-based modelling. Technol Forecast Soc Chang 77:1014–1036
35. Shapiro JN (2008) Terrorist organizations' inefficiencies and vulnerabilities. In: Giraldo JK, Trinkunas HA (eds) Terrorism financing and state responses: a comparative perspective. Stanford University Press, Stanford
36. Yang CC, Sageman M (2009) Analysis of terrorist social networks with fractal views. J Inf Sci 35:299–320
37. Acharya A (2009) Targeting Terrorist Financing. Routledge, New York
38. Shapiro JN, Siegel DA (2007) Underfunding in terrorist organizations. Int Stud Quart 51:405–429
39. Burt RS (2004) Structural holes and good ideas. Am J Sociol 100:349–399
40. Burt RS (1992) Structural holes: the social structure of competition. Harvard University Press, Boston
41. Burt RS (2000) The network structure of social capital. In: Sutton RI, Straw BM (eds) Research in organizational behavior. JAI Press, Greenwood, pp 345–423
42. Horne C, Horgan J (2012) Methodological triangulation in the analysis of terrorist networks. Stud Conflict Terrorism 35:182–192
43. Morselli C (2008) Assessing vulnerable and strategic positions in a criminal network. J Contemp Crim Justice 26:382–392
44. Morselli C, Roy J (2008) Brokerage qualifications in ringing operations. Criminology 46:71–98
45. Associated Press (2009) Third man pleads guilty in Somali terror case. 12 Aug 2009
46. Asal V, Nassbaum B, Harrington DW (2007) Terrorism as transnational advocacy: An organizational and tactical examination. Stud Confl Terrorism 30:65–74
47. Chen H, Qin J, Reid E, Zhou Y, Sageman M, Weinmann G (2008) Uncovering the dark web: case study of Jihad on the web. J Am Soc Inform Sci Technol 59:1347–1359
48. Joseph J (2011) Terrorists move to Skype, frustrate eavesdroppers. The Economic Times, India
49. Ogun MN (2012) Terrorist use of internet: possible suggestions to prevent the usage for terrorist purposes. J Appl Secur Res 7:203–217

50. Sageman M (2008) The next generation of terror. Foreign Policy 165:36–42
51. Schmidle RE (2010) Positioning theory and terrorist networks. J Theor Soc Behav 40:65–78
52. United States District Court, District of Minnesota (2010) Indictment, United States of America vs. Amina Farah Ali and Hawo Mohamed Hassan
53. Dougherty JM (2006) Hawala: how terrorists move funds globally. Corp Fin Rev 10:333–335
54. de Goede M (2003) Hawala discourses and the war on terrorist finance. Environ Plann D: Soc Space 21:513–532
55. Looney R (2003) Hawala: the terrorist's informal financial mechanism. Middle East Policy 10(1): 164–167
56. Tupman WA (2009) Ten myths about terrorist financing, J Money Laundering Control, 12(2):189–205
57. Bakker RM, Raab J, Milward HB (2012) A preliminary theory of dark network resilience. J Policy Anal Manage 31:33–62
58. U.S. Attorney's Office, District of Minnesota (2009) Terror charges unsealed in Minnesota against eight defendants. Justice Department Announces (23 Nov 2009)
59. Kron J (2011) American identified as bomber in attack on African Union in Somalia. New York Times, USA
60. Yuen L (2011) Family IDs Minn. man allegedly behind Somali suicide bombing. Minnesota Public Radio News, USA
61. Razavy M (2005) Hawala: an underground haven for terrorists or social phenomenon? Crime, Law, Soc Change 44:277–299
62. Eck JE (2006) When is a bologna sandwich better than sex? A defense of small-n case study evaluations. J Exp Criminol 2:345–362
63. van Duijn MAJ, Vermunt JK (2006) What is special about social network analysis? Methodology 2:2–6
64. Bowyer-Bell J (2000) The IRA 1968–2000: An analysis of a secret army. Frank Cass, London
65. Bruce S (1992) The Red Hand: Protestant Paramilitaries in Northern Ireland. Oxford University Press, Oxford
66. Coogan TP (1995) The IRA. Harper-Collins, London
67. Della Porta D (1995) Social movements, political violence and the state. Cambridge University Press, Camridge
68. Fair CF (2008) Who are Pakistan's militants and their families? Terrorism Polit Violence 20:49–65
69. Hegghammer T (2006) Terrorist recruitment and radicalization in Saudi Arabia. Middle East Policy 13:39–60
70. Jamieson A (1989) The heart attacked: terrorism and conflict in the Italian state. Marian Boyers, London
71. Taylor M (1988) The terrorist. Brassey's, London
72. White RW (1993) Provisional Irish republicans: an oral and interpretive history. Greenwood Press, Westport
73. LaFree G (2008) Generating Terrorism event databases: results from the global terrorism database, 1970 to 2008. In: Lum C, Kennedy LW (eds) Evidence-based counterterrorism policy. Springer, New York, pp 41–64
74. Chermak SM, Freilich JD, Parkin WS, Lynch JP (2011) American terrorism and extremist crime data sources and selectivity bias: an investigation focusing on homicide events committed by far-right extremists. J Quant Criminol 28:91–218
75. Leuprecht C, Hall K (2013) Networks as strategic repertoires: functional differentiation among Al-Shabaab terror cells. Global Crime 14:287–310
76. Adams S (2012) 'Just been shot in neck by assassin': American jihadist live-tweets attack in Somalia and shows picture of his bloody neck. The Daily Mail, UK
77. BBC (2012) Danish–Somali terror suspect accused of training five weeks in Al-Shabaab camp (June 4)

78. Danish cartoonist hid in 'panic room' during attack (2010, Jan 2) CNN
79. Hiiraan Online (2006) Ontario municipal election: Somali Canadian prospective (Dec 10)
80. Freisen J, Freeze C (2009) Are Somali–Canadians fighting for the shadowy al-Shabab?. The Globe and Mail, Ontario
81. O'Toole M (2009) Terror suspect Mohamed Hersi to head directly to trial. The National Post, Ontario
82. Greenwood C, Bates D (2012) Seven held over smuggling of a banned stimulant from England to U.S. to fund terror. The Daily Mail, UK
83. Hartley A (2012) This is khat: The natural high available on British streets and suspected of funding terrorism. The Daily Mail, UK
84. Abdulle A (2010) Dutch police arrest Al shabaab commander's father over foiled blasts. The Suna Times, Somalie
85. Associated Press (2010) Dutch arrest 12 Somalis on terror suspicions. CBS, New York
86. Meryhew R, Walsh J (2009) Young men: searching for something better. Minneapolis Star Tribune, USA
87. Yuen L (2012) Terrorist pipeline continues to flow from Minn. to Somalia. Minnesota Public Radio 26 Oct.
88. Shirwa's Journey (2009) Minneapolis star tribune
89. United Sates District Court, District of Minnesota (2009) Indictment, United States of America vs. Mahamud Said Omar (20 Aug 2009)
90. Yuen L (2009) Man killed in Somalia may have recruited others to the cause. Minneapolis Public Radio, USA
91. Elliot A (2009) A call to Jihad, answered in America. New York Times, USA
92. Elliot A (2009) Charges detail road to terror for 20 in U.S. New York Times, USA
93. Meryhew R (2009) Minneapolis Somali man killed in homeland. Minneapolis Star Tribune, USA
94. District of Minnesota, state of Minnesota (2011) Affidavit of Michael N. Cannizarro Jr.', United Sates of America vs. Abdow Munye Abdow (Oct 9)
95. United States District Court, District of Minnesota (2010) Government's sentencing positioning paper, United Sates of America vs. Abdow Munye Abdow (13 July 2010)
96. Jones BT, Cooney JF (2011) Man extradited from Netherlands appears in federal court on charges of supporting terrorists. United Sates Department of Justice, USA
97. United Sates District Court, District of Minnesota (2009) Information, United Sates of America vs. Adarus Abdulle Ali (27 Oct, 2009)
98. United Sates District Court, District of Minnesota (2012) Plea agreement and sentencing stipulations, United States of America v. Ahmed Hussein Mahamud
99. United States District Court, Southern District of California (2011) Plea agreement, United States of America v. Nima Yusuf
100. United Sates District Court, District of Minnesota (2001) Government's total brief, United Sates of America vs. Omer Abdi Mohamed (14 July 2001)
101. Fife-Yeomans J, Vollmer T (2009) Rant in court by one of four man who appear in court accused of terrorism offences. The Daily Telegraph, Australia
102. The Supreme Court of Victoria, Melbourne Criminal Division (2011) VSC 681, R. vs. Wissam Mahmoud Fattal, Saney Edow Aweys and Naser El Sayed, 16 Dec 2011
103. An uncle's regret (2009, Aug 6) A flight from Somalia gone wrong. Sydney Morning Herald, Australia
104. Warne-Smith D, Wilson L (2009) Somali terror suspects 'new to mosque'. The Australian, 6 Aug 2009
105. Aulakh R (2009) Did five Torontonians join jihad in Somalia?. The Toronto Star, Canada
106. Shephard M (2008) Canadian insurgent 'Asparo' killed in Somalia. The Toronto Star, Ontario
107. Bell S (2009) Imam tries to calm fears about missing Somali-Canadians. The National Post, Ontario

108. Freeze C, MacArthur G (2011) Man jailed in new kind of terrorist case for Canada. The Globe and Mail, Ontario
109. Bell S (2013) Canadian linked to terrorist group was killed in suicide attack in Somalia: community source. The National Post, Ontario
110. Bell S (2010) Martyr video claims Toronto man 'succeeded'. The National Post, Ontario
111. United Sates District Court, Eastern District of Missouri, Eastern Division (2010) Indictment, United Sates of America v. Mohamud Abdi Yusuf, Duane Mohamed Diriye and Abdi Mahdi Hussein (21 Oct 2010)
112. United States District Court, Southern District of California (2013) United States' trial memorandum, United States of America v. Basaaly Saeed Moalin, Mohamed Mohamed Mohamud, Issa Doreh and Ahmed Nasir Taalil Mohamud (28 Jan 2013)
113. Piazza JA (2009) Is islamist terrorism more dangerous?: an empirical study of group ideology, organization, and goal structure. Terrorism Polit Violence 21
114. Al Jazeera (2013) French agent 'executed' by al-Shabab (Jan 17)
115. Maclean W, Khamis N, Ahmed M (2012) Special report: in Africa, a militant group's growing appeal. Reuters, 30 May 2012
116. The Amsterdam Herald (2013) First Dutch militia fighter killed in Syrian conflict. http://www.amsterdamherald.com/index.php/rss/748-20130320-first-dutch-militia-fighter-killed-syrian-conflict-delftmilitia-delft-aivd-intelligence-netherlands-dutch-security. Accessed 20 March 2013
117. Radio Netherlands Worldwide (2012) Al-Shabaab plotting attacks in Netherlands. http://www.rnw.nl/english/bulletin/al-shabab-plotting-attack-netherlands. Accessed 28 May 2012

Social Network Analysis Applied to Criminal Networks: Recent Developments in Dutch Law Enforcement

Paul A. C. Duijn and Peter P. H. M. Klerks

Abstract This article examines the application of social network theory in Dutch law enforcement. Increasing amounts of information about habitual lawbreakers and criminal networks are collected under the paradigm of Intelligence-Led Policing. Combined with data gathered from open sources such as social media, such resources allow criminal analysts trained in social network analysis (SNA) at the Police Academy of The Netherlands to apply advanced network analysis methodology and crime scripting. This in turn helps the police to identify crucial weak spots in illicit arrangements and criminal business processes. A case study of the 'Blackbird' crime network, involved in the wholesale cultivation of cannabis is presented to illustrate the power of SNA when combined with crime script analysis. Using a mix of quantitative and qualitative analysis, the topology of the 86-strong Blackbird network is laid out and its substructures and key individuals exposed. In detailing the network's social embeddedness, the authors clarify the importance of female actors for the flexibility and efficiency of the network structure and thereby for the continuity of criminal business. Applying SNA is already helping criminal intelligence units of the Dutch police in identifying intelligence gaps and potential informants. Working in symbiosis, analysts and informant handlers develop a better understanding of strategic targeting and access points to relatively unknown criminal communities and –markets. To be delivered in a timely way to be useful in ongoing criminal investigations, SNA products require even faster data processing. Also, when applied to dark networks SNA should be tailored to better take network dynamics into account, in particular regarding the adaptability to network disruption.

P. A. C. Duijn (✉)
Department of Research and Analysis, Dutch Police Unit the Hague, The Hague, The Netherlands
e-mail: pacduijn@gmail.com

P. P. H. M. Klerks
Board of Procurators-General, Public Prosecutor's Office, The Hague, The Netherlands

A. J. Masys (ed.), *Networks and Network Analysis for Defence and Security*, Lecture Notes in Social Networks, DOI: 10.1007/978-3-319-04147-6_6,

Keywords Social network analysis · Organized crime · Script analysis · Crime logistics · The Netherlands · Intelligence led policing

1 Introduction

Criminal networks can be defined as networks operating outside the boundaries of the law, for which network achievements come at the cost of other individuals, groups or societies [1, 2]. Across the globe criminal networks have a significant impact on national defense and security. Criminal networks try to infiltrate legal businesses and governments, infecting economic branches with violence and corruption. Moreover, upcoming threats like cybercrime, child pornography, maritime piracy, match fixing, illegal logging and identity theft, cause substantial harm to society and require proactive interventions to target the criminal networks underlying them [3, 4].

Government and law enforcement agencies therefore seek ways to effectively disrupt criminal network structures, preferably at an early stage. Since criminal networks face a constant threat from government agencies as well as aggressive criminal competitors, network members tend to evade detection and intervention [1]. This makes it difficult to assess and collect reliable criminal network data. Therefore criminal network structures remain largely unknown as compared to other types of empirical networks [5, 6, 7]. Consequently, little empirical knowledge concerning the impact of different disruption strategies on criminal networks is available to policymakers and law enforcement agencies.

The Netherlands has a relatively long tradition in controlling and studying organized crime. Over the years different paradigms have shaped the way in which law enforcement strategies against organized crime were applied. In the late 1980s Dutch law enforcement agencies thought of criminal organizations as hierarchical structures, leading to prolonged and extensive investigations targeting the presumed 'Capo di Tutti Capi' at top of the pyramid. Only recently have criminologists acknowledged organized crime from a different perspective through the social network paradigm [2, 8–11]. Gradually this paradigm is being adopted within the Dutch law enforcement intelligence community, now leading to innovative ideas about law enforcement strategies.

Two important opportunities for network analysis have contributed to this development. First the Dutch Police have invested in the process of Intelligence-Led Policing, leading to an increased focus on collecting information in the frontlines of law enforcement. Consequently, more detailed data about network members and their illicit activities now become available within different police databases. Secondly research shows that more and more information about criminal network members can be found in bright networks, e.g. within online communities [12, 13]. Dutch law enforcement agencies are experimenting with the retrieval of such open source intelligence for operational purposes.

The increased availability of data on criminal networks enables network analysis on two levels. First, it offers opportunities for scientific research in network analysis, aimed at revealing how these criminal networks operate and how they react following network disruption. These insights offer a better understanding for law enforcement decision makers in estimating the effectiveness of different criminal disruption strategies in general (e.g. [14–16]). Secondly, network analysis is becoming an important method within operational intelligence projects, leading to more strategic ways of targeting criminal networks. To stimulate this development, the Dutch Police Academy currently offers police analysts special training in Social Network Analysis (SNA), aimed at applying this additional analysis tool in both criminal investigations and strategic intelligence projects. Which lessons can be learned from this application of network analysis in crime control? What are the practical implications of applying network analysis in targeting criminal networks and strategic intelligence gathering? What does dynamical network analysis research tell us about the effectiveness of criminal network disruption? How does the network paradigm connect with law enforcement decision-making? These are the questions this chapter aims to address.

The aim of this chapter is thus twofold: First, to inform the reader about recent developments of the application of network analysis in controlling crime in the Netherlands. Secondly, to offer insight into the practical application of network analysis in law enforcement, specifically applied to effectively target criminal networks.

The remainder of this chapter is as follows: Sect. 2 describes the evolution of three different paradigms for organized crime and how this shaped control strategies across time. Section 3 describes the results of a case study of SNA used to understand the structure and resilience of a cannabis cultivation network. Following the limitations and challenges of this case study, Sect. 4 describes the recent progress and developments within overcoming these challenges within the application of SNA in Dutch law enforcement. Section 5 ends of this chapter with an overall conclusion.

2 Three Paradigms of Organized Crime

This section discusses the evolution of three different paradigms of organized crime, as well as their impact on control strategies. This is illustrated by developments within Dutch law enforcement over the last 30 years.

Organized crime was first recognized as a relevant phenomenon for Dutch law enforcement in the mid-1980s, when narcotics traffickers were found to engage in worldwide smuggling operations connections dozens of operators and making enormous profits [17]. When the first crime analysts began to draw up their reports on criminal gangs around 1988, they portrayed mostly hierarchical groups in which often dozens of criminals worked under a division of labor on (most often) the import and distribution of hashish, heroin and cocaine. Every group had a clearly identified leadership, and the strategy by which the police attempted to tackle them was mostly to intercept and confiscate drug shipments and arrest those involved. The idea was to build up pressure on a group's business, and thus force the supposed organizers in the background to expose themselves and show their hand. 'Dismantling criminal structures' and sentencing ringleaders to long prison sentences, so it was thought, would counter organized crime. This strategy resulted in major confiscations and some prison sentences of 10 + years, but it soon became clear that the intercept rate never reached more than about 20 % of the estimated total drug markets.

In the early 1990s, while most academic researchers still shied away from studying organized crime, the police and public prosecutor's office began to understand the Dutch narcotics underworld through such metaphors as a 'monkey rock' or 'octopus': a more or less integrated and hierarchical criminal conglomerate in which markets were divided and coordinated through negotiations and occasional conflicts. This called for a 'war on crime', including drastic measures such as the deployment of criminal informants who were allowed to grow into a position where they would be able to provide incriminating information on the supposed premier league of criminal masterminds. Such 'growing informants' could not always be kept under control. One such resourceful informant was permitted by his police handlers to bring shipments of thousands of kilos of cannabis, hashish, ecstasy and cocaine on the market while customs authorities conveniently looked away. When this came to light in 1994, a traumatizing scandal erupted of which the shock waves are noticeable even today. A parliamentary inquiry followed, and by the year 2000 Dutch criminal investigation procedures had become as strict as anywhere in the world.

This hierarchical pyramid or bureaucracy model of organized crime, while appealing to enforcement practitioners and some journalists, has never attracted interest or support from Dutch academic researchers. Their interest was raised substantially however when, in the wake of the 'IRT affair', four of the leading Dutch criminologists were tasked with writing a thorough and comprehensive study of organized crime in The Netherlands. In 1996 their authoritative report changed the perception of organized crime. Criminal gangs and entrepreneurs were found to have gained footholds in several inner-city areas and branches such as prostitution and parts of the catering industry. There was however no sinister master mind at work: most criminal markets were relatively open for competition, with varying sets of illegal entrepreneurs often profiting from lax or gullible branches of local government. Thus, organized crime became conceptualized as mostly entrepreneurial in nature, with fluid criminal groups working in clandestine logistical arrangements to overcome the challenges of serving illegal markets. Consequently, crime control strategies became more sophisticated with an explicit responsibility for administrative bodies to impose tighter controls on permits in vulnerable branches and city areas [18]. Also, attempts were made to increase the financial investigative capacities of the police and more interventions were aimed at reducing criminal opportunity structures and controlling chokepoints in criminal business processes. Criminal 'facilitators' were targeted that bridge the gap between illegal entrepreneurs and their legitimate environment providing them with financial and logistical services, thus blocking money laundering channels and the acquisition of apparatuses for producing narcotics [19].

From the mid-1990s onwards, dozens of researchers in the Netherlands became involved in empirical studies of organized crime phenomena [8, 20, 21]. With the police now more open to outside scrutiny and public debate, it had become much easier for serious scholars and their students to gain access to police files. Many detailed and extensive studies appeared, allowing for a more knowledge-based crime control policy. Both the police and the justice department commissioned their own periodical crime monitors and threat reports and gradually, a mild consensus formed on the approximate size and shape of the organized crime problem [22–30]. Since around 2002 the social network approach to organized crime has become increasingly popular, initially among academic students of organized crime and through their teaching and involvement in crime control projects, also among a new generation of analysts and investigators. Where initially the concept of 'criminal networks' was amply defined, serving rather as an antithesis to the traditional paradigm of rigid hierarchical organizations, researchers like Klerks [9, 17], Kleemans [8] and Spapens [10, 27] advanced this to include the micro-level of networks (criminal groups or 'collectives') forming and operating in the context of criminal macro- and meso-networks. The macro network in theory is worldwide and connects all able and willing (potential) offenders through criminal relations. In practice this macro network clusters into smaller meso-networks, thus establishing criminal opportunity structures located in specific periods times and areas.

From the academic literature through teaching and intellectual osmosis, the social network model gradually began to permeate police reports. In 2005, the Board of Chiefs of Police brought out their strategic vision paper '*Politie in ontwikkeling*' (Police in development) which contained concepts like the 'nodal orientation' inspired by the thinking of Manuel Castells [31]. This nodal orientation implied that the police can only be effective in a network society if they organize surveillance and intervention capabilities on the 'nodes' through which streams of people, products, money and information flow such as airports, seaports, highway interchanges and the Internet. A popular handbook on net-centric strategies in law enforcement, distributed for free among policymakers and practitioners, further helped to familiarize these audiences with networking concepts [32]. Police researchers and analysts gradually became acquainted with more technical network applications through social network analysis courses given at the Police Academy and from internal reports such as Neve [33], and they began to use them in their work. Currently, nearly one hundred law enforcement analysts have passed the Police Academy exams in SNA and an increasing number of them apply their SNA skills in either operational or strategic intelligence work. Some of them publish their experiences in articles and internal reports, such as Bosveld [34] on the application of forensic SNA in cold case investigations, Visser [35] on post-intervention readjustments in the modus operandi of a criminal network, and Van der Horst et al. [36] on the time-saving application of SNA for targeting criminal youth gangs.

Current conceptualization of organized crime in the Netherlands centers on the notion of illegal entrepreneurship serving illegitimate markets including narcotics, human trafficking, stolen vehicles, illegal arms trade and irregular waste disposal. Holland being a trade and transit rather than a production economy, its mirroring illegal economy also has a predominant character of international trade with the important exception of producing marihuana and synthetic drugs. Four conceptual dimensions are now considered important in combating organized crime:

1. the criminal business processes and logistics (and the ways to disrupt them);
2. the physical infrastructures enabling illegal entrepreneurs to unobtrusively produce and ship their merchandise;
3. the social networks that spawn criminal cooperation and conflicts;
4. the financial streams that provide energy and motivation to the 'underworld', connect it with the 'upperworld' and allow investigators to link discrete organizers and their social entourage to the repellent crimes from which they profit [37].

All four of these dimensions profit from the tools and techniques of social network analysis, and the insights they can provide.

3 A Case Example: Unraveling the Blackbird Network with SNA

The previous section explained how network theory is getting increasing attention within both criminology and Dutch law enforcement agencies involved in organized crime control. The interest for this paradigm among intelligence analysts also influences the way some police commanders and public prosecutors in the Netherlands think about the current control strategies for organized crime. As described above, the development of a practical SNA training course for intelligence analysts at the Dutch Police Academy has stimulated this development and leads to a growing number of explorative case studies. Besides empirical knowledge of criminal network problems, these case studies offer great lessons for the operational application of SNA. In this section one of these case studies is presented, specifically demonstrating the advantages, limitations and challenges of the practical implementation of SNA within Dutch law enforcement.

3.1 Description of Operation Blackbird and Research Questions

In the autumn of 2007 an investigation team within a regional police department in the Netherlands started criminal investigation Blackbird against a criminal group involved in organized cannabis cultivation. This operation was part of project Umbrella, the goal of which was to target a regionally active but extensive criminal network involved in multiple forms of organized crime, such as cannabis cultivation, ecstasy production, cocaine trade, extortion, violence and even first degree murder. The objective of operation Blackbird was twofold: (1) to target the core members ('big fish') of the criminal network that were specifically involved in organized cannabis cultivation; (2) to retrieve additional intelligence about criminal network members within in the embedding criminal network. Operation Blackbird lasted for nine months in total, leading to the initial arrest of eleven suspects and a final conviction of the three presumed core members (the big fish) for involvement within a criminal organization.[1] The operation was considered a success, because the first goal of catching the big fish was achieved.

However analysts and detectives working within project Umbrella soon retrieved signals that although the three important network members were convicted and detained, this didn't stop the remainder of the criminal network to continue their cannabis cultivation activities. It showed that the cultivation network was highly resilient against network disruption and didn't fall apart as implicitly expected at the start. Therefore public prosecutors and law enforcement managers evaluating the case asked themselves how the network's resilience could be explained. An answer to this intelligence question might contribute to the adjustment of future law enforcement strategies.

A team of three analysts involved in the Social Network Analysis (SNA) training program at the Dutch Police Academy started a search for the answers using SNA methodology. The primary question of the analysis was twofold: what is the structure of the Blackbird network and how did it contribute to the observed resilience against law enforcement interventions? An implicit third question was: can SNA methodology help in targeting these criminal network structures more effectively from the start?

3.2 Data Sources and Research Design

As usual, the analysis process started with the identification of possible data sources and collection of data. In conventional SNA research, data are mostly collected by taking surveys from all members of the studied social network, including questions about the origin and nature of their mutual social relationships [38]. Unfortunately most criminal network members don't like to be asked questions about their criminal activities, nor are they easily approached to discuss their mutual criminal relationships [39]. Therefore the first challenge in the criminal network analysis field is collecting substantive relational data with

---

[1] Within Dutch Penal law, 'participation within a criminal organization' is an independent misdemeanor punishable under article 140 of the Dutch Penal Code.

enough content to interpret the nature of mutual relationships. In general most SNA practitioners within criminology therefore turn to criminal investigations as a main source of data. These investigations involve wiretap data, eyewitness and suspect statements and surveillance data over a certain period of time, containing valuable clues about the nature of social and criminal relationships, specific language used, ways of communication and participation in (criminal) activities of members in a criminal network [2, 8, 9, 40].

3.2.1 Data Collection

Operation Blackbird was part of a larger operation Umbrella, covering multiple criminal investigations aiming at different criminal hotshots at the same time. Additional relational data about the Blackbird network members could therefore also be retrieved from four other operations during that same time period. This strengthened the validity of our observations, because every network representation based on criminal investigations data is biased to a certain extent toward its initial objectives [2]. Because these four operations had different objectives, this validity problem could be confined. Initial relational data were therefore retrieved from extensive wiretap data sets, eyewitness statements, suspect statements and surveillance data from these four investigations, to be combined into one dataset.

In addition to these primary data sources data from social media were also obtained. A quick scan within different Internet communities such as Facebook and Hyves, showed that several members of the Blackbird network were actively participating in these social network sites.[2] These data contained additional relational information revealing other (social) relationships within the embedding Blackbird network.

3.2.2 Data Processing

The data where processed using different actor-by-actor matrices and graphs as described by Scott [41] and Hanneman and Riddle [38]. In addition we used the UCINET 6 and Netdraw software package [42]. One of the central research questions was to reveal features from the network structure that contribute to its resilience. Some detectives of the investigative team pointed already at the importance of family—and affective relationships for its social structure. Following this hypothesis, it was decided to distinguish between the following types of connection according to the type of relationship between actors, including:

1. Criminal ties
2. Kinship ties
3. Affective ties.

[2] www.Hyves.nl and www.netlog.nl were popular Internet communities in the Netherlands in 2007 and 2008.

For every type of relationship another actor-by-actor matrix was processed, in order to compare the different networks at a later stage.

3.2.3 Combing SNA with Crime Script Analysis

In addition to exposing the criminal network structure through the social network analysis method, 'crime script analysis' adds insight into the individual positions of actors within a criminal network. Cornish [43] was one of the pioneers describing criminal markets in terms of crime scripts. Following this method a crime script is a systematical blueprint of the different phases of a criminal business process, that each consist of different facets. The permutation (possible combinations to pass all phases) is an indicator of its flexibility. In other words, the more options (facets) built into the crime script to pass the different phases, the more resilient the crime script is against disruption. Sparrow [11] already emphasized that this method could be very useful in intelligence analysis to identify actors with unique roles.

Bruinsma and Bernasco [44] combined crime script analysis and social network analysis to describe the flexibility within the criminal markets of heroin trade, trafficking in women and car theft. They found some evidence that the structure of criminal networks was shaped according to the features of the criminal activities and illegal markets, for instance the possible legal and economic consequences of the specific criminal activities. Additionally, Morselli and Roy [45] integrated crime scripting with Social Network Analysis methodology in labeling different actors within a criminal network according to their involvement in the different phases and facets of the crime script of organized car theft. They identified the importance of brokers between the different roles in the crime script. According to Sparrow [11] these actors have low 'substitutability' and are therefore interesting targets for network disruption, because this means that most of the criminal network depends on just a few actors for a successful outcome of the criminal business process. Sparrow [11] emphasizes that disruption of actors with specific skills might have major consequences for the criminal network, as compared to actors involved in more general tasks or roles. Crime script analysis is therefore an essential additive to contemporary social network analysis methods in the criminal intelligence toolkit.

In accordance with the previous studies, crime script analysis was also used to unravel the structure of the Blackbird network. One of the selection criteria for actors to be included into the Blackbird network is involvement in organized cannabis cultivation business. Cannabis cultivation is a complex and delicate criminal business, involving many roles and tasks. Based on observations in the data and studies by Morselli [46], Spapens et al. [28] and Emmet and Broers [47],

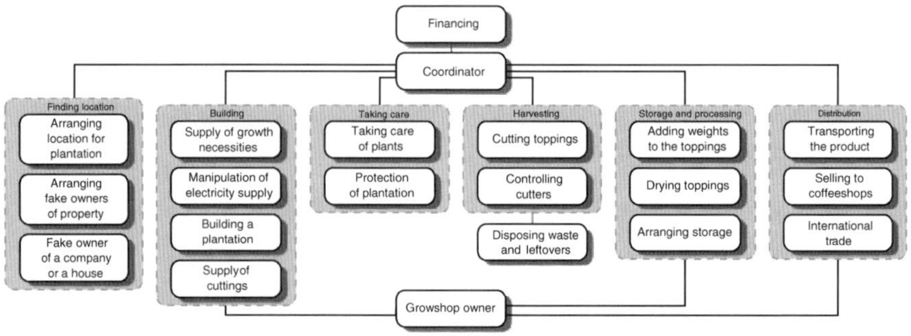

Fig. 1 Crime script of cannabis cultivation [47, 28, 46]

Table 1 Example of integrating crime script analysis within an actor by variable matrix

Arranging location	Building plantation	Taking care of plants	Harvesting	Storage processing	Distribution	.Etc
1	0	0	0	0	1	
0	1	0	0	0	0	
0	0	1	1	1	0	
0	0	1	1	1	0	
0	0	1	0	1	0	

the configuration of the crime script of organized cannabis cultivation was retrieved (see Fig. 1).[3]

To integrate the crime script analysis in the social network analysis framework an actor-by-variable matrix was used to assess the participation of each actor in the different phases of the cannabis cultivation business process (See Table 1). In this way tasks or roles could be identified that are thinly populated within the overall network structure. Actors representing these roles might be difficult to substitute [11].

3.2.4 Consideration About Data Validity

All data-sources (wiretaps, statements, reports) were scored on these variables by the three analysts at the same time. This required intensive discussion of interpretations of language used by the actors in their communication. As a form of counterstrategy the actors within the Blackbird network often used coded language when referring to (specific) illegal activities, accomplices or locations, which would be susceptible to multiple interpretations [2, 9]. This constituted a major

[3] Unfortunately a fully detailed description of the different phases and facets of the organized cannabis cultivation process is out of scope of this chapter. For a more detailed description see Spapens et al. [28] or Potter et al. [48].

risk to data validity, as conversations between actors had to be interpreted in a similar way by all three analysts. To ensure this consistency, fifty wiretap conversations were scored by all three analysts separately and checked on differences in interpretation. This check was often repeated.

Sometimes coded language was easily overlooked. In some conversations the actors talked about *'getting a cup of coffee'* for example. In the context of conversations later on in time, it was found that this was code language for ordering specific cannabis growth necessities. Therefore processed conversations had to be rechecked in order to preserve data validity. In addition to coded language the use of nicknames made it difficult to identify individual actors. In the end it seemed two presumed different actors were actually one and the same individual. Fortunately, this could be corrected afterwards within the actor-by-actor matrix by merging the two identities. To keep track of such changes and make these considerations transparent, every decision was noted in a log file.

3.2.5 The Boundary Specification Problem

After all raw network data were processed into the different matrices, the question arose which actors to include or exclude from the analysis [11, 39, 49, 50]. This 'boundary specification problem' might affect the structure and scope of the criminal network. Therefore selection criteria for boundary specification have to be set prior to the analysis process [41]. These selection criteria might be based on theory or practical considerations derived from the principal research questions.

Police reports and wiretap data on the Blackbird network showed that actors were involved in more than one criminal activity. Some actors combined cannabis trade with the production of synthetic drugs and firearms trade. Our analysis was focused on actors from the Blackbird network involved in organized cannabis cultivation. Therefore the decision was made to apply the criteria of 'involvement in cannabis cultivation' for inclusion or exclusion from the final dataset. This meant that actors involved in one or more facets of cannabis cultivation process were included, leading to a network consisting of 88 identified actors. As two actors were recognized as isolates, the final network representation consisted of 86 actors in total. This was the starting point for answering the intelligence questions.

3.3 Quantitative Analysis of the Blackbird Network

For simple networks network visualizations are often useful for analyzing the features of network structure. However, for bigger networks these visualizations soon resemble plates of spaghetti, in which individual positioning is difficult to identify with the naked eye. To overcome this problem the SNA toolbox contains numerous algorithms that can be used to calculate individual actor features within the densest of networks. SNA practitioners, like some intelligence analysts, are often confronted with the dilemma of which algorithms to use to answer a specific research question. Choosing the right algorithms for a specific research question requires a high level of understanding of all possibilities and their implications. In order to decide which SNA algorithms are suitable for answering the research questions Roberts and Everton [51] introduce an analytical framework that divides network structure into three levels:

1. Whole network level
2. Subgroup level
3. Individual level

Within SNA methodology distinctive measures are associated with these different levels of network structure. This classification gives SNA practitioners a good reference for choosing the right algorithms to use. Roberts and Everton [51] point out that in order to fully understand network structure it's essential to understand all three levels. Baker and Faulkner [52], Robins [53] and Morselli [2] all emphasize that features of individual positioning and subgroups within illegal networks can only be interpreted properly, if the overall network topology is understood in the first place. Robins [53] even points specifically at the symbioses of individual psychological features and properties of network topology. These practitioners call for an integrated analysis of the different levels of criminal network structure, to understand the way these criminal networks operate. Elaborating from these considerations, this same analytical framework was used to unravel the structure of the Blackbird network in relation to its network resilience. In this next section the results of this quantitative analysis are described.[4]

3.3.1 Blackbird Network Topology

SNA offers many measures to analyze network topology. According to Hanneman and Riddle [38] and Everton [54], the five most important measures for network topology are: centralization, density, average degree, average path length and network diameter. Table 2 shows the results of applying these algorithms on the overall Blackbird Network (N = 86).

In SNA terms, a degree centralization of 58.49 % and betweenness centralization of 41.47 % indicate that the network gravitates around a few central actors who have relatively more direct connections in the network than the rest of the network. It means that there is a distinction between a core and periphery in the network. This implies that peripheral actors depend on a few central actors for their information and resources flowing through the network. According to network theory this gives the central actors a powerful and influential position in the network [38].

[4] A full description and explanation of all possible SNA measures would go beyond the scope of this chapter. For an extensive overview of these measures, see Hanneman and Riddle [38].

Table 2 Topology measures of the Blackbird network (N = 86)

Measures of network topology	Score
Degree centralization	58.49 %
Betweenness centralization	41.47 %
Density	0.0663
Average degree	6.442
Average path length	2.249
Network diameter	5.0

The third important metric for unraveling network topology is density. This metric is defined as the total number of ties within a network divided by the total possible number of ties, which means that network density measures range from 0 to 1. Density gives insight in the speed at which information diffuses among the actors and the extent to which the actors in general have high levels of social capital (many connections) [38, 54]. Density within the Blackbird network is relatively low (0.0663). In network theory this means that information doesn't spread effectively through the network. This again emphasizes that there are actors who depend on other actors for their information about network activity and are therefore not that well connected.

This interpretation is further underpinned by the results for average degree centrality. This algorithm represents the average number of direct connections of actors in the network. A high score for average degree means that actors are very well connected. Actors in the Blackbird network have an average of 6.44 direct connections. The total network consists of 86 actors. This means in theory that every actor can have a maximum of 85 (N − 1) connections. In this sense 6.44 average direct connections per actor is quite low.

Finally, the average path length and network diameter were calculated. Average path length is calculated by finding the shortest path between all pairs of nodes, adding them up, and then dividing by the total number of pairs. This shows, on average, the number of steps it takes to get from one actor within the network to another. Network diameter is equal to the longest of all the calculated shortest paths in a network. Table 2 shows that actors in the Blackbird network can reach each other in an average of 2.2 steps with the longest distance between two actors in the network being 5 steps apart.

In sum, analysis of network topology showed the network is centralized (58 %), but that it has a rather low density (0.066) and average degree (6.44).This means that the network gravitates around (a few) central actors and that there are less connections throughout the network. However, the average distance between actors in the network (2.2) indicates information flows fast, meaning that most information and resources has to pass through the central actors to become available to other actors within the network. This suggests that a substantial part of the Blackbird network is dependent on these central actors for their information and resources. A further understanding of these mathematical results follows from the sociogram of Fig. 2. It reveals that a high number of single link actors form a

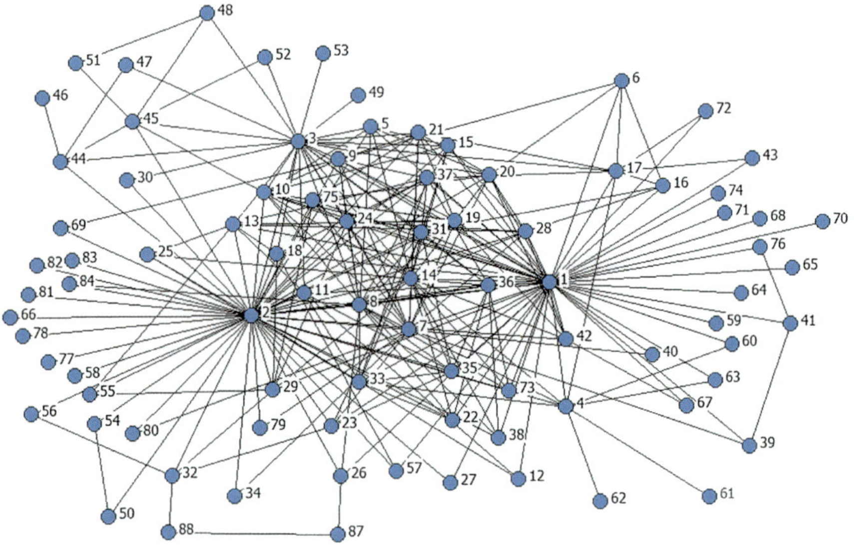

Fig. 2 Sociogram of total Blackbird network

"loose" periphery, that is connected by a just few actors with the dense core of the network.

3.3.2 Substructures Within the Blackbird Network

In the previous section we analyzed the network as a whole. In fact this was a top-down approach to unravel its structure. In this section we will analyze the Blackbird network from a bottom up approach, as we seek for substructures that keep the network together. SNA methodology covers many different measures for identifying substructures of groups within a social network [38].[5] Here we apply two of the most common algorithms for identifying substructures: K-core analysis and clique analysis.

The K-cores metric uses degree centrality to identify clusters of actors that are tightly connected. This approach doesn't pay attention to the degree of individual actors in the network but to the degree of all actors within a cluster [54]. A cluster is called a K-core, for which K indicates the minimum degree of each actor within the cluster. This means a 3-core cluster contains all actors that have three or more ties to other actors.

The results of the K-cores analysis of the Blackbird network are shown in Fig. 3a, b. Figure 3a shows that the network gravitates around a highly connected

[5] See Hanneman and Riddle [38] for a complete overview of all possible algorithms to identify subgroups.

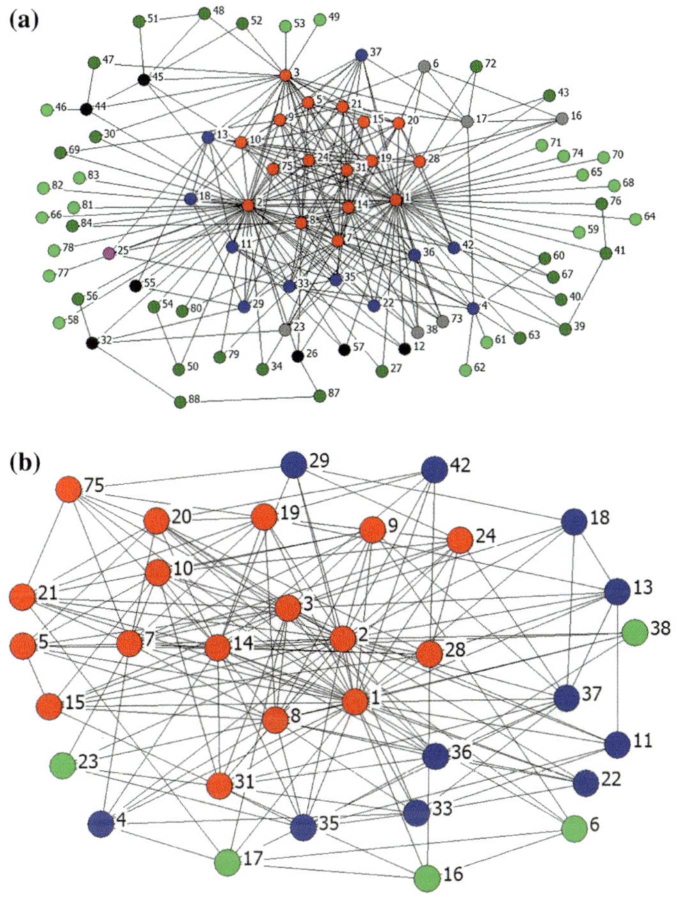

Fig. 3 The K-core distribution is visualized as part of the total network (3a). Secondly its core structure is visualized in 3b, depicting different K-core levels: 6-Core (*red*), 5-Core (*red* + *blue*) and 4-Core (*red* + *blue* + *green*)

6-Core, represented as the red actors in the graph. Figure 3b zooms in on this core, representing the 6-Core (in red), 5-Core (in blue and red combined) and 4 Core (in green, blue and red combined). The 6-core sub-network consists of 17 actors in total. Calculation of the topological measures of the core network of 3b, reveals that density is higher (0.3) then for the overall network and diameter is shorter (2.0) (Table 3). This supports the hypothesis that the network is highly centralized around and gravitates around a tightly knit core.

Within SNA every measure has its own approach. Therefore it's essential to combine different measures in order to draw any conclusions about network structure [54]. Based on the results for the K-Core analysis, another important method to identify subgroups in the overall network is clique analysis. In essence,

Table 3 Topological features of the core of the Blackbird network (N = 32)

Measures of network core (minimum of 4–6 connections)	Score
Degree centralization	34.09 %
Betweenness centralization	22.39 %
Density	0.3087
Average degree	10.91
Average path length	1.700
Network diameter	2.0

a clique is a sub-set of a network in which the actors are more closely and intensely tied to one another than they are to other members of the network. In a clique all nodes are connected to every other node [38]. Clique analysis offers a second "bottom up" approach to understanding network structure. It focuses attention on how solidarity and connection of large social structures can be built up out of small and tight components [38]. The result for the clique analysis of the Blackbird network are depicted in Fig. 4a, b.

Figure 4a shows the number of cliques in relation to its members. Clique analysis thus confirms that the Blackbird network is built up out of a total of 64 tightly knit coalitions (cliques). It also shows that some actors are part of many different cliques (actor 1, 2, 3). Although there are many cliques identified within the Blackbird network, they tend to stay small in size. Figure 4b shows the biggest clique identified (N = 8) as part of the total network (N = 86).

In sum, the K-Core analysis shows that the Blackbird network is built around a tightly connected core of actors. This might be an explanation for its network resilience. Hence, when actors 1, 2 and 3 were arrested the rest of its core members were mutually well-connected to prolong the cannabis production process. Furthermore the clique analysis shows that the Blackbird network is built up out of numerous small coalitions (cliques) within both its core and its periphery. In theory this adds more flexibility to the network's structure and strengthens the chance that ties between the core members and (essential) peripheral actors become restored. In short, the particular structure identified offers resilience against network disruption. If one coalition falls apart due to arrests, there is a great chance that remaining actors can fall back on other coalitions and re-establish the lines of production.

3.3.3 Individual Positioning Within the Blackbird Network

The previous section offered one explanation for the network's flexibility against disruption caused by the criminal investigation. In order to find additional evidence for this hypothesis, we need to zoom in on the individual level of the network. Networks consist of individuals that have different influence and power

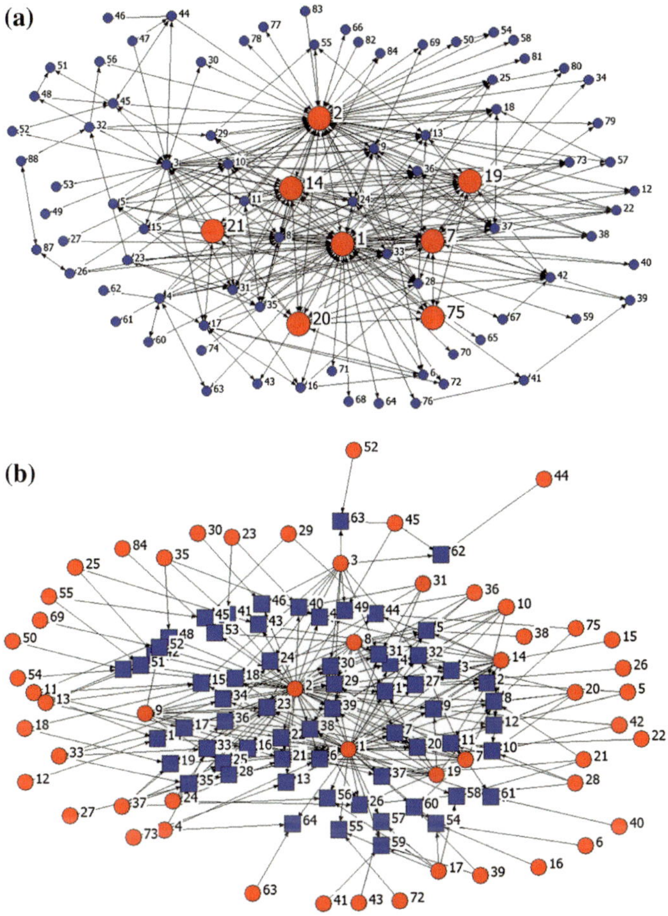

Fig. 4 Results for clique analysis of the Blackbird network with **a** Biggest clique **b** Involvement of actors (*red circle*) in the identified cliques (*blue square*)

within the network. Understanding individual actors' properties in terms of influence and power is important for understanding overall network structure. One of the most significant measures related to influence and power is actor centrality [38]. There are many different measures to estimate centrality. The most commonly used centrality measures are listed and explained in Table 4.[6]

Table 5 shows all scores for the top 15 actors on these centrality measures. Not surprisingly these different measures for centrality reveal that actor 1 and 2 are in highly central positions within the network. Although they might have a lot of influence within the network, the relatively low scores for Bonacich Power (32 %)

[6] For a full description of these measures see [38].

Table 4 Common measures to estimate centrality [38]

Degree centrality	Number of direct contacts that an actor has
Bonacich power	The extent to which an actor is connected to other actors that score high in degree centrality
Closeness centrality	Indicates how close each actor is to all others
Betweenness centrality	The number of paths that connect pairs of nodes that pass through a given node

Table 5 Distribution of different actor centrality measures (top 15)

Actor	Degree	Bonacich power	Closeness	Betweenness
2	55	32.29	73.913	0.444
1	53	31.81	70.833	0.388
3	36	20.37	57.432	0.132
14	23	21.92	56.291	0.018
8	21	20.41	55.921	0.019
7	20	18.26	55.195	0.034
10	13	15.70	52.469	0.005
19	13	15.18	52.147	0.002
31	13	14.41	51.829	0.006
20	12	13.58	51.829	0.005
9	12	13.52	51.515	0.008
24	11	13.48	51.515	0.002
36	11	12.78	51.205	0.002
4	11	10.16	50.595	0.056
13	11	9.97	51.205	0.007

reveal that they might in fact not be all that powerful in network terms. Bonacich [55] argued that being connected to others that are not well connected makes someone powerful, since such actors are dependent on you—whereas well-connected actors are not. So according to Bonacich an actor's power in networks depends not solely on their own connections, but mostly on the connections of their direct neighbors.

We can explain this further by looking at the graph of Fig. 4a representing the size of the nodes according to the scores for degree centrality. Although actor 1 and 2 score high on degree centrality, a representative part of their direct neighbors within the core of the network, are well-connected themselves. This means that these neighbors aren't solely dependent on actor 1 and actor 2 for their resources or information. According to network theory this reduces the power that actor 1 and 2 have over the core members, as they are self-sufficient for their resources and information. However, Fig. 2 reveals that the actors in the periphery of the network are often dependent on actor 1 or 2 for their participation in the cannabis cultivation process. This might give actors 1 and 2 a strategic advantage and an opportunity to apply power to these peripheral actors.

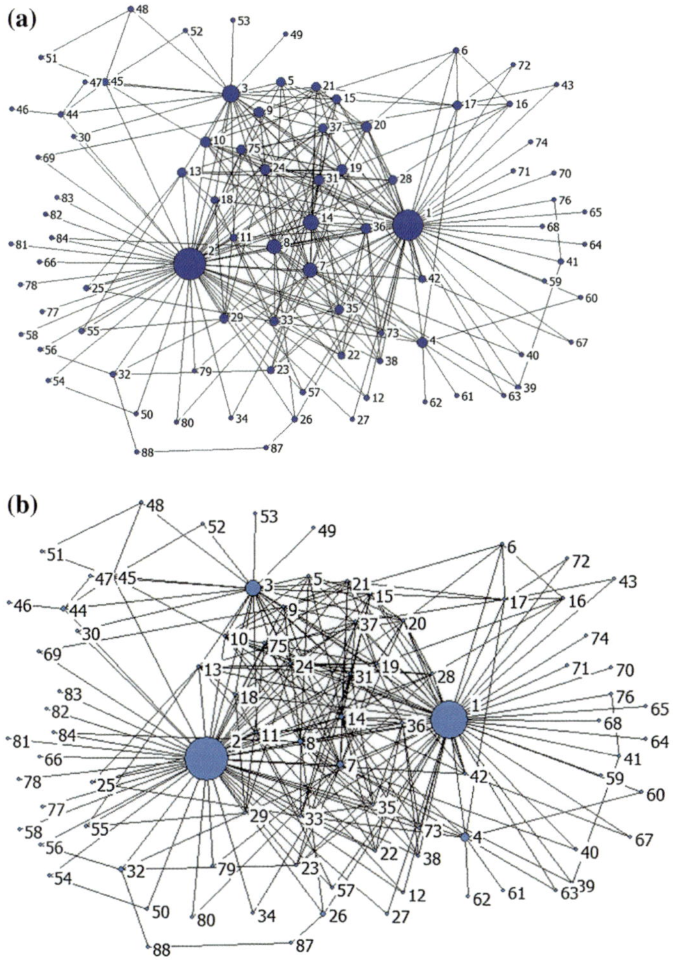

Fig. 5 The size of the nodes corresponds to the scores on **a** Degree centrality and **b** Betweenness centrality

More evidence for this assumption can be found in the individual scores for betweenness centrality (Table 5). The size of the nodes in Fig. 5b represents the score on betweenness centrality. It reveals that actor 1, 2 and 3 score high on betweenness as opposed to the remainder of the network. The graph indicates they form a bridge between the network's periphery and core. According to network theory, betweenness centrality is associated with strategic advantage. Burt [56] offers a theoretical framework for understanding this phenomenon and found that having quick access to information offers some actors abilities to fill positions that allow them to seize rewarding opportunities. The entrepreneurial opportunity that follows from the position of 'bridge' between two separated parties is called a *structural hole*. According to Burt [56], actors with the capacity to enrich their personal network with a proportionally higher set of structural holes may come to control other actors in the network. Morselli [46] specifically studied *brokerage positioning* in criminal networks of career criminal Howard Marks in the international cannabis trade. Morselli found that the key to Marks's successful criminal career was that he structurally stayed in between different criminal groups. This brokerage position contributed to his reputation and strategic advantage in the worldwide criminal macro network.

In accordance with Burt [56] theoretical framework, brokerage positioning within the Blackbird network was analyzed. First, the results of betweenness centrality analysis reveal that actors 1, 2 and 3 occupy score high on potential brokerage positions. In addition, structural hole analysis reveals that actors 7, 8, 9 and 14 often occupy a structural hole position. As these specific actors were left out of the scope of the investigation and final arrests, this could be an additional explanation for the observed network resilience after intervention. These results might offer another explanation for network resilience. Hence, this analysis reveals that the structural holes that are left behind by the arrested actors 1, 2 and 3 could be easily be occupied by actors 7, 8, 9 and 14.

In sum, analysis of the individual positions within the Blackbird network, revealed that the network is built around two highly connected actors. Betweenness centrality (Table 5) shows that these actors are also important hubs for the flow of information and resources throughout the network. Figure 5b shows that this brokerage role connects the 'loose' periphery with the tight core of the Blackbird network, making actors 1 and 2 influential actors. However, the results for the Bonacich power analysis (Table 5) reveals that their power might be reduced, because they are connected to well-connected others. Figure 5a shows that these well-connected others (high degree) are part of the network core (N = 32). This can be acknowledged by the structural hole analysis, which revealed that actors 7, 8, 9 and 14 are often in structural holes position themselves. Referring to network resilience this means these actors might play an important role in network recovery, in case actor 1, 2 and 3 become arrested.

The aim of this analysis was to unravel the structure of the Blackbird network and how it managed to continue with cannabis cultivation after law enforcement interventions. To search for the answers, we used quantitative social network analysis to unravel the Blackbird network structure on three levels: network topology, substructures and individual positioning. Through analysis on all network levels, different network properties could be identified that might have contributed to the observed network resilience against a major enforcement intervention. Although this gives us some answers, it also leaves us with a lot of additional questions: What causes this network to be highly centralized? How can the high level of redundancy within the core of the network be explained, compared to the non-redundant periphery? What are the characteristics of the central actors and their 'independent' well-connected neighbors? Answers to these question are essential for making any meaningful recommendation can be made for law enforcement tactics. This is the point where mathematical methods stops and more qualitative methods of social network analysis enter into play. The next section describes how qualitative features of the Blackbird network can place the observed Blackbird network structure into a different perspective.

3.4 Qualitative Analysis of the Blackbird Network

Many SNA scholars have emphasized the importance of integrating individual characteristics within the study of criminal networks (e.g. [2, 53, 57, 58]. Although quantitative methods offer us direction of interesting network features, qualitative methods are essential for placing these results in the right context or even revealing other aspects of network structure that cannot be calculated. As described above, detectives who worked many hours on the Blackbird investigation, pointed in the direction of an embedding social structure of kinship and affective relationships as an important explanation for the observed flexibility within the criminal network structure. Based on this hypothesis kinship and affective relationships were scored in different matrices in addition to all observed criminal relationships. The graph of Fig. 6 shows the results for combining these different networks matrices.

As in the previous graphs, criminal relationships are visualized by gray lines. More interesting in this graph is the way in which the network structures of affective ties (red lines) and family ties (green lines) are intertwined in the core of the criminal network. Another interesting feature in relation to this network's structure is revealed by visualizing male (blue nodes) and female (pink nodes) actors. Figure 6 shows, that women are an essential part of the core, suggesting they might hold influential and powerful positions within the Blackbird network. Additional evidence for this hypothesis was already found in the quantitative results on individual centrality measures (Table 5), for which five of the most central actors are female. But how do these women end up in these influential positions in the network? The answer can be found in the social network in which the criminal activities were embedded.

3.4.1 The Social Embeddedness of the Blackbird Network

Figure 7a displays this embedded social network of combined affective (red lines) and kinship (green lines). It reveals that actor 2 is most important for introducing women in the network. Evidence for this is found in many wiretap conversations over the six-month time period and in statements made by some of these women after the final arrests. These police reports show that actor 2 was a skilled networker who applied his organizational skills not only in his criminal environment but also in his social life, as he managed to maintain affective relationships with different women at the same time. Furthermore it shows that as these women were introduced into the network by actor 2 over time, they became accepted within the

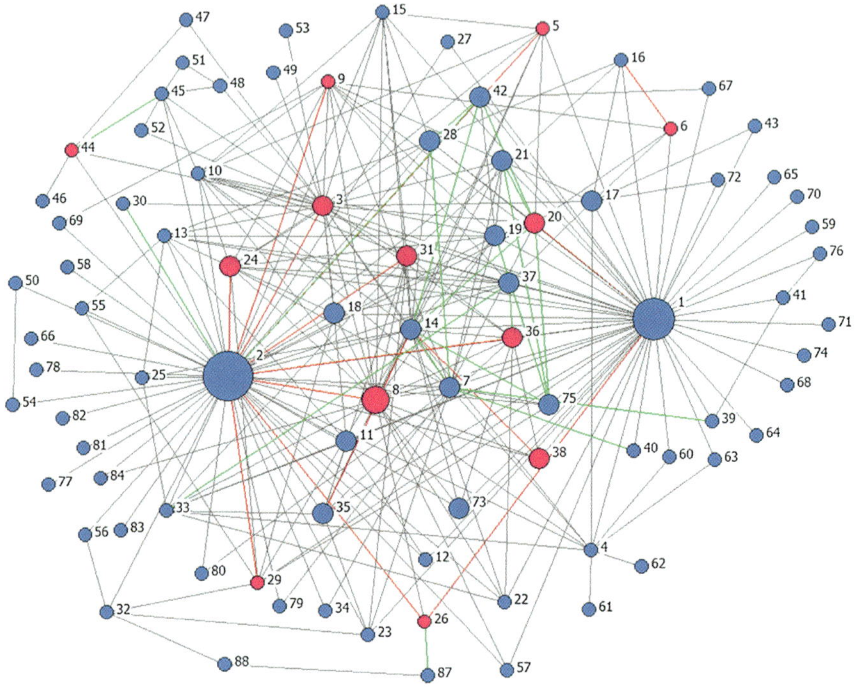

Fig. 6 The Blackbird network including affective ties (*red lines*) and kinship ties (*green lines*). Furthermore the blue nodes represent males and the pink nodes represent women in this graph

tight social core surrounding actor 1 and 2. Some women even started new love affairs within this social network, after their relationship with actor 2 had ended. An important factor in this respect, deriving from the surveillance reports and wiretap data, is that all activities within this social structure seem to gravitate around a small geographic infrastructure of cafés, hangouts and restaurants.

Furthermore, as these women became members of this social core, they became connected to other women in the network (see Fig. 7b). Various wiretap conversations and surveillance reports show that besides their progressing influence in the social network, most women were introduced to the illegal activities of cannabis cultivation. The development of trust seemed to play an important role in this. Different suspect statements reveal that as these women proved themselves to be reliable and loyal members of the embedding social network, they were allowed to participate in criminal activities. Moreover, newly-introduced women began to establish mutual criminal relationships between themselves. Figure 7b presents the criminal relationships observed between female actors.

Although some of the women played important roles in coordinating different phases of the cannabis cultivation process, they were ignored as serious suspects in the Blackbird operation. These findings are consistent with a study of Kleemans and Van de Bunt [59] on the '*social embeddedness*' of organized crime. Based on

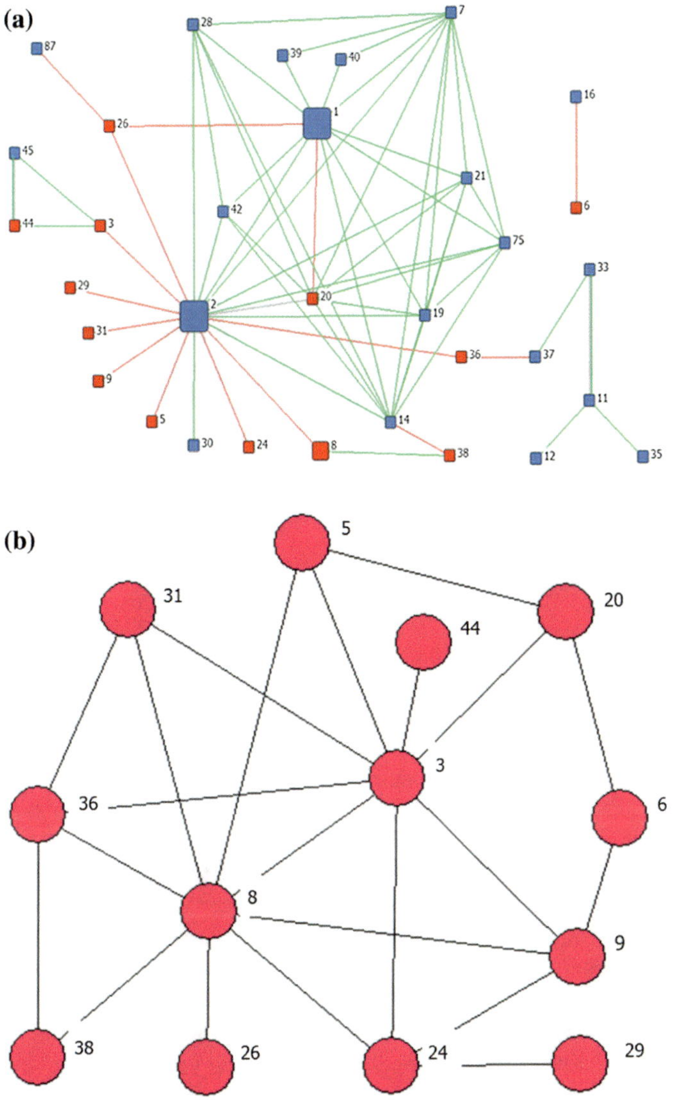

Fig. 7 **a** Visualization of the embedded social network of combined affective (*red lines*) and kinship (*green lines*). Red nodes correspond to females and blue nodes correspond to males in the Blackbird network. **b** Criminal relationships between females after they had entered the embedded social network

their analysis of 40 cases of organized crime, they found that women were not only important for maintaining and establishing contacts between different parts of the criminal network, but were in some cases in charge of a whole criminal association. They concluded that the importance and influence of women in terms of social embeddedness in organized crime is often a blind spot in law enforcement control strategies.

In addition to these affective relationships, another important part of the embedding structure of the Blackbird network was formed by kinship ties, as indicated by green lines in Figs. 6 and 7a. These graphs show that Kinship ties play an important role for the observed redundancy within the network's core. In addition, based on quantitative measures for network positioning, it was already concluded that actor 14 is an influential actor often positioned on a 'structural hole' position within the network. In fact, qualitative analysis shows that actor 2 is his father. These findings support the idea that actor 14 might have inherited his father's criminal achievements (and possibly reputation) and therefore his social and criminal capital. This observed generational heritage in criminal carreer opportunities, is consistent with observations made by Spapens [10] based on his study of criminal ecstasy networks operating in the south of the Netherlands.

The importance of social ties for criminal network development was first addressed by Granovetter [59]. He introduced the theoretical concept of '*the strength of weak ties*'. According to Granovetter strong ties are important for illegal as well as legal transactions, because trust is built between like-minded actors. Especially within the hostile and uncertain environment of organized crime, strong ties of family, friendship and even love often offer a necessary fundament of trust. Different studies show that trust in criminal networks is often found in an embedded network of social ties [2, 10, 17, 46, 59, 61]. This is in accordance with our findings in the Blackbird network. The tight core of this network is formed not only through criminal relationships, but more importantly through affective and family relationships. Additional empirical evidence is found in the analysis of a high number of wiretap conversations between core members. Most conversations concern a mixture of social and criminal activities. In general it can be concluded that strong ties of affective and kinship ties form an important framework of trust, from which criminal activities in the Blackbird network originated.

Granovetter [60] also emphasizes the importance of *weak ties* for expanding business opportunities. Weak ties are connections between people who are not intimate or close. In these relations mutual trust is not easily attained. But precisely because of this, these ties are not redundant and are therefore essential for access to new resources and information. Weak ties can therefore offer new opportunities, especially within an illegal enterprise [17, 46, 59].

Besides strong ties, the Blackbird network also consists of a high number of weak ties. The introduction of women in the Blackbird network is an example of this 'strength of weak ties' principle. In the beginning they are recognized as weak ties, but as the number of connections with the redundant core increases over time, these actors become trusted and serious participants in the criminal activities. Another example of this mechanism within the Blackbird network is the difference in observed redundancy between the core of the network and the embedding periphery. Most actors in the periphery of the Blackbird network are connected with the core of the network through a single tie with actors 1, 2 or 3 (Fig. 5b). These connections are in fact non-redundant, but part of the network because of their direct involvement within the cannabis cultivation process. Qualitative analysis of these weak ties in the Blackbird network reveals that most of these actors are not 'isolated' freelancers, but often representatives or even brokers between the Blackbird network and other criminal networks that are connected with actor 2 for expanding their criminal business.

For instance, based on additional content analysis of wiretap data, it was recognized that actor 87 is an important representative of a foreign mafia organization and an important buyer of cannabis from actor 2 from the Blackbird network. Besides his criminal relationship with actor 2, he also had a short love affair with female actor 26 (Fig. 5a). Female 26 seemed to have had an intimate relationship with actor 2 in the past. Although the investigative data doesn't allow drawing a timeline of the initiation of these criminal and affective relationships, it's evident that this social connection played an important role in the initiation of an export route of cannabis from the Blackbird network to Italy. The weak ties between the Blackbird network and the Italian Mafia network that offered new opportunities for the Blackbird network to market their illegal product, developed into a stronger tie over time due to social embeddedness. Based on these findings it can be concluded that the social embeddedness the observed within Blackbird network structure offers another explanation of its flexibility and resilience against disruption.

3.4.2 The 'Division of Labor' Within the Blackbird Network

In the beginning of this chapter the application of crime script analysis was explained to identify unique roles in the cannabis cultivation process. In Fig. 7 the actor-by-variable matrix of involvement in cannabis cultivation is visualized. Every link represents the involvement of an actor in a specific phase of the cannabis cultivation process. Actors of which role specific information was missing were left out in the final representation. However, even without the missing data this visualization shows a highly redundant division of labor, in which every task is covered by multiple participants. Based on crime scripting analysis, it was assumed that actors responsible for 'manipulation of electricity supply' would be thinly populated within the network, because of the specific skills and knowledge needed to complete this task. Figure 8 on the contrary shows that this 'specialized' task is covered by no less than five actors within the network. Qualitative analysis of the wiretap conversations reveals that these specific skills were learned in the network by actors through experience. This is a form of *differential association* described in classic studies of learning criminal networks [61, 62]. In this way the network could efficiently replace these actors in case of arrest or other external interventions from its own redundant core network.

In addition, Fig. 8 also reveals that the most central actors 1 and 2 are involved in many specific tasks themselves instead of delegating from a distance. On the one hand, this gives them a lot of control over the criminal business process, but on the other hand it increases their visibility and therefore their vulnerability. This could probably be one of the explanations for their final arrests, which raises the

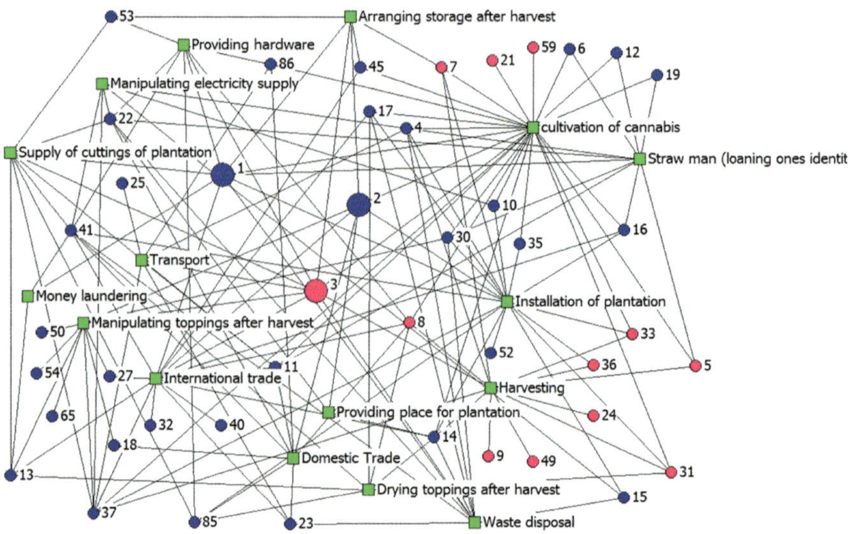

Fig. 8 Division of labor within the Blackbird network concerning illegal cannabis cultivation. The most central actors 1 and 2 are involved in many specific tasks themselves instead of delegating from a distance

question: might there be an external 'supervisor' from the periphery who was missed in this investigation? Unfortunately, answers to this question are out of the scope of the dataset.

Another interesting aspect of this division of labor is the position of women within the criminal process. Figure 8 shows that most females are involved in simple tasks, such as '*harvesting*' and '*helping with the installation*' of the plantation. The specific task of harvesting requires that these women were working together in a small and closed room for several hours. Qualitative analysis of the surveillance-, wiretap- data and eyewitness statements reveals that this was one of the reasons these women developed mutual social and criminal connections. Moreover it seemed that females 3 and 8 were also involved in tasks of coordination and leadership within the whole network. This is also observed within the '*women network*' of Fig. 7b, in which these females occupy a central position. Although female 3 got arrested in the end, female 8 might have played an important role in network recovery and reestablishing the division of labor within the whole process.

These findings for 'division of labor' are in line with research associated with the tradeoff between efficiency and security that is revealed within previous research [5, 7, 52, 63]. On the one hand illicit networks try to keep their illegal activities concealed from the government or criminal competitors. This means that direct communication between co-conspirators concerning illegal activities needs to be restricted to a minimum. On the other hand risks have to be taken in times of action, often demanding highly efficient communication and trust among its participants [5, 63]. This tradeoff shapes the way illicit networks are structured. For instance, criminal networks demanding high levels of action and therefore efficiency are often characterized by high levels of redundancy. Terrorist networks on the contrary often demand just one successful action to reach its network objectives. These network structures are therefore characterized by high levels of non-redundancy and compartmentalization in different cells [16, 50]. Terrorist networks use this strategy to decrease the risk of becoming detected by the arrest or detection of a single actor. Criminal networks also try to built in security, but as times-to-task are much shorter efficiency often predominates.

Analysis of the Blackbird network according to this theoretical framework reveals that a certain level of compartmentalization can be found in its network structure. Hence, there seems to be a clear separation between the network's core and periphery, which might offer some security to core members if peripheral actors become arrested. However the crime script analysis in combination with SNA reveals that tasks are divided in a highly redundant way, which is typical for 'action-minded' criminal networks (e.g. [16]). On the one hand this gives the network the advantage of flexibility in replacing actors after arrests or seizures, which is also observed in our analysis of the Blackbird network. On the other hand, this increases the risk for exposing the network as a whole if a single actor gets arrested. Hence, due to the high level of redundancy, chances are substantial that a single arrested actor is directly connected to the central actors 1 and 2. This increases the risk of exposing these important actors, for instance by tracking previous telephone calls. Apparently the low level of security within their network structure is exactly what ultimately caused the arrest of actors 1 and 2.

3.5 Conclusion

By combining quantitative and qualitative methods of SNA, the structure of the Blackbird network was unraveled. Quantitative analysis revealed that its overall network structure gravitates around a few central actors. These actors form a redundant core that is connected with a non-redundant periphery by just a few highly connected actors. These actors occupy strategic positions, but because they are connected to well-connected others their positions are not irreplaceable. Qualitative analysis reveals that the core of the criminal network is embedded in family and affective relationships. Women and children play an important role in this embedded and criminal network, as they add to overall network redundancy and fulfill coordinating tasks that become specifically important after their husbands and fathers become arrested. Crime script analysis revealed that this redundancy is also translated to the division of labor, for which all tasks can be fulfilled by multiple actors. In part this offers an explanation for the observed flexibility and resilience against disruption. On the other hand, it has been shown that this redundancy increases network visibility and offered opportunities for arrests. However, it can be concluded that these control strategies were ineffective, as the process of cannabis cultivation continued due to the flexibility and efficiency that is built into its network structure. Based on these conclusions it could be recommended in search for effective control strategies in the future to take notice of the active and important participation of women and direct relatives in the organization of criminal activities.

3.6 Discussion

As described in the beginning, this case example demonstrates how the application of Social Network Analysis (SNA) could be of value in understanding the effects of current control strategies and creating and adjusting future strategies aimed at these complex criminal network problems. This case study also demonstrates that applying SNA on criminal networks demands a twofold approach, integrating quantitative and qualitative methods. As was demonstrated in this case study, this is essential in understanding not only the answers to the 'what' questions, but also the important 'why' questions. For instance, an additional qualitative interpretation was crucial to understanding why women occupied central positions in the core of the network. Answers to these 'why-questions' are therefore the key to really understanding the 'covert' mechanisms associated with criminal networks and for the translation of such insights into concise recommendations for law enforcement control strategies. This case-study therefore shows that the application of quantitative and qualitative methods of SNA together with crime script analysis constitutes a powerful tool for agencies confronted with criminal network problems. However, in addition to such advantages, the case study also revealed some important limitations.

First, practitioners should realize that the final representation of the criminal network is to a large extent a product of the boundary specification criterion and available data [2, 11, 39]; (Krebs 2002). For instance, we observed based on qualitative analysis, that some 'isolated' actors in the periphery that were selected for their involvement in the cannabis cultivation process, were in fact representatives of other criminal groups. This emphasizes the fact that our observed Blackbird network is in fact part of a bigger macro-network [10]. Another important point to address is missing data. Our observations are solely based on investigative data. This naturally filters the data collection process and therefore the network representation to a certain degree according to the initial goal of the investigation [2].This becomes especially important when interpreting the results from quantitative measures, such as centrality and individual positioning. The fact that actors occupy strategic positions in the local setting of the Blackbird network, doesn't necessarily mean that they are powerful or influential in general. Placing quantitative results in a qualitative context is therefore crucial when using this method. The challenge for the practical application of SNA would therefore be to combine different sources of relational data, for instance intelligence data, street cop data, arrest records and 'online' data. Every source has its own filter through which we observe 'reality'. Combining 'filters' might increase the reliability of the final network representation.

Secondly, one of the critical success factors within intelligence-led-policing is timely intelligence products. Time in this context is related to law enforcement demanding swift decision-making. Decision makers in law enforcement settings therefore want fast, reliable and concise advice [64]. Social network analysis on the contrary is a time-consuming exercise. As shown in this case example, data have to be collected and processed in a structured way. Additionally, results need to be interpreted in the right context. The application of SNA within an operational law enforcement environment might therefore become problematic. The final results might only come available too little too late. The challenge for the practical application of SNA within organized crime control is therefore to find a way of processing data in a faster way.

Thirdly, like any social network, criminal networks aren't static but dynamic [57]. The structure of the network as well as its activities is ever-changing. Our case study focuses on the network configuration before the final arrests. This offers unique insights into the properties of network structure that explains its flexibility, but it doesn't offer us insight in the way the network really adapted to the arrests. The challenge for the practical application of SNA in organized crime control would therefore be to find ways of observing these network dynamics and the network's adaptability to network disruption.

These are tough challenges that are not easy to translate into practice. However, in the Netherlands these challenges are slowly becoming reality. In the next section the progress and developments in the Dutch Police in answering these challenges will be discussed.

4 SNA and Recent Developments in Dutch Law Enforcement

As described above, this case study is one of the recent experimental examples of the practical application of SNA in current Dutch law enforcement. However, the issues and challenges that were addressed are no novelty, as they were already recognized before by Sparrow [11]. It can therefore be concluded that the implementation of SNA within the operational law enforcement environment is a major challenge, as two decades after Sparrow introduced and addressed these issues they are still topical in Dutch law enforcement. Yet this isn't a lost cause, as there are promising developments that might help to translate network theory into SNA practice: the increasing availability of data on criminal cooperation and advances in SNA methods from computational science.

4.1 ILP and the Increasing Availability of Criminal Network Data

One of the important challenges for the practical application of SNA within criminal intelligence that Sparrow [11] identified is creating an automated data-management system for parallel processing technologies in which different databases can be linked together in a structured way. Klerks [9] was confronted with this challenge in his SNA-based study of Dutch criminal networks involved in international drugs smuggling. The initial coded data within the police databases turned out to be unreliable for SNA practice. For instance, specific persons were registered multiple times. The data had to be recoded all over again, requiring a lot of time and effort.

One reason for these data validity problems in Dutch police databases is that information gathering and processing aren't always recognized as one of the primary tasks of law enforcement officers in the frontlines of police work. This often results in poor quality of data, especially about the more circumstantial features of observed criminal cooperation and communication which are important for SNA. For instance, within the Blackbird operation 'social' conversations between women in the network were labeled 'irrelevant' by detectives, but SNA of the Blackbird network revealed that these conversations specifically emphasized the importance of these women as 'mediators' in case of internal conflict. Therefore, the practical application of SNA within law enforcement in general depends for an important part on the '*information-mindedness*' of police officers and detectives.

Still, compared to 10 years ago the general information quality and quantity in Dutch law enforcement shows progress. One reason is the introduction of the concept '*intelligence led policing*' (ILP), which increased the general awareness of specific intelligence tasks involving daily police work [64]. This has resulted in the introduction of specified intelligence tasks during police surveillance, aimed at retrieving information from the direct observation and registration of '*local heroes*' or '*hot spots*' (e.g. bars, restaurants) associated with local organized crime. In practice, this has led to the recognition of ties between high profile criminals that weren't observed before. Information collection is therefore increasingly recognized within Dutch law enforcement as a primary task of regular police work. It also helps that the 25 regional police forces in Holland were merged into one National Police in January 2013, facilitating the implementation of shared doctrine, ICT etcetera.

Another important development associated with the introduction of ILP is an increased awareness of the importance of powerful ICT tools for 'user-friendly' data processing in criminal investigations. Although data quality in general is still a concern, these tools are already showing increased uniformity and quality in data processing. These developments are still in their infancy, but they are promising for the effective application of SNA in the law enforcement environment. A more specific aspect of this development is the effective use of human intelligence (HUMINT) and social media intelligence (SOCMINT) for proactive law enforcement intelligence purposes. These developments will be discussed in the next section.

4.1.1 Human Intelligence

Every regional police unit in the Netherlands maintains a Criminal Intelligence Unit (CIU). These CIU's are specifically tasked with retrieving 'human intelligence' from criminal informants and have primarily been focused on assisting ongoing criminal investigations with supporting evidence. More recently, it is recognized that the CIU's are also important for delivering proactive intelligence products aimed at discovering strategic trends in illegal criminal markets and the translation of such trends in effective operational targeting of subjects at the start of investigative operations. The growing symbiosis between analysts and handlers in the criminal intelligence process has strengthened this development. This leads to a more goal-oriented intelligence collection process. For many years the search for criminal informants has been rather opportunistic, often the result of sudden opportunities following arrests in criminal investigations or conflicts between known criminal rivals. Although this remains a fruitful tactic for recruiting motivated informants, it mostly leads to more information on already familiar actors and well-known criminal markets. The biggest challenge for CIU's is therefore to find potential informants in criminal networks or criminal topics that are still relatively unknown to the police, for instance cybercrime networks or human trafficking rings.

Following these considerations, SNA is increasingly recognized as an important method for intelligence analysts in profiling such potential informants and identifying opportunities for approaching them. For instance, SNA helps to identify criminal brokers within criminal networks that might function as potential '*points of access*' to relatively unknown criminal communities and –markets. It needs no further explanation that these brokers might be high-potential sources of criminal intelligence. In this way SNA stimulates strategic thinking about proactively shaping intelligence positions according to novel trends in the criminal environment, as opposed to the traditional opportunistic selection of informants based on ongoing operations. Moreover, as this increases the validity and reliability of human intelligence databases, this offers chances for a more concise application of SNA with the aim of targeting criminal networks effectively.

4.1.2 Social Media Intelligence

A second development that offers improved opportunities for SNA in law enforcement is the increasing usage of open source intelligence in Dutch law enforcement. A growing number of studies reveal that Internet communities such as Facebook, Twitter and Google + are not only used by criminals for ordinary social reasons, but also for expanding their criminals markets or even threatening criminal rivals [12]; (Decker and Pyrooz 2009). Social Media Intelligence (SOCMINT) is therefore recognized as an indispensable source of operational intelligence about criminal network structures [65].

In practice, social media intelligence offers an opportunity to peek behind the social network structures embedding the criminal cooperation. For instance, it was found that some members of the Blackbird network shared the same hobby: sport fishing. Combining police information with the pictures they posted on social media posing with their fishing trophies, some new (criminal) ties in the criminal network could be revealed that had not been observed before. SOCMINT therefore offers a different perspective on underlying social network structures, often unknown to law enforcement. Still, the application of SOCMINT is associated with some difficulties. First, it takes some time to adjust legislation for the use of intelligence to encompass such exponentially expanding technological developments. This problem leads to tension between public goods of security on the one hand and citizens rights to the rule of law, liberty and privacy on the other [65]. Secondly, Intelligence analysts are confronted with *big data*, which is near-impossible to analyze using traditional SNA analysis methods. Therefore computational methods are essential in addition to traditional SNA for mining these big sources of data for relevant information [65].

A recent study by Dijkstra et al. (2013, in press) explores the application of computational text-mining models for identifying an illicit network of drugs users within the online social network *Livejournal*, covering 2,6 million Russian users. A '*webcrawling*' computational algorithm was used to find patterns of textual terminology associated with drug use within the enormous amount of Livejournal user blogs. In this way a network of bloggers interested in drug related subjects could be identified. Further analysis of these user-profiles revealed that this network showed some similarities in their interests, such as specific hobbies or beliefs. Although this study reveals promising future opportunities of the computational approach for identifying illicit networks within social media, these models need further elaboration before application in an operational intelligence environment is possible. For instance, based on the results the possibility cannot be excluded, that the identified network also contains actors that are anti-drugs activists rather than drug users [13]. Omand et al. [65] therefore emphasize the importance of validation of the results of these data mining methods on the Internet by comparing them with other criminal intelligence sources and placing it in the context of the specific characteristics of the target (criminal) subgroup and social media platform. By using this data and knowledge to 'train' these computational models, validity and reliability of the results will increase over time.

4.2 Towards a 'Real-time' SNA Approach to Organized Crime

Following these previous considerations it can be concluded that HUMINT and SOCMINT are in themselves important pieces for constructing the criminal network representation. However, this intelligence puzzle cannot be completed based on these data sources alone. Moreover, as criminal networks are dynamic in nature the developments within criminal networks following from these data-sources need to be monitored continuously. In Dutch law enforcement, these considerations have lead to a 'real-time' SNA approach for analyzing criminal networks. In essence, this approach consists of the structural integration and analysis of multiple data sources into one relational database. This method offers the opportunity of assessing 'missing data' in the criminal network representation [2]. This leads to the identification of *'intelligence gaps'*, which can be translated into topical intelligence collection plans [66]. For instance, an identified social tie between two known criminals that is identified based on SOCMINT can be translated into concise intelligence questions for criminal informants from a HUMINT approach to assess the nature of this relationship. Although verifying information in this way is already part of everyday police work, continuously and structurally combining multiple sources in the context of previously collected intelligence aimed at identifying criminal networks, is not always a matter of course. Because new information is continuously interpreted in the light of previous developments in the criminal environment, opportunities for identifying recent change in such networks arise. Identifying these changes is an important part of effectively targeting criminal networks at a certain point in time. However, it needs no further explanation that continuously monitoring different information sources by hand would be very time-consuming. Powerful ICT tools are therefore essential to this approach, because different data sources with varying data formats have to be integrated and merged automatically into one relational database. State of the art database analysis tools, such as IBM's i2 iBase, offer these ICT solutions with integrated visualization and SNA applications.

This approach has some important advantages for the application of SNA in law enforcement:

1. Data from different data sources can easily be validated with other data sources, leading to a more strategic approach for data collection.
2. Because data is collected and processed in an ongoing and automated process in a structured format suitable for SNA methodology, time is saved for actual SNA practice and making recommendations. This results in timely intelligence products that find the connection with the time-dependent decision-making cycle that characterizes operational law enforcement management.
3. Because new data are continuously analyzed, changes in criminal networks or criminal markets can be identified. The flexibility offered by this approach is important for recognizing chances for effective interventions at a certain point

in time and offers the possibility to analyze criminal networks as dynamic structures instead of static snapshots.

4.2.1 Combining Computational Methods with SNA

Besides the application in retrieving relevant data from social media, computational methods are increasingly important for understanding the complex dynamics of criminal networks. Appreciating these dynamics may have major consequences for the way we think about the effectiveness of control strategies aimed at criminal networks. SNA scholars agree however that capturing network dynamics is one of the most difficult challenges in criminal network research (e.g. [2, 11, 67]). The limited number of studies on this topic identifies four methods for capturing network dynamics: descriptive, statistical, simulation and visualization methods (Doreian 1997); [67]. Descriptive methods are focused on structural changes in social networks by comparing structural properties across time. These structural changes are associated with changes in nodes, links or groups within the network. The statistical approach is not only focused on structural changes but also involves an evaluation of the reasons for such changes, for instance the effect of gender for preference in social bonding. Simulation methods rely on multi-agent technology, for which actors are modeled as agents making decisions based on specific criteria. These criteria are translated into algorithms. Visualization methods aim at comparing network maps at certain points in time through visual inspection (Doreian 1997); [67].

An example of the application of descriptive and visualization method was recently presented by Bright and Delaney [14]. These authors studied the evolution of a drug trafficking network and found that participants change their specific role in the crime script based on needs, as opposed to simply recruiting replacements to fill those needs. Secondly, they found that these changes have a direct impact on the centrality of single actors in the network. Bright and Delaney [14] emphasize that law enforcement needs to respond flexibly to these changes in network composition. However, one of the important limitations of this study was that the observed changes in the network could be artifacts of intelligence collection methods.

Capturing network dynamics with simulation modeling is less sensitive to this type of bias, as network behavior is not empirically observed but simulated with multi-agent technology. This method offers the opportunity to perform "what-if" scenarios to study how social networks adapt to different external shocks [67]. This simulation methodology was used in a recent study by Duijn et al. (in press), aimed at unraveling the dynamics of criminal network resilience against disruption. For this study a unique dataset consisting of multiple data sources, including criminal intelligence data, investigative data, police reports and arrest records was used to reconstruct a criminal cannabis cultivation network (N = 793 actors), including variables on specific roles within the cannabis cultivation crime script (Fig. 1) and its embedded criminal macro network (N = 29,346 actors). These unique features offer the opportunity to simulate the interaction between a criminal micro network and the embedding macro-network as a result of network disruption and recovery. Based on previous research it can be assumed that the embedding macro network, plays an important role in finding replacements after network disruption (e.g. [2, 10, 59]).

To simulate different control strategies, five algorithms for network disruption were applied to the criminal cannabis network, associated with attacks against *social capital* (e.g. degree, betweenness) and attacks against *human capital* (e.g. specific crime script roles) within criminal networks. In addition to these network disruption strategies, network recovery (replacement of nodes and links) was also simulated at the same time. In this way criminal network resilience against network disruption could be estimated. The results of these simulations show that criminal networks become more efficient as a result of network recovery. More specifically, network density increased due to network recovery, resulting in higher levels of general redundancy. As criminal network structures apparently become more efficient following intervention, this result might be disturbing for law enforcement agencies trying to control these network structures. However, criminal networks face a constant tradeoff between efficiency and security as previously explained in Sect. 3.4 [52, 16]. This means that the increased efficiency has a negative effect on its general security, as on average network members become more visible within the overall network structure due to increased redundancy. From a law enforcement perspective, this offers new chances for further law enforcement surveillance and criminal investigation. Based on these results, disrupting these flexible criminal networks seems to require a long-term effort.

Although the Duijn et al. (in press) study aims to understand these criminal network dynamics in general, this multi-disciplinary method might have direct relevance to the operational law enforcement environment. As these models can be 'trained' and adjusted over time by the increased availability of empirical criminal network data, this approach might become a powerful method for pro-actively experimenting with "*what-if*" scenarios and strategically thinking about intervening effects on live criminal networks. Secondly, these models might contribute to the identification of 'trigger events' often hidden in the data, which might function as an early warning for upcoming criminal activity, travel movements, unusual financial transactions or changes in criminal network structures. These early warnings might be translated into proactive, well-timed and specifically targeted control strategies. Combining SNA and practical law enforcement knowledge with simulation methods and the increased availability of data may therefore become an integral part of proactive organized crime control in the near future.

5 Conclusion

The aim of this chapter is to inform about recent developments of the application of network analysis in controlling crime in the Netherlands. It offers insight into the practical application of network analysis in Dutch law enforcement, specifically applied to effectively targeting criminal networks. Based on the developments described in the chapter, some conclusions can be drawn about the practical implementation of SNA: (1) It can be concluded that SNA is a useful method for unraveling the structure of criminal networks. It offers renewed understanding of hidden social structures that might be of direct relevance to strategic planning within organized crime control. (2) The strength of SNA within law enforcement becomes most evident if quantitative and qualitative methods are combined. This places the quantitative results in the necessary context. (3) The biggest limitations of traditional SNA methodology (as applied in the case study) are that it's time consuming, static and often too little too late in the eyes of law enforcement decision makers. (4) Due to an increasing 'information mindedness' within Dutch law enforcement in general and availability of advanced ICT applications, new opportunities arise for a data driven approach to SNA in law enforcement. This makes it possible to combine multiple data-sources, which can be connected and integrated automatically. (5) The progress with this approach is strengthened by strategic planning in the field of human intelligence (HUMINT) and social media intelligence (SOCMINT) gathering. The ultimate goal of this approach is to establish a 'real-time' intelligence position on organized crime, from which topical changes in criminal network structures, compositions and activities can be monitored and identified. This offers timely opportunities for proactive control strategies. (6) Simulation methods from computational science might play an important role in understanding these complex criminal network dynamics in the near future. Not exclusively in contribution to the field of science, but also towards operational organized crime control.

In resemblance with the fluid dynamics observed in such criminal network structures, these developments show that the practical application of SNA in Dutch law enforcement is not at all static. Moreover, as the net-centric doctrine of organizing law enforcement cooperation between various agencies and partners becomes more accepted and implemented, flexible criminal networks and law enforcement networks begin to show increasing similarities. While government agencies will always be restrained by legal requirements and subject to budget restrictions, they appear to become somewhat more attuned to the fluid and opportunistic tactics of illicit entrepreneurs. Aided by advanced analytical methods such as SNA, they may become increasingly effective in tackling vital elements of the criminal machinery.

For the near future, law enforcement organizations will at least formally continue to resemble the geometric hierarchy that every civil servant knows as the line-and-block chart, while criminal entrepreneurs will operate in the fluid, random and seemingly chaotic environment that we have come to conceptualize as networks. It is not hard to comprehend that agile and flexible entities unrestricted by laws will often succeed in outsmarting rigid and policy-obese government agencies, even though the latter have the law on their side. Social network analysis provides the guardians of society with a better understanding of the mechanics of criminal networks. As they gradually learn to appreciate some of the benefits of networking, law enforcement and intelligence organizations may become more effective at their core business of safeguarding society.

References

1. Milward HB, Raab J (2006) Dark networks as organizational problems: elements of a theory. Int Pub Manage J 9(3):333–360
2. Morselli C (2009) Inside Crim Netw. Springer, New York
3. Europol (2011) OCTA: EU organized crime threat assessment. European Police Office, The Hague
4. United Nations Office on Drugs and Crime (2010) The globalization of crime: a transnational organized crime threat assessment. United Nations Publications, Vienna
5. Morselli C, Giguère C, Petit K (2006) The efficiency/security trade-off in criminal networks. Soc Netw 29(1):143–153
6. Xu KS, Kliger M, Chen Y, Woolf PJ, Hero III AO (2009) Revealing social networks of spammers through spectral clustering. In: Proceedings IEEE conference communications
7. Lindelauf R, Borm P, Hamers H (2009) The influence of secrecy on the communication structure of covert networks. Soc Netw 31(2):126–137
8. Kleemans ER, Brienen MEI, Van de Bunt HG et al (2002) Georganiseerde criminaliteit in Nederland: tweede rapportage op basis van de WODC-monitor. WODC, Den Haag
9. Klerks P (2001) The network paradigm applied to criminal organisations: theoretical nitpicking or a relevant doctrine for investigators? recent developments in the Netherlands. Connections 24(3):53–65
10. Spapens ACM (2010) Macro networks, collectives, and business processes: an integrated approach to organized crime. Eur J Crime, Crim Law, Crim Justice 18(2):185–215
11. Sparrow M (1991) The application of network analysis to criminal intelligence: an assessment of the prospects. Soc Netw 13:251–274
12. Décay-Hétu D, Morselli C (2011) Gang presence in social network sites. J Cyber Criminol 5(2):876–890
13. Dijkstra LJ, Yakushev AV, Duijn PAC, Boukhanovsky AV, Sloot PMA (2012) Inference of the Russian drug community from one of the largest social networks in the Russian Federation. arXiv:1211.4783v2
14. Bright DA, Delaney JJ (2013) Evolution of a drug trafficking network: mapping changes in network structure and functions across time. Glob Crime 14(2–3):238–260
15. Duijn PAC, Kashirin V, Sloot PMA (submitted 2014) The relative ineffectiveness of criminal network disruption
16. Morselli C, Petit K (2007) Law enforcement disruption of a drug importation network. Glob Crime 8(2):109–130
17. Klerks PPHM (2000) Groot in dc hasj: theorie en praktijk van de georganiseerde criminaliteit. Samsom Kluwer Rechtswetenschappen, Antwerpen
18. Van de Bunt HG, Van der Schoot C (eds) (2003) Prevention of organised crime: a situational approach. Boom Juridische Uitgevers, Den Haag
19. Nelen H, Lankhorst F (2008) Facilitating organized crime: the role of lawyers and notaries. In: Siegel D, Nelen H (eds) Organized crime: culture, markets and policies. Springer, New York, pp 127–142
20. Kleemans ER, Van den Berg EAIM, Van de Bunt HG et al (1998) Georganiseerde criminaliteit in Nederland: rapportage op basis van de WODC-monitor. WODC, Den Haag
21. Van de Bunt HG, Kleemans ER et al (2007) Georganiseerde criminaliteit in Nederland: derde rapportage op basis van de Monitor Georganiseerde Criminaliteit. Boom Juridische Uitgevers, Den Haag

22. Boerman F, Grapendaal M, Nieuwenhuis F, Stoffers E (2012) Nationaal dreigingsbeeld 2012 Georganiseerde criminaliteit. Dienst IPOL, Zoetermeer
23. Korps landelijke politiediensten (2008) Nationaal dreigingsbeeld 2008: Georganiseerde criminaliteit. KLPD Dienst IPOL, Zoetermeer
24. Kruisbergen EW, Van de Bunt HG, Kleemans ER et al (2013) Georganiseerde criminaliteit in Nederland: vierde rapportage op basis van de Monitor Georganiseerde Criminaliteit. WODC, Den Haag
25. Nationaal dreigingsbeeld zware of georganiseerde criminaliteit: Een eerste proeve (2004). Dienst Nationale Recherche Informatie, Zoetermeer
26. Soudijn M (2006) Chinese human smuggling in transit. Boom Juridische uitgevers, Den Haag
27. Spapens AC (2006) Interactie tussen criminaliteit en opsporing: De gevolgen van opsporingsactiviteiten voor de organisatie en afscherming van xtc-productie en -handel in Nederland. Intersentia, Antwerpen
28. Spapens AC, Van de Bunt HG, Rastovac L et al (2007) De wereld achter de wietteelt. Boom Juridische uitgevers, Den Haag
29. Staring R, Engbersen G, Moerland H et al (2005) De sociale organisatie van mensensmokkel. Zeist, Kerckebosch
30. Zaitch D (2002) Trafficking cocaine: colombian drug entrepreneurs in the Netherlands. Kluwer Law International, Den Haag
31. Projectgroep Visie op de politiefunctie, Raad van Hoofdcommissarissen (2005) Politie in ontwikkeling: visie op de politiefunctie. NPI, Den Haag
32. Roobeek AJM, Van der Helm M (2010) Netwerkend werken en intelligent opsporen: een meervoudige uitdaging voor de Nederlandse Politie. Free Musketeers, Zoetermeer
33. Neve R (2010) Netwerken op de stromen. KLPD-IPOL, Zoetermeer
34. Bosveld M (2010) Van je vrienden moet je het hebben… Een verkennend onderzoek naar de toepassing van Forensisch Sociale Netwerk Analyse in Cold Case onderzoeken. Student paper, Politieacademie, Apeldoorn
35. Visser R (2013) Effecten van politie-interventies: Onderzoek naar de ontwikkeling van een crimineel network na een politie-interventie. Student paper. Politieacademie, Apeldoorn
36. Van der Horst PH, Sutmuller AD, Vredengoor S (2013) Real-time op alle niveaus: Snel tot de kern! Politieacademie, s.l
37. Verantwoording aanpak georganiseerde criminaliteit 2012 (2013) Openbaar Ministerie & Politie, s.l
38. Hanneman RA, Riddle M (2005) Introduction to social network methods. University of California, Riverside, (published in digital form at http://faculty.ucr.edu/~hanneman/)
39. Van der Hulst R (2009) Introduction to social network analysis (SNA) as an investigative tool. Trends organ crime 12:101–121
40. Natarajan M (2006) Understanding the structure of a large heroin distribution network: a quantitive analysis of qualitative data. J Quant Criminol 22(2):171–192
41. Scott J (2000) Social network analysis: a handbook. Sage, Newbury Park
42. Borgatti SP, Everett MG, Freeman LC (2002) Ucinet for windows: software for social network analysis. Analytic Technologies, Harvard
43. Cornish DB (1994) The procedural analysis of offending and its relevance for situational prevention. Crime Prevention Stud 3:151–196
44. Bruinsma G, Bernasco W (2004) Criminal groups and transnational illegal markets. Crime, Law, Soc Change 41:79–94
45. Morselli C, Roy J (2008) Brokerage qualifications in ringing operations. Criminology 46(1):71–98
46. Morselli C (2001) Structuring Mr. Nice: entrepreneurial opportunities and brokerage positioning in the cannabis trade. Crime, Law Soc Change 35:203–244
47. Emmet I, Broers J (2008) The green gold: report of a study of the cannabis sector for the national threat assessment of organized crime. KLPD-IPOL, Zoetermeer
48. Potter GR, Bouchard M, Decorte T (2011) The globalization of cannabis cultivation. In: Decorte T, Potter G & Bouchard M (eds) World Wide Weed. Ashgate. pp 1–20

49. Coles N (2001) It's not what you know, but who you know that counts: analyzing criminal crime groups as social networks. British J Criminol 41:580–594
50. Krebs VE (2002) Mapping networks of terrorist cells. Connections 24(3):43–52
51. Roberts N, Everton SF (2011) Strategies for combating dark networks. J Soc Struct 12(2):1–32
52. Baker WE, Faulkner RR (1993) The social organization of conspiracy: illegal networks in the heavy electrical equipment industry. Am Sociol Rev 58:837–860
53. Robins G (2008) Understanding individual behaviours within covert networks: the interplay of individual qualities, psychological predispositions, and network effects. Trends Organ Crime 12:166–187
54. Everton S (2010) Tracking, destabilizing, and disrupting dark networks with social network analysis. Dark Networks Course Manual
55. Bonacich P (1987) Power and centrality: a family of measures. Am J Sociol 92(5):1170–1182
56. Burt RS (2002) Structural holes versus network closure as social capital. In: Lin N, Cook KS, Burt RS (eds) Social capital: theory and research. Transaction, New Brunswick, pp 31–56
57. Carley KM, Ju-Sung L, Krackhardt D (2002) Destabilizing networks. Connections 24(3):79–92
58. Varese F (2012) How Mafias take advantage of globalization the Russian Mafia in Italy. Br J Criminol 52:235–253
59. Kleemans ER, Van de Bunt HG (1999) The social embeddedness of organized crime. Transnatl Organ Crime 5:19–36
60. Granovetter M (1983) The strength of weak ties: a network theory revisited. Sociol Theory 1:201–233
61. McCarthy B, Hagan J (1995) Getting into street crime: the structure and process of criminal embeddedness. Soc Sci Res 24:63–95
62. Sutherland E (1937) The professional thief- by a professional thief: annotated and interpretated by Edwin Sutherland. University of Chicago, Chicago
63. Erickson B (1981) Secret societies and social structure. Soc Forces 60:188–210
64. Ratcliffe JH (2008) Intelligence-Led Policing. Willan Publishing, Cullompton
65. Ormand D, Bartlett J, Miller C (2012) Introducing Social Media Intelligence (SOCMINT), Intelligence and National Security, pp 1–23
66. McDowell D (2009) Strategic intelligence: a handbook for practitioners, managers and users. Scarecrow professional intelligence education series, no 5
67. Xu J, Marchall B, Kaza S, Chen H (2004) Analyzing and visualizing criminal network dynamics: a case study. In: Intelligence and security informatics, vol 2072. Proceedings of the ISI 2004 Second Symposium, Tucson, pp 359–377

The Networked Mind: Collective Identities and the Cognitive-Affective Nature of Conflict

Manjana Milkoreit and Steven Mock

Abstract Using a cognitive approach to the study of conflict that conceptualizes the mind as a network of mental representations, we make three arguments about the role of collective identities in the emergence, persistence and resolution of conflict. Collective identities are subsystems of larger networks of mental representations that make up an individual mind. Because they manifest the group within the mind of an individual, but also connect and align the individual mind with that of other group members, collective identities are an essential element of a complex, multilevel process that constitutes the group in the first place—they are necessary for the emergence of the social group phenomenon. Finally, collective identities are "sticky" in the sense that they are more resistant to change and trigger stronger—more emotional—defensive responses than other mental representations when challenged.

Keywords Cognition · Emotion · National identity · Conflict

1 Introduction

Network analysis is enjoying increasing popularity among defense and security scholars, policy-makers and practitioners engaged in conflict management, conflict resolution and disaster response, and also international relations scholars more generally [1]. So far most network analyses in security studies focus on the structure and vulnerability of physical networks, such as critical infrastructure [2],

M. Milkoreit (✉) · S. Mock
Waterloo Institute for Complexity and Innovation, University of Waterloo, Waterloo, Canada
e-mail: manjana@mac.com

Global Institute of Sustainability, Arizona State University, Phoenix, USA

S. Mock
Balsillie School of International Affairs, University of Waterloo, Waterloo, Canada
e-mail: sjmock@uwaterloo.ca

A. J. Masys (ed.), *Networks and Network Analysis for Defence and Security*, Lecture Notes in Social Networks, DOI: 10.1007/978-3-319-04147-6_7,

or on the harder-to-detect characteristics of social networks and their internal information flow [3]. Social networks are treated either as a threat factor, as in the case of terrorist networks [4, 5], or as a resilience factor, for example, the role of local communities in counterterrorism or disaster preparedness [6]. We apply a network-based analysis to an area that suffers from even greater empirical challenges than social network analysis: the human mind.

Working with the assumption that all human behavior has cognitive origins this chapter has three aims. First, we make the case for the relevance of a cognitive approach to defense and security studies, exploring areas of application and potential insights. Second, we use cognitive theory (emotional coherence) and a complex systems approach to explore the role of specific cognitive elements—collective identities—in the emergence and resolution of conflict. Third, we introduce cognitive-affective mapping as a tool to apply our theoretical framework to specific empirical cases and walk the reader through a specific case study, the international climate negotiations, as an example of non-violent political conflict implicating collective identities at the multilateral level.

At the heart of this chapter are three distinct arguments regarding the role of collective identities in social conflict. Collective identities are subsystems of larger networks of mental representations that make up an individual mind. Because they manifest the group within the mind of an individual, but also connect and align the individual mind with that of other group members, collective identities are an essential element of a complex, multilevel process that constitutes the group in the first place; they are necessary for the emergence of the social group phenomenon. Finally, collective identities are "sticky" in the sense that they are more resistant to change and trigger stronger, more emotional defensive responses than other mental representations when challenged. This last insight has important implications for understanding conflict dynamics and efforts at conflict resolution.

2 A Cognitive Approach to Security Studies

Cognitive analysis—the attempt to identify, describe and understand the content, structure, and dynamics of systems of mental representations—can address important lacunae in the field of security studies, in particular when analyzing and managing conflicts. Building on recent advances in the cognitive sciences and rapidly evolving technological support tools for studying the mind, a cognitive approach to theorizing political behavior can answer four major questions: How does the mind represent the world? How do people make decisions? How do people's minds change over time? How can we understand the relationship between individual and collective beliefs and decisions? In this chapter we touch upon all of these questions, but focus on collective decisions related to conflict.

Explaining the human capacity for collective action may well be the fundamental problem of political science [7–10]. Understanding how and why people regularly function as groups to create public goods or engage in conflict is important not just to understanding political behavior in the present, but may prove crucial to the wellbeing of societies as we confront increasingly complex and intractable problems on a global scale. The dominant paradigms of international relations have not been fully successful in this task. The rational choice and equilibrium models still prevalent in economics and political science have proven inadequate for explaining such unpredictable collective phenomena as economic crises and social revolutions. Social-constructivist approaches have sought to correct this excessively materialist understanding, but their insights remain unintegrated, and at times tend to the opposite problem of mystifying collective entities by attributing to them a manner of agency analogous if not equivalent to that of the individual.

Identity is a key variable in explaining collective action, but also an inherently problematic concept. Originating in the study of psychology, it refers by definition to what makes the individual distinct; to the perception of the autonomous self as separate from the outside world. But in the social sciences it is most commonly used in a seemingly opposite sense: to refer to the properties characteristic of a collectivity such as a nation, class or culture. Given the extent to which collective belonging is crucial to the manner in which the individual negotiates his or her place in the world, these disparate common-sense uses of the term between different disciplinary categories are not as contradictory as they might appear; the question "who am I?" is often answered in significant part through the question "who are we?" Still, until a means is available to represent the precise dynamics of the relationship between individual and collective identity, the casual conflation of these concepts has the potential to generate methodological confusion.

2.1 The Need for a Cognitive Approach to Security and Defense

This chapter makes the case for a cognitive approach to security studies, and in particular to the study of the role of collective identities in conflict, based on five general arguments. First, the explanatory power of cognitive frameworks potentially exceeds that of rational choice models or constructivist theories because it is not constrained by their respective conceptual categories. Identifying cognitive processes at the individual and collective actor level can create new insights into the content of group identities, associated emotions, the interaction between identity and other cognitive elements, and the change of collective identities over time.

Second, cognitive theory can build on and potentially integrate existing theories in international relations. Rational decision-making and normative beliefs can be considered part of the potentially numerous, parallel cognitive processes that motivate conflict participants.

Third, a cognitive approach is able to integrate emotions into theory and methodology.

Fourth, past methodological limitations for studying the human mind are beginning to be relaxed with the development of new technologies and techniques that allow the researcher to investigate, depict and simulate thought processes related to group identities.

Finally, a cognitive approach offers not only empirically driven analytical insights, but also opportunities for shaping political interventions, for instance, in attempts to frame or reframe contentious political issues, or by supporting conflict resolution and efforts to 'humanize' parties in violent conflict.

2.2 Chapter Goals

In this chapter, we use a cognitive approach to theorize about the role of group identities in the emergence and dynamics of social and political disputes. We focus in particular on how cognitive systems facilitate the relationship between an individual group member and the collective, and the special role played by mental representations of group identities.

Our argument in favor of a cognitive approach to security studies is relevant to network analysis given our conceptualization of the human mind as a network of mental representations, and consequently a phenomenon that is open to both a network analytic approach and complex systems analysis.

In the following section we review the existing literature on identity in international relations scholarship (Sect. 3). After a brief outline of basic concepts of cognition and cognitive-affective mapping as a tool (Sects. 4–5), we develop our central arguments about the function of collective identity concepts in social conflict, especially its bridging function between the individual and social scales, using the example of national identity (Sects. 6–7). In Sect. 8 we apply these insights to the global climate change negotiations, demonstrating their utility for the analysis of real-world conflict and negotiation.

3 Identity as a Variable in International Relations

The development of constructivist approaches since the 1980s has triggered new theorizing and empirical work on identity as a variable in international relations and conflict. In contrast to rational choice-based theoretical frameworks, which assumed self-interested rational actors (states) as constant and exogenously given [11], constructivists began to explore the possibility of changing identities and interests, driven either by domestic or international influences [12]. Wendt argued that shared identities among states could emerge at the international system level, leading to political outcomes structural theorists could neither expect nor explain [13]. He suggests that a shared identity induces actors to take other group members' interests into consideration and consequently leads to different interest definitions and political behavior than a purely egoistic approach.

A significant body of research has explored when and how social identities affect the foreign policy behavior of states and decision-making elites [14]. Many scholars recognize the importance of individuals' beliefs, belief system dynamics and perceptions regarding their own group or other groups when conducting foreign policy. However, most of the existing empirical research in this area focuses on individuals (e.g., US presidents) and their belief systems to analyze specific policy decisions (e.g., initiating war with another country). While collective identities tend to play a role in these belief systems, they have not been singled out as a variable that might play a special role in political decision-making or as a factor that interacts with individual beliefs.

Seeking greater conceptual clarity and empirical tractability in the study of social identity, Abdelal et al. [15] present a framework that distinguishes between content and contestation of identity dimensions and outline four identity content types: constitutive norms, social purposes, relational comparisons with other social categories, and cognitive models. While this is a useful framework to explore the substance of a particular social identity, it does not provide any guidance on the role of identity in conflict situations. It also disregards the individual-group relationship.

Peace and conflict studies have built upon work in sociology, in particular Tajfel's theory on inter-group relations. Concepts like in-group preference and out-group discrimination have become the foundation of theories of ethnocentrism [16] and ethnic conflict [17, Chap. 4, 18, pp. 33–71]. Important theories of conflict have explored the phenomenon of stereotyping [19, Chap. 16] combined with perceptions of injustice, in other words, an individual's assessment of the unfairness of the in-group's disadvantage in comparison with the out-group becomes the central source of conflict. The out-group is blamed for the current injustice, and (violent) conflict is considered the best strategy to remedy this situation [19, Chap. 15, 17]. Finally, relative deprivation theory (Berkowitz in [20], Davies in [20]) uses similar categories of comparative assessments of the wellbeing of in- and out-groups, leading to violence only when the difference between the groups has become unjustifiably large to the disadvantage of the in-group.

While none of these theories addresses the role of collective identities directly, they lay the foundation for identity research by pointing to the importance of identity groups, group membership, and perceived differences between groups that facilitate the identification of an other that eventually becomes the enemy and target of violence.

4 The Fundamentals of a Cognitive Approach

Cognitive science is the multidisciplinary study of mind and intelligence. Conceptualizing the mind as a complex network of mental representations, cognitive theories deal with the elements, structures and processes of thought and emotion.

Individual cognitive elements are network nodes that can be activated by links between them. The central process for problem solving or decision making in a network of mental representations is coherence [21]. Cognitive change requires changes to several nodes and links simultaneously, thereby restructuring the network in a manner that maintains coherence at the system level.

The basic objects of analysis in cognitive science are mental representations, structures and processes. Types of mental representations include concepts [22, Chap. 4], beliefs, and goals or motivations [23, Chap. 6]. Cognitive structures are the configurations of linkages or relationships between mental representations. Clusters of concepts form belief systems, images, issue frames, ideologies or other structural entities. Processes include decision-making, problem solving, and risk assessment. Meaning emerges from the connections between multiple cognitive elements, structures and processes as much as from their relationship to entities in the material and social worlds [24].

Recognition of the inexorable link between cognition and emotion—that feeling is integral to knowing—has become increasingly relevant to recent cognitive science research [25–28]. The emotional values associated with concepts, far from being hindrances to rational processes as often assumed, are in fact indispensable elements of human perception, understanding, and rationality. Thagard argues that earlier views of cognition as computational processes of deliberative coherence are incomplete, putting forward instead a theory of cognition as a process of emotional coherence [27, Chap. 2, 29]. Concepts, beliefs and goals all come with emotional valences that shape coherence assessments as much as does logical reasoning. Emotions are also involved in cognitive processes such as the rejection or revision of beliefs.

4.1 Cognitive-Affective Mapping

What is needed then is a method for representing individual minds as complex cognitive systems, and tools to model how interactions between multiple agents and their cognitive systems generate emergent social patterns and properties. Thagard developed *Cognitive-Affective Mapping* as a method for graphically diagramming cognitive systems as networks of mental representations [30, 31]. This approach satisfies traditional methodological individualism, as it is predicated on a positivist rejection of mind/body dualism that would otherwise tend toward mystifying ideas and emotions as abstract or intractable aspects of the human condition. Instead, it frames them in cognitive science terms as mental representations that are the product of brain processes—patterns of neural firing. However, it also acknowledges that interactions between multiple minds or cognitive systems can yield non-linear emergent patterns independent of those that characterize individual minds.

The products of this method—cognitive-affective maps (CAMs)—represent an individual's concepts and beliefs about a particular subject, such as another

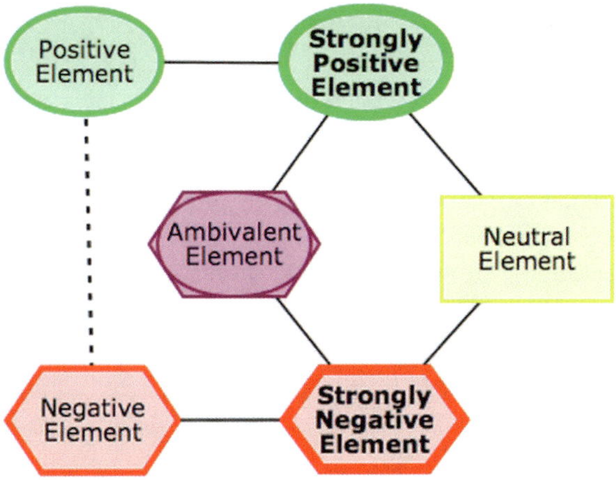

Fig. 1 Basic cognitive-affective map (http://cogsci.uwaterloo.ca/empathica.html)

individual or group or an issue in dispute. Particularly valuable is the way that CAM incorporates emotion directly into the representation of belief systems, in recognition of the principle that issues of cognitive and emotional coherence are intertwined in processes of rational decision-making; an insight accepted in social psychology and cognitive science over recent decades that has yet fully to penetrate the study of politics [32–34].

The CAM approach adopts the following conventions. Shapes and colors depict network elements. There are four different types of nodes indicating discrete cognitive elements with different emotional valences. Green ovals represent emotionally positive elements and red hexagons emotionally negative ones. Yellow rectangles represent elements that are emotionally neutral. A superimposed oval and hexagon (purple) indicates ambivalence; a single element that can have either positive or negative emotional valence depending on context. The thickness of the shapes' lines represents the relative strength of the positive or negative valences associated with them.

Links (edges) between shapes depict relations between cognitive elements. Solid lines represent relations between elements that are emotionally compatible or mutually supportive—if you like one you also like the other. Dashed lines represent relations between elements that are emotionally incompatible with each other. The thickness of links indicates the strength of the emotional connection between two elements.

The product amounts to a network of interconnected elements, as illustrated in Fig. 1.

The CAM method, and the assumptions on which it is based, allow for a cognitive system to be depicted in a manner equivalent to a scale-free network. This opens the possibility for modeling emergent patterns and properties that might result from the interaction between network elements within an individual mind, as well as the multi-level interactions between multiple individual cognitive networks or between an individual and a group through systems of social communication.

5 Collective Cognition?

Any effort to apply cognitive science principles at the level of the social sciences will run into the problem of agency at the individual and collective level. Cognition takes place in the brain of an individual. Yet when the units of analysis are communities such as nations or institutions such as states, these units are often conceived as thinking, acting and feeling entities: "America invaded…", "banks reacted…", and so on. Does this reflect fallacious thinking, "metaphorical pointers" [23], or a convenient way of speaking to a more complex reality? Collective meaning-making is a fundamental process for human societies, and it is at the heart of political decision-making. But can we really attribute beliefs to social groups? And if not, how can the processes and results of collective sense making be conceptualized? It is precisely this impasse between the psychological and social levels of analysis that we aim to resolve with a special focus on identity-related cognitions.

Cognition is a process that consists of interacting mechanisms on multiple levels, including the neural, molecular, psychological, and social [35]. Given that groups do not have brains, and collective mental representations are therefore no real entities, we argue, following Thagard, that the individual-group problem can also be conceptualized in terms of multilevel interacting mechanisms. The interaction between individual-level mechanisms (molecular, neural and psychological processes) and group-level mechanisms (e.g., communication, sensory interaction, emotional contagion) create the bonds that hold a group together. The key to collective cognition is the individual who thinks about himself as a member of the group [35, p. 274]. Individuals acquire and change group-related beliefs through interactions with other people and with other elements of the group, for instance, certain spaces and office buildings, the use of collective resources and property, or the experience of events. This process of social communication and physical-sensory interaction works both ways: a group member not only receives information about the group and develops an understanding of the group as a collective entity, she also contributes to other people's mental representations and experiences of the group. The nature of the group depends on this recursive process between individual cognition of and social interaction between group members. Writing about conflict Ellemers confirms this view by exploring the conditions under which "the group self"—thinking about oneself as a member of a group—becomes more important than the individual self [36].

Understanding collective beliefs therefore requires first of all an understanding of how individuals envision themselves as group members and the emotions, values and meanings they attach to this membership ([23]; see also Tajfel, Introduction in [37, pp. 2–3]). Further, communication and physical interactions between group members and with the social-material environment are important processes toward creating shared understandings in a group and enabling individual and collective cognitive change.

6 The Relationship Between Individual and Collective Identity

To begin with, we must divide the notion of "the group" conceptually into two distinct aspects: the social and the cognitive-individual. It is easy to conceive of the group as larger than the individual; as a collection of individuals and the networks of social communication between them. But at the cognitive level the group is smaller than the individual, insofar as it is a subset of the mental representations that make up an individual mind. The group cannot exist without these two necessary conditions in place: (1) a collection of individuals with networks of social communication between them, and (2) a cognitive construct common to the minds of member individuals according to which the group is conceived and defined. Neither condition can exist without the other, hence the group is the product of a multi-directional feedback effect between these two systems at different levels of analysis. Networks of social communication are required to create the cognitive construct of the group in the minds of member individuals; yet those networks are themselves generated and formed by the presence and content of the cognitive construct.

Conceptualizing the group as a sub-system of mental representations as well as a super-system of individuals opens a new approach to the problem of collective cognition, emotion and agency. Thought and emotion are brain states, but groups do not have group brains. Only individual minds are capable of retaining mental representations with emotional valence. Therefore, the group, as such, cannot think, feel and act. However, the individual mind that contains a part reflecting the group—the subset of mental representations that amount to the internalized group construct—is capable of thought, emotion and agency. The systems of mental representations communicated to the individual mind through interaction with the group contain within them not only objects and connectors but an inherent emotional logic and coherence to them. In addition to symbols and narratives group communication also conveys feelings. Thus even though groups, as such, cannot think and feel, the common adoption of complex identity constructs that include emotional valences creates a dynamic that can be considered or at least usefully labeled collective cognition and emotion. These clusters of mental representations related to identity are the sources of collective identity and agency as emergent properties of cognitive-social systems. They are the origins of the group as a social phenomenon.

However, the interdependence between systems at the social and cognitive levels will cause the sub-system of mental representations that amount to the group identity to function according to different rules than do other 'ordinary' mental processes. Though like all brain processes they are internal to the mind, because they are received via consensual social signals external to the individual they are experienced as unitary, reified and external facts, objectively perceived as part of the 'natural' environment. The form that these sub-networks take is not organically determined according to the individual's rules of cognitive-affective coherence.

Instead content and structure are determined by external social processes over which the individual experiences no control, because (1) the identity construct is perceived to be a finished product, not an open process, (2) they adhere to their own internal rules of coherence, driven in part by their utility to the social system, and (3) they are received by the individual via social communication. Communication, in this context, must be understood in broad terms to include not just verbal and written signals transmitted directly to the individual by other people, but also more implicit social processes such as mechanisms of child-rearing, value-laden behaviors observed and imitated, exposure to prevalent myths and symbols, the experience and use of certain places and sites associated with the group, and even the ambient sights, sounds, smells, tastes and textures established as familiar to the group. These experiences will vary between member individuals, but certain elements will be tacitly established as common and definitive, incorporated into a shared understanding of what the group *is*.

In short, these hypotheses summarize our understanding of the cognitive constructs that constitute collective identity:

1. They are sub-sets of larger cognitive networks (clusters of concept nodes).
2. They are shared in more or less identical form between a limited set of individuals distinguished at least in part according to possession of these networks.
3. They are tacitly agreed upon by the group so formed, and received by the individual via social communication.
4. Because of the specificities of this process of formation and transmission, they display different behaviors than other cognitive elements like concept nodes.
5. This difference stems from their *reification*—they are experienced as unitary objects, conflated with external laws and objects in nature.
6. Because the concepts and valences that compose these reified networks are set by collective consensus and imparted to the individual via external networks of social communication, the individual experiences them as beyond his or her agency to change, rendering them significantly more resistant to change than other cognitive elements.
7. Therefore, if challenged or threatened, emotions associated with collective identity constructs intensify, potentially elevating the challenge to an existential threat.

This last point is key: stimuli that challenge the perceived reality of concepts that define the group consensus tend to provoke strong negative emotions of fear, anger and disgust [38], which are both cause and effect of their "stickiness" relative to other mental representations. This is because shared systems of representations are needed for the group to exist as a group and to thereby generate the benefits, material as well as ideational, that accrue from collective behavior. These constructs must be experienced as natural and invariable in order to secure the stability of the group on which the individual relies upon for goods of existential significance; not just identity and meaning, but comfort, security, perhaps even bodily survival. It is only when these constructs are commonly experienced as real that each individual member can be plausibly reassured that the beliefs and behaviors of all other group members will predictably coincide and co-ordinate around them in perpetuity. Any input that threatens to expose the conventional and ephemeral nature of the group consensus exposes the group and its member individuals to the very real threat of social breakdown and the costs such breakdown will incur.[1]

One could therefore hypothesize various explanations at different levels of analysis—cognitive, social, and even evolutionary—behind this reification phenomenon. At a cognitive level, one could theorize that awareness, conscious or unconscious, of the importance of the group to the wellbeing of the individual causes the individual to attach strong emotion to shared symbols associated with the group. At the level of social evolution, it could be argued that groups built around ideational constructs that are coherent, emotionally salient and easily conflated with natural objects and laws are more likely to survive, perpetuate, and compete by commanding greater loyalty and sacrifice. And at the level of biological evolution, it could be proposed that those individuals psychologically more inclined to reify the group and feel strongly toward symbols of group belonging were more likely to engage in co-operative social behavior that furthered their survival and ability to perpetuate their genetic material.

This is not to suggest a sharp dichotomy between group-associated mental representations that are reified, and non-group-associated mental representations that are not. Nearly all mental representations are formed in response to external stimuli, and not all concept nodes associated with a group identity will be reified to an equal extent. It is better to understand this as a continuum, with fully reified mental representations and identity constructs at one end, and other constructs more flexible to individual idiosyncrasies on the other. What's more, it is not to be assumed that all group members will reify the same collective concepts and beliefs to precisely the same extent, though group membership might be tested and measured—in whole or in part, whether explicitly or tacitly—by the extent to which individuals do. Some members might retain the ability to question or alter elements of the reified group construct, and we could further hypothesize that such individuals are likely to be found at the periphery of the group—those with only partial or dubious commitment to the group identity—but also potentially at the group's core in the form of charismatic leaders of social movements.

Indeed, one could divide the mental representations that compose an individual's cognitive-affective map into three categories according to origin—the universal, the specific, and the particular. Universal representations are ones that can be expected to be common to all of humanity, present in some form in every

---

[1] This view coincides with the theories of Durkheim [39] and Geertz [40] on the meaning and function of religion in society. In constructivist international relations literature, it could relate to Mitzen's argument that in addition to physical security, political actors seek "ontological security" in the form of the preservation of rigidly routinized relationships with others to which they have become attached, and that this tendency may contribute to the perpetuation of even dangerous routines such as protracted conflict [41].

cognitive-affective matrix due to their being an inherent or at least standard aspect of the human condition (say, "water" or "mother"). Specific representations are the product of the individual's unique experience of the world. But the particular is particular to the group(s) into which the individual is socialized, therefore held in common by members of the group and only by members of the group. Of course, this is not a clean typology; any set of mental representations will inevitably be a product of a mixture of universal, specific and particular influences. After all, each individual experiences the universal and the particular on his or her own specific terms. Nonetheless, our focus on group-level cognitive elements concerning group identity starts from the hypothesis that representations will interact with one another differently within a given cognitive-affective system depending on their universal, specific and particular character. Therefore, there may be times when it is beneficial to tease out the particular, isolating it as a component for separate analysis.

CAM is a method suited to this task. It can be used to distinguish the concepts, beliefs and values that come to be reified as part of a collective identity, and model how these interact with other elements of a cognitive-affective system. It thereby offers a means to represent testable theories as to what concepts and connections could trigger ontological crisis in a given population or hypothetical individual member of said population; in other words, to represent the impact of a collective identity on individual behavior.

7 Nation, Cognition and Emotion

We illustrate this approach with reference to what is arguably the most pervasive form of identity in the modern international system: the nation. In the study of national identity and conflict the causal role of symbolic and emotional content tends to be subsumed in a wider debate on agency between schools of thought that could be broadly termed primordialist, instrumentalist, and constructivist.

Primordialism views identities and their related symbolic attachments as in some way embedded if not in the very nature of the human species than at least in a long history during which the identity group can be recognized as a continuous protagonist.[2] This often leads to the assumption that conflicts between groups that implicate symbolic attachments are the consequence of "ancient hatreds", that essential characteristics of distinct groups place them in conflict, and that these

[2] For example, sociobiologists such as van den Berghe [42] who view nations as extensions of the evolutionary mechanism of kin selection explained by, among others, Dawkins [43]; or culturalists such as Grosby [44], drawing from the works of anthropologists like Clifford Geertz who saw groups as forming around perceived a priori "givens" such as descent, language or religion [40]. *From the standpoint of our argument, we could also include in this category the "ethnosymbolist" approach of Smith* [45, 46] that frames the nation as a modern social construct nonetheless dependant on continuity with durable pre-modern ethnic communities.

antagonisms are the very properties of an enduring group identity. While such theories are best at taking emotional attachments to symbols seriously as causal factors in identity construction and conflict, in doing so they tend to mystify and essentialise them.

Instrumentalists challenge this tendency, drawing from the rational-choice paradigm to construe such attachments as tools in the service of ultimately material ends, with a focus on the interests and agency of elites.[3] Indeed, it is not difficult to find empirical support in cases, throughout history and in current situations of conflict, where national symbols have been manipulated intentionally by elites seeking to mobilize populations further to the standard imperatives of power politics. But it is implausible that this is the whole story, since for every instance where elite manipulation of national symbols is successful in altering identities or mobilizing animosities, there are many more in which attempted manipulations fail to resonate. Elites are constrained in which symbols will generate mass emotional response, and instrumentalism alone cannot explain why one symbol will resonate and another not.

A third category of theory, which could be termed constructivist, frames the nation not as a thing in nature, nor as an instrumental fabrication, but rather as an emergent social construct, the product of a distinctly modern convergence of norms and instrumentalities. While this is a sensible premise, beyond the truism that national identities are social constructs, general theories as to how and why they are constructed prove notoriously difficult to verify or falsify.[4]

A cognitive theory of identity offers the potential to break this impasse, suggesting a means and method by which the questions raised by each existing approach can be addressed simultaneously: What are the emotional attachments at the core of a given national identity; the network of myths, symbols, values and animosities that are experienced as felt realities? How could the manipulation of elites or other perturbations be expected to affect the system and in what circumstances? And what are the essential elements of the emergent social construct that becomes the nation, enabling it to function as a coherent system of binding norms and shared mental representations? Specifically, any model that reconciles these disparate positions on the basis and origins of national identity must account for the fact that, although national identities ultimately reside in individual human minds, they are perceived and experienced as external facts equivalent to objects in nature, and this perception is crucial to their function in affecting collective

---

[3] A view most forcefully articulated by Brass [47, pp. 40–41], but also evident in the works of Laitin [48]; as well as historians Breuilly's [49] notion of nationalism as a form of politics geared toward control of the state and Eric Hobsbawm's concept of nationalism as the product of "invented traditions" engineered by elites to mobilize masses in the age of mass politics [50, 51].

[4] Anderson [52], for example, saw the nation a consequence of the decline of universal religions, leading to the formation of territorial states formed around the vernaculars generated by the market demands of print capitalism. Gellner [53] saw them as the product of the social changes needed to maintain a modern growth economy, accelerated by the impacts of industrialization on the relationship between imperial cores and their culturally distinct peripheries.

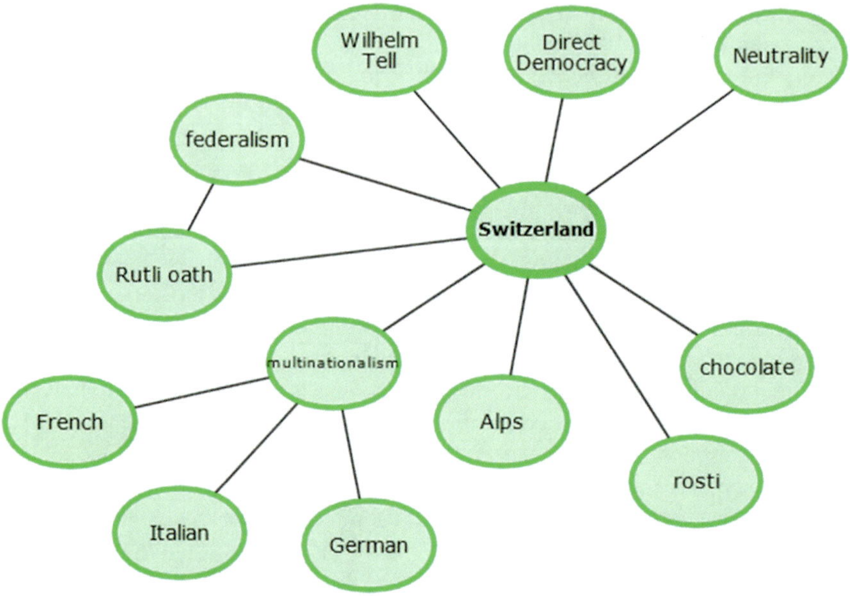

Fig. 2 Simplified representation of Switzerland

behavior and social cohesion. As Ernest Gellner evocatively observed, though the nation may be a fundamentally modern construct, having one is considered akin in the modern world to having a nose and two ears [53]. It may be possible for one to lack any of these things, but unnatural, and experienced as the consequence of some extraordinary tragedy.

To illustrate our approach by way of example, consider the figure below to be a rudimentary cognitive-affective map of the concept of "Switzerland" (Fig. 2):

These are the concepts that compose the socially determined consensus of what "Switzerland" is. Of course, it is dramatically oversimplified for the purpose of the illustration. A thoroughly researched CAM depicting any ideational system as multi-faceted as a national identity will be more than simple nodes connected to positively associated objects, but rather will involve intricate networks of interactions along with negative associations—those things that the nation *is not* in addition to what it *is*. This can also concern negative elements the group has to grapple with, for example, Germany's world war history. But this simplified CAM does show that these constructs will include a wide variety of types of mental representation: principles for establishing group boundaries (language, ethnicity, territory), in-group organization (federalism, democracy), inter-group relations (neutrality); founding myths (Rutli Oath), symbols and stories (Wilhelm Tell), territory and physical objects (Alps), cultural traits, even seemingly superficial associations like characteristic foods. Indeed, a truly comprehensive CAM should be able to incorporate images, sounds, smells and tastes.

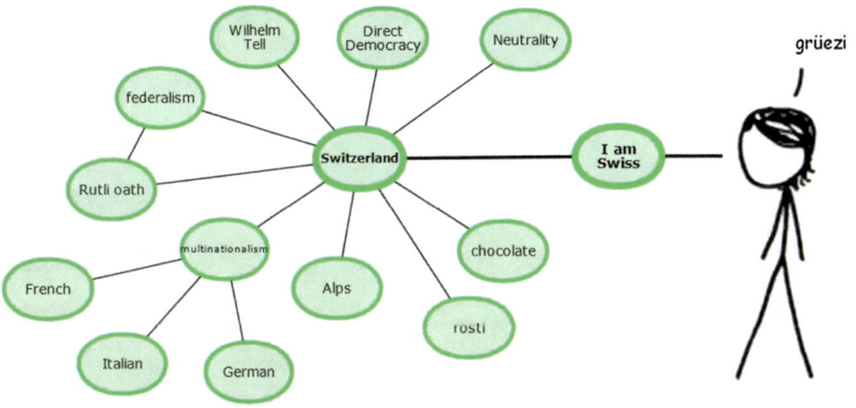

Fig. 3 Representation of Swiss identity

But establishing the reified construct in the mind of a group member is not all that it takes to create a group identity. After all, the same construct of "Switzerland" will be present in the minds of non-members as well, at least partially, depending on the extent of that non-member's contact with the concept. The incorporation of an externally received and reified set of mental representations into identity is more of a process, depicted as follows[5]: (Fig. 3)

This process leads not just to the adoption of a set of mental representations, but also the activation of a characteristic pattern of emotional valences that impact how these mental representations will interact with others within the individual mind. The concept of "Swiss" is inherently connected to the concept of "Switzerland". Therefore, once an individual's belief system includes the proposition "I am Swiss", the mental representations and valences associated with "Switzerland" are imported as a whole coherent system into the individual's cognitive-affective network. They may be imported imperfectly, according to the specificities of the individual's unique experience of the construct. And once imported, they are subject to interaction with other elements of the individual's cognitive-affective map. This can account for differences between group members with regard to their mental representations of and relationships to the group. But the individual is not unconstrained in choosing what mental representations are adopted along with the identity association, at least concerning a certain number of common essentials.

The received representations are communicated to the individual via social signals and therefore experienced by the individual as external and given, similar to natural objects and laws. This is not to say that they exist outside of the individual. Everything depicted in these diagrams is internal to the mind/brain. The point is that the internalization of the concept of "Switzerland" involves the

[5] With the stick-figure, borrowed from xkcd.com, standing in for the sum total of remaining mental representations, structures and processes that make up the human individual.

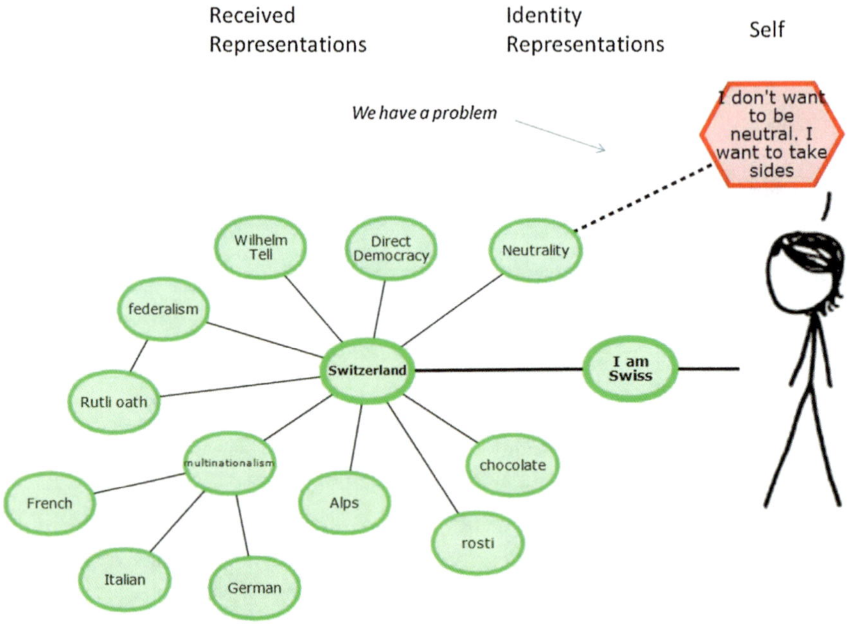

Fig. 4 Challenging the identity concept "neutrality"

internalization of a network of representations determining what "Switzerland" means; and, though internal to the mind, the boundary mechanisms and symbolic content of this network are determined prior to their importation, the product of processes of social communication, and therefore experienced as originating and residing outside of the self, beyond control of the self. Anyone can see a mountain, and, with a rudimentary set of representational tools, can form a mental representation of a mountain. This much can be classed as "universal". But it takes the internalization of a particular set of external social signals to know that a mountain is the Matterhorn (and it's not just *called* the Matterhorn; it *is* the Matterhorn), and it takes a mechanism for the importation of that set of social signals into one's identity to respond emotionally to the Matterhorn as a national symbol.

What then happens if an external stimulus (e.g., scientific reasoning, personal experience, the influence of other potentially conflicting group constructs, etc.) generates individual mental representations in conflict with the ones that have been imported from the social environment? To pick a random example, let's say the individual is shown a reason (how and what is unimportant for now) to disapprove of the notion of neutrality (Fig. 4).

The result is the introduction of conflict between concepts into the CAM that must be resolved in order for the system to remain stable. There are three types of solution. One can alter the conflicting mental representation (Fig. 5), the mental representation associating the individual with the identity construct (Fig. 6), or the contents of the received identity construct (Fig. 7). These solutions become more

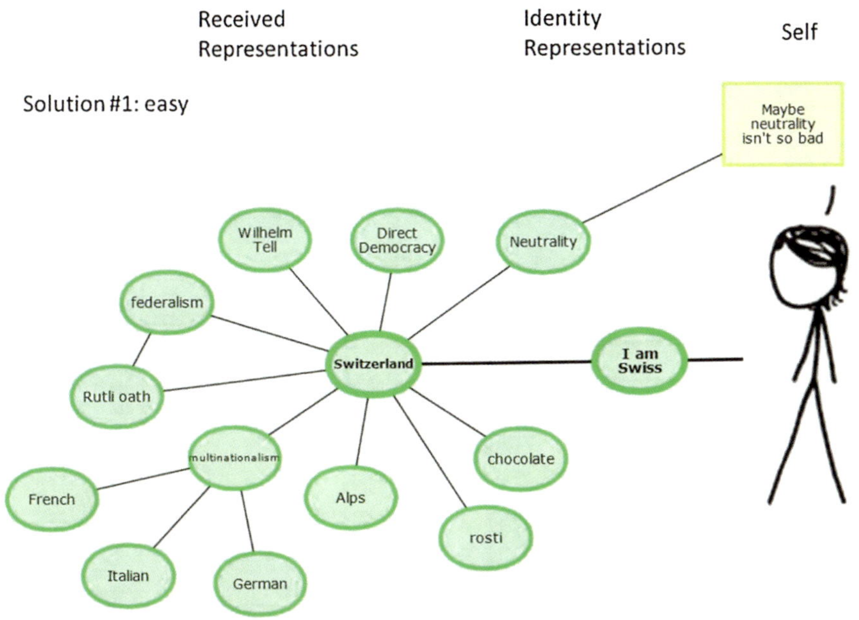

Fig. 5 Solution 1 to offending identity concepts

difficult the more they involve the received identity construct. Since this construct is perceived as an unchanging natural object or law, its permanence is something each group member depends upon for social stability.

The easiest thing for the self to do is therefore to change her mind about, reason her way around, or at least soften her emotional commitment to the offending concept in the face of the seemingly unmoving influence of the imported and shared social construct (Fig. 5).

The individual depicted above still could not be said to *believe* in neutrality. Nonetheless, she has imported the concept and thereby come to accept it along with its positive valence as part of her system of identity representations. This illustrates how attachment to a collective construct of identity can compel a person to alter his or her behavioral and emotional environment around a concept that he or she does not personally believe in. This might explain, for example, how individuals who are not religious believers can nonetheless be sincerely moved to mobilize, fight and sacrifice for symbols and objects of a religion associated with their group identity.

There may be situations where the individual is unable or unwilling to adjust the emotional valence attached to the offending representation. In that event, the next easiest approach is to weaken one's association with, or alter the emotional valence of the entire identity category that has caused the conflicting representation to be included in one's belief system (Fig. 6).

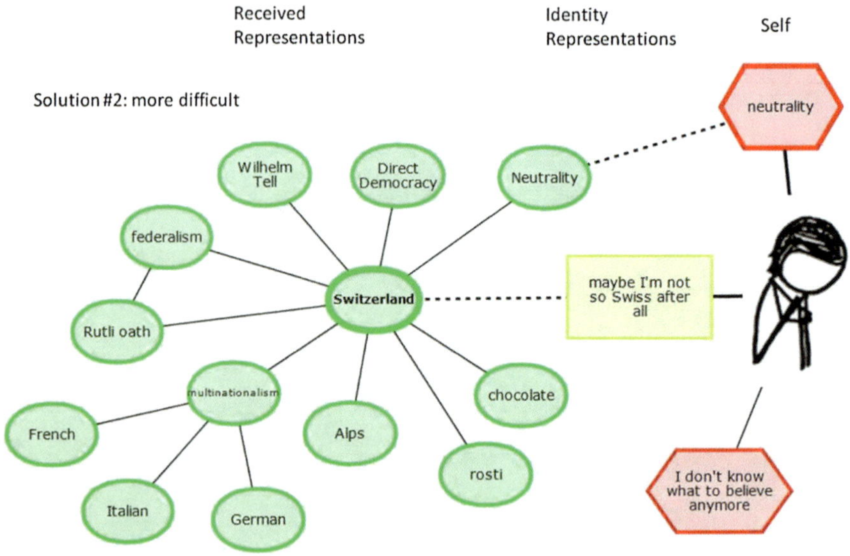

Fig. 6 Solution 2 to offending identity concepts

If one simply cannot reconcile oneself to a representation received as definitive of a given identity construct, one may question whether one really has a right to consider oneself part of that identity group. This is more difficult because it magnifies the overall impact of the solution on the rest of the individual's belief system. When one distances oneself from an identity group like the nation one is born into, one's perception of the reality of all other representations associated with that group come under question as well, exposing the individual to a demoralizing *anomie*.

This dynamic illustrates a potential threat to social cohesion. If the environment of a significant number of group members changes such that these individuals start to receive continuous and undeniable signals that conflict with an essential element of the reified group identity, the resulting cognitive-affective incoherence may cause the group to collapse and effective in-group co-operation to degenerate; at least if it is unable to conspire collectively to alter its received identity constructs accordingly through solution number three.

This third solution, however, is the most difficult: keep the individual opinion, keep the identity association, but remove the offending representation(s) from the received identity construct in order to restore cognitive-affective coherence not just in the individual mind but for the entire group (Fig. 7).

If we were to treat received representations the same as other mental representations, this would appear to be the most straightforward solution. But the special nature of collective identity constructs causes these sub-systems of mental representations to function according to different rules. Because the mental representation of Switzerland with the essential component of neutrality is received

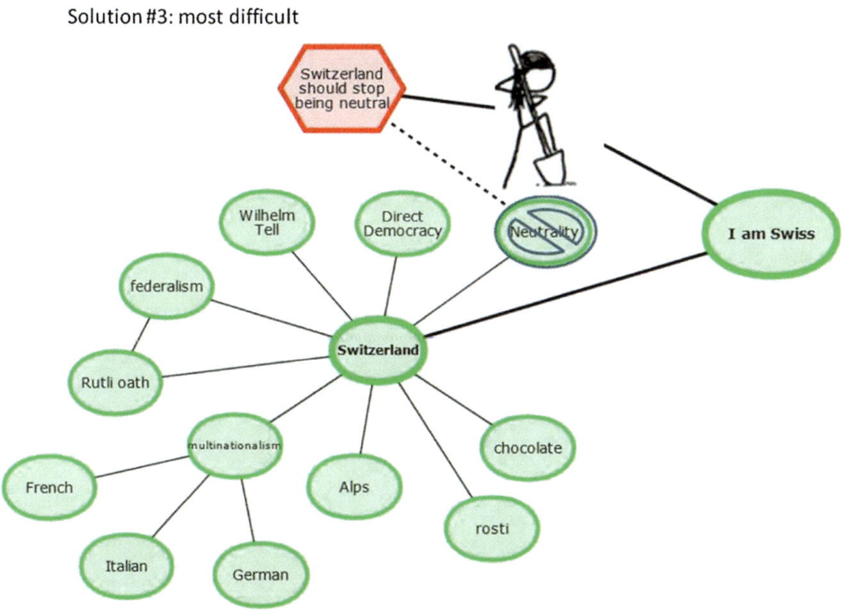

Fig. 7 Solution 3 to offending identity concepts

and validated via external signals, it cannot be altered without the connivance of those external, collective forces; namely, the whole (or at least a critical mass) of the group (Fig. 8).

Neutrality is a social convention. But to someone who has internalized the Swiss identity construct it is a natural and inherent attribute of an actual thing called Switzerland. In principle, there is no reason why the social convention of neutrality cannot be removed from the concept of Switzerland, save for the fact that if it were really that easy, the larger concept would not seem real enough to secure the loyalty and submission of group members in the first place.

This has significant implications to the way one would approach situations of ethnic or national conflict. When a group's symbolic attachments are the cause of conflict intractability, such as when disputed territory is imbued with religious or national significance, this third solution—removing the offending concept from the group's construction of identity—is often presumed by outsiders to be the easiest and preferred approach. So long as one rejects the primordialist assumption that such attachments are embedded in the group's essential nature, the simplest course of action would appear to be to convince disputants to forego this attachment by exposing its irrational, inessential and historically contingent character. Symbolic attachments are not built into our DNA, so if they are the cause of strife we should have the sense to change or abandon them.

But symbolic attachments do not have to be primordial in order to be deeply felt, or for there to be dramatic and unforeseen consequences to their disruption. And each of the coping mechanisms described above points to a corresponding

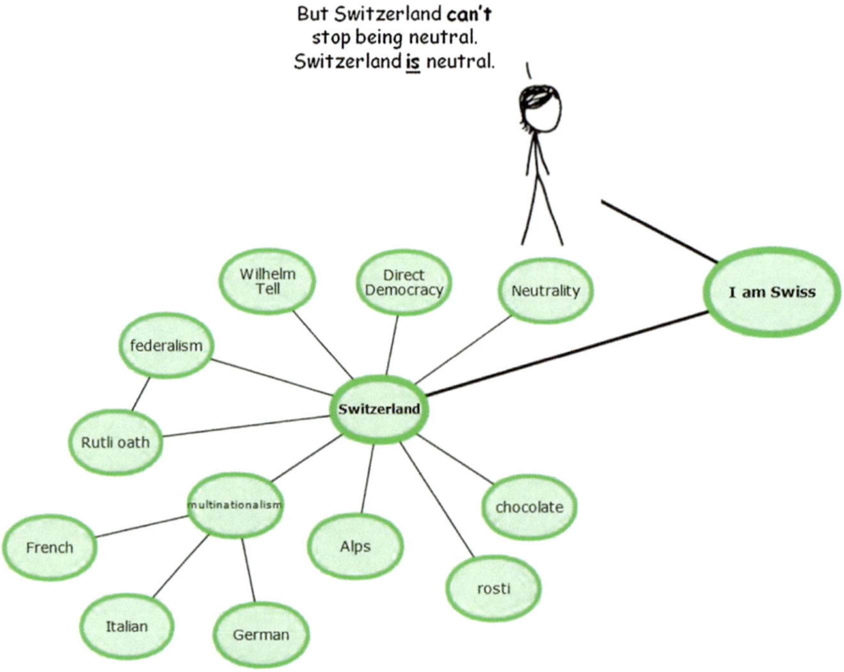

Fig. 8 Resignation in the face of offending identity concept

danger to the simplistic assumption that a group can or should be talked out of its symbolic attachments when these become inconvenient. An individual does not have to be personally convinced of the religious significance or even historical accuracy of a myth or symbol in order to continue to experience and be moved by it as part of the reified group construct. Whereas efforts that successfully cause members of a group to question or discard a concept integral to their group construct can have unforeseen effects on other connected components, possibly threatening the equilibrium of the construct as a whole and weakening bonds of social communication and norms of social constraint. Fear of such breakdown, and the very real dangers of violence and suffering that accompany it, generates fierce emotional resistance in the face of such change against anyone—insider or outsider—who would threaten it.

Our cognitive approach demands that we treat reification not just as a fallacy, but as a cognitive-emotional process that can be integral to the maintenance of group cohesion and the continuation of group behavior. A challenge to any mental representation that is an integral part of a reified collective identity construct threatens to upset the coherence of the system as a whole. To the extent that identity concepts are linked to material realities under contestation—for example, territory that two groups deem to be objectively theirs—a cognitive lens on the structure of identity can reveal more about the conflict and possible paths and pitfalls toward resolution than any rational approach.

8 Case in Point: Global Climate Change Negotiations

In this final section we apply the concepts developed above to the case of global climate change governance, illuminating non-violent conflict dynamics in the UNFCCC negotiations with the help of identity concepts. The same analytic framework can easily be applied to more obvious and clear-cut cases of conflict, including ethnic violence, revolutions or genocide. Here, we focus on the beliefs of two types of climate change negotiators, who tend to take strongly contrasting positions: representatives of small island states and of the United States (US). We present and analyze cognitive-affective maps of the private belief systems of one individual from each group. Using *Empathica* we generated these CAMs based on semi-structured interviews conducted in 2012 as part of Milkoreit's research project on cognition in global climate politics. The CAMs were shared with the research participants, who were encouraged to provide feedback and request changes to first drafts. We support our arguments with evidence from multiple additional interviews not depicted as CAMs.

The interview-based CAM of a US diplomat is complemented with a CAM based on multiple text sources, including speeches and press conferences given by prominent negotiators like Todd Stern and Jonathan Pershing. Rather than representing the beliefs of an individual, this supporting CAM represents the collective position of the US delegation to the negotiations that is perceived by other negotiation participants as the will or beliefs of the United States. This reflects the fiction of collective cognition at the level of a state and the unitary actor (and brain) assumption common in international relations scholarship.

Conflicts can challenge group identities in different ways and to varying degrees. Climate change presents very different identity challenges to various political actors, triggering in turn different political demands and concerns. In highly simplified terms the contrasting negotiation positions of small island states and the US can be explained by the differences in their diplomats' and likely citizens' respective threat perceptions regarding climate change. Islanders perceive climate change impacts as a fundamental threat to their collective identity, even to their existence as a group, and consequently desire urgent and significant collective action to address this problem. Americans do not feel threatened by the impacts of climate change itself, but by international demands to take economically costly mitigation action in response to climate change. The negotiation positions presented by these different groups reflect their attempts to protect reified national identity constructs (Figs. 9, 10).

These CAMs demonstrate that each individual's belief system contains not just one but multiple concepts that specify group identities, including those groups the person identifies with—the in-groups—and those perceived as other—out-groups. The most important in-group for negotiators is usually the country they represent. Others include their own negotiation alliance (e.g., AOSIS) and other negotiation blocks, "poorer countries" or even the "human community". Individual negotiators envision themselves as members (or outsiders) of these groups and have a

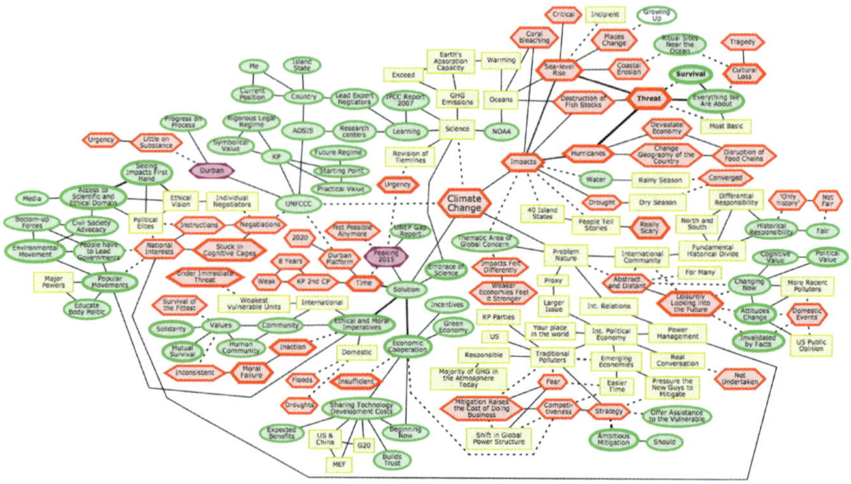

Fig. 9 CAM of small island state representative

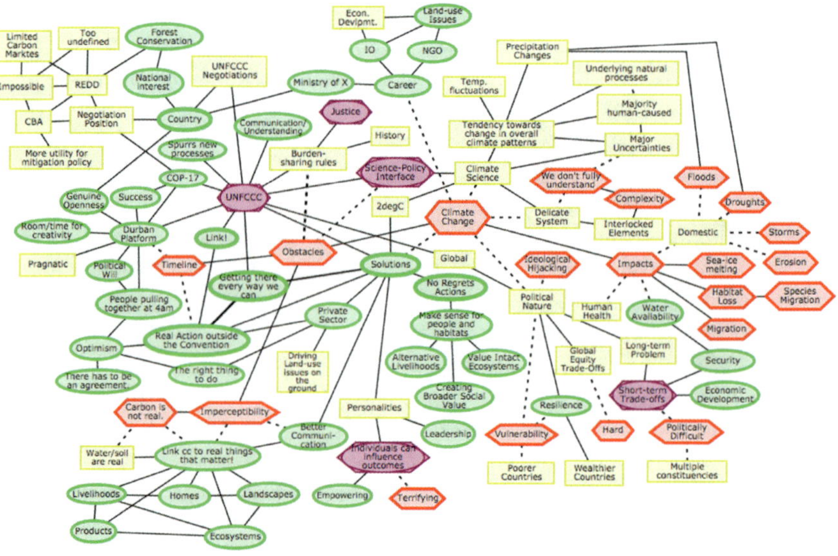

Fig. 10 CAM of US representative

range of ideas regarding the nature of these collectives. Only very few of these ideas are depicted here because interview questions did not inquire about these groups directly. They include concepts like "island state", "vulnerable," or "welfare system."

One of the most important differences between the CAM of the small island state representative and the US diplomat concerns the connection between climate change impacts and national identity. The small island state representative is concerned about specific climate change impacts, including sea-level rise, hurricanes and droughts. These are perceived as a significant threat not only to the physical integrity of the country—eroding and inundating cost lines, flooding roads and disrupting food and water supplies—but also the cultural identity and existence of the island nation because they are destroying cultural sites near the ocean and changing the place in a way that affects all social activity. Concepts like "Everything we are about", "Cultural Loss" and "Tragedy" indicate that the person conceptualizes climate change as a threat to the most basic, fundamental elements of island society. Other CAMs offer similar clues, including concepts like "fishers become farmers," "change to the social fabric," and "my poor little country will be no more."

A specific subset of concepts related to the issue of identity loss concerns place-based identity [54]. The concept has received little attention in international relations scholarship, but as the CAMs demonstrate place-based identity elements are an important feature of belief systems about climate change and help understand how material features of a person's or group's environment—parts of the system structure in a neo-realist sense—can be linked to individual and group identity. These elements can shape conflict dynamics because threats to the physical features, for example ritual or burial sites near the ocean that are being eroded by sea-level rise, become threats to the person's or group's identity.

The negotiation positions of small island states are an effort to protect current national and cultural identity concepts from the threat of climate change. At the same time some island states have begun the hard work of changing previously reified identities—akin to removing "Neutrality" from the Swiss identity—forced by the physical realities of sea-level rise and extreme weather. Kiribati is starting to relocate its population off their island territory, leaving behind national symbols, cultural sites and practices. Along with these social and physical realities they abandon previously reified parts of their collective identity.

The belief system of the US negotiator also acknowledges significant impacts of climate change, including domestic ones, but it lacks threat perceptions that link climate *impacts* to group identity. Instead US positions often reflect a concern about the material-economic costs of climate *policy*, especially of immediate mitigation. The individual CAM depicted above does not contain any concepts that directly confirm this argument, but the CAM of the US negotiation position (Fig. 11) offers some evidence in this respect. Solution-oriented concepts evolve around the idea of mitigation effectiveness, which requires complete legal parity of all parties to a multilateral agreement, in other words, no differentiation of rules for emerging or developing countries, and large emission coverage of such an agreement. Those concepts are based on the assumption that any economic cost imposed on the US must be equally imposed on all other countries to maintain a 'level economic playing field'.

Underlying this rather abstract reasoning about multilateral rules is a deep domestic concern about economic competitiveness with rising powers coupled with strong conservative ideological attachment to neoliberal values including free

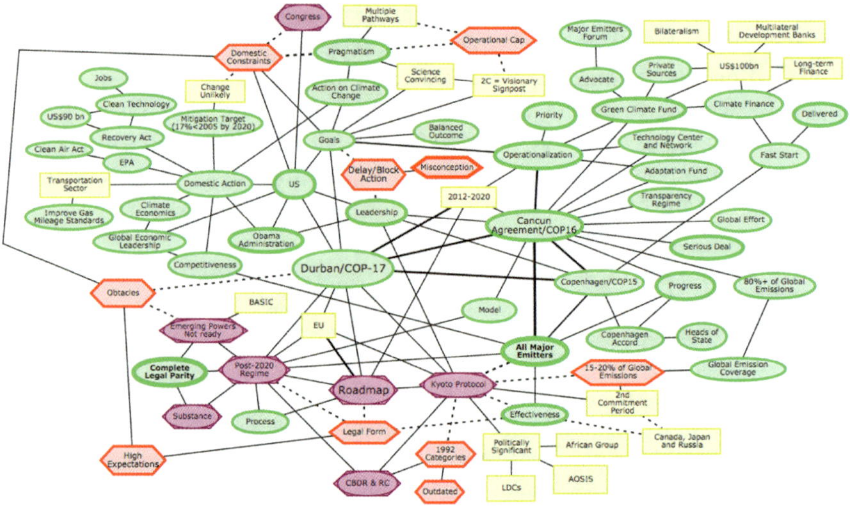

Fig. 11 CAM of the US negotiation position in 2012

markets and enterprise, small government, protection of private property and the right to exploit natural resources for profit. Those values form the center of the national identity of many Americans, and are threatened by proposals that the US should lead global mitigation efforts, should act earlier than others and do more than others because of its large emission share and economic capacity. Such mitigation leadership would require major government intervention ('big government') curbing the free market by regulating production processes, and maybe even limiting individual freedom by constraining consumption processes. These ideas pose fundamental challenges to the American way of life in a society that is increasingly polarized over the role of government and markets for social wellbeing.

In this case it is not climate change but climate change action that challenges national identity constructs and is the target of resistance in the US negotiation position. While this resistance takes the form of climate change denial in the domestic political sphere it is expressed in terms of economic reasoning at the multilateral level.

In short, much of the contestation within the climate negotiations can be traced to different threat perceptions of two distinct groups of individuals: those who perceive climate impacts as a threat to their national identity and even existence as a country, and those who perceive climate action as a threat to their national identity and way of life.

This key difference in threat perception has important implications for a range of dispute dimensions, for instance, the importance and meaning of justice and fairness, or more generally the relevance of norms. They also affect concepts of time and urgency. Perceived long-term existential threats to small island states create an acute awareness of the importance of time, a focus on the systemic linkages between action today and effects tomorrow and therefore a strong sense of urgency. Consequently AOSIS members plead for immediate action and ambitious temperature targets. No sense of urgency exists in the belief systems of US representatives. Concepts of time are almost absent or very vague; linkages between today's and future generations are not important although they are occasionally acknowledged.

Apart from these differences in content, one can also observe very different levels of emotional involvement between island representatives and US diplomats. The interventions of the former are often infused with heightened emotions: they give passionate speeches and referring to the present and expected loss and suffering in their home countries they seek to mobilize the international community with pleas for immediate action on climate change. Emotions matter for Americans as well, especially when a concept that is linked to group identity is challenged. This was observable in Durban (COP 17, 2011) when an intervention by an NGO representative challenged the self-perceptions of the US delegation regarding their leadership role in the negotiation process. The challenge triggered a statement by a visibly shaken Todd Stern during the subsequent press conference, seeking to "correct the misconception" that the US was blocking a legally binding agreement in Durban and wanted to delay action until 2020 (Stern, US Press Briefing 12/08/2011). Self-perceptions in line with the US national identity as a responsible global leader and problem solver contrasted starkly with the perceptions of other negotiation parties regarding the role of the US in the negotiations. But the notion of US leadership is fully coherent with their overall view of the political history and situation today. More generally, the American negotiation position and interventions are affectively muted or 'cool' in comparison to the statements and arguments brought forward by islanders.

This observation offers provisional evidence for our hypothesis that collective identity challenges intensify the emotions associated with threatened identity elements and trigger a defensive response.

9 Conclusion

We have argued that understanding group identities and the individual-group relationship are key for making sense of social disputes, whether they are violent ethnic conflicts or contentious climate change politics. A cognitive approach is particularly well suited to tackle this challenge theoretically and methodologically. Introducing cognitive theories like emotional coherence and tools like CAM into security and conflict studies also opens up new prospects for the empirical exploration of the multifaceted belief systems of political actors and the emotional logics that hold them together.

The mind can be conceptualized as a complex system consisting of a vast number of networked mental representations whose interactions create emergent patterns of thought and meaning. Identity constructs are sub-systems of the mind—

clusters of mental representations at a meso scale that define and give rise to social groups. Collective identities are synchronized mental constructs that exist in the minds of individual group members and specify the nature of one's in-group. Due to their special social stabilizing functions and their dependence on external networks of communication, collective identity constructs adhere to special rules within the mind's overall tendency toward emotional coherence. Most importantly, they are highly resistant to change. If elements of a collective identity are challenged the individual is most likely to resist that challenge and to protect the identity status quo. Consequently, group members experience actions or events that stand to threaten the perceived reality of collective identity as attacks that provoke emotional reactions difficult for outsiders to predict or comprehend. It is in this way that they can become catalysts for misunderstanding and conflict, while internal challenges—for instance, group members seeking to remove or change identity concepts—can lead to social destabilization.

The methods we have proposed, predicated on an understanding of identity constructs and their associated emotions as brain processes, offer a means to model these "sticky" mental representations in a given case and thereby to identify actions or perturbations that might be experienced by group members as threats to the group's ontological security. At the same time, our theoretical argument challenges not only international relations scholarship but also cognitive science to explore the role of cognitive identity constructs and our hypotheses outlining the special cognitive rules that apply to them due to their reification and special social functions. Finally, our focus on emotion contributes to an emergent literature on the role of emotion in political decision-making within international relations, calling for more sophisticated multi-disciplinary approaches to studying this phenomenon.

References

1. Hafner-Burton EM, Kahler M, Montgomery AH (2009) Network analysis for international relations. Int Org 63(03):559–592
2. Lewis TG (2006) Critical infrastructure protection in homeland security: defending a networked nation. Wiley, New Jersey
3. Scott J, Carrington PJ (2011) The SAGE handbook of social network analysis. SAGE Publications, California
4. Ressler S (2006) Social network analysis as an approach to combat terrorism: past, present, and future research. Homel Secur Aff 2(2):1–10
5. Enders W, Jindapon P (2010) Network externalities and the structure of terror networks. J Conflict Resolut 54(2):262–280
6. National Research Council (2009) Applications of social network analysis for building community disaster resilience: workshop summary. National Academies Press, Washington, D.C
7. Dietz T, Ostrom E, Stern PC (2003) The struggle to govern the commons. Science 302(5652):1907–1912
8. Hardin G (1968) The tragedy of the commons. Science 162(3859):1243–1248

9. Homer-Dixon T (2001) The ingenuity gap: can we solve the problems of the future? 1st edn. Vintage Canada
10. Ostrom E (1990) Governing the commons: the evolution of institutions for collective action. Cambridge University Press, Cambridge
11. Olson M (1965) The logic of collective action: public goods and the theory of groups. Harvard University Press, Cambridge
12. Onuf NG (1989) World of our making: rules and rule in social theory and international relations. University of South Carolina Press, South Carolina
13. Wendt A (1994) Collective identity formation and the international state. Am Polit Sci Rev 88(2):384–396
14. Kaarbo J (2003) Foreign policy analysis in the twenty-first century: back to comparison, forward to identity and ideas. Int Stud Rev 5(2):155–202
15. Abdelal R et al (2006) Identity as a variable. Perspect Polit 4(04):695–711
16. LeVine RA, Campbell DT (1972) Ethnocentrism: theories of conflict, ethnic attitudes, and group behavior. Wiley, New Jersey
17. Horowitz DL (1985) Ethnic groups in conflict, 1st edn. University of California Press, California
18. Ignatieff M (1998) The warrior's honor: ethnic war and the modern conscience. Holt Paperbacks, New York City
19. Brown R (1986) Social psychology, 2nd edn. Free Press, New York
20. Davies J (1971) When men revolt and why, 1st edn. Transaction Publishers, New Jersey
21. Thagard P (2000) Coherence in thought and action. MIT Press, Cambridge, Mass
22. Thagard P (2005) Mind: introduction to cognitive science, 2nd revised edn. MIT Press, Cambridge
23. Thagard P (2010c) The brain and the meaning of life, 1st edn. Princeton University Press, New Jersey
24. Markus HR, Hamedani MG (2007) Sociocultural psychology: the dynamic interdependence among self systems and social systems. In: Kitayama S, Cohen D (eds), Handbook of cultural psychology. Guilford, New York pp 3-39
25. Damasio AR (1995) Descartes' error: emotion, reason, and the human brain, 1st edn. Harper Perennial, New York City
26. Loewenstein GF et al (2001) Risk as feelings. Psychol Bull 127(2):267–286
27. Thagard P (2006) Hot thought: mechanisms and applications of emotional cognition. MIT Press, Cambridge, Mass
28. Vohs KD, Baumeister RF, Loewenstein G (2007) Do emotions help or hurt decision making?: a hedgefoxian perspective. Russell Sage Foundation, New York City
29. Thagard P (2008) How cognition meets emotion: beliefs, desires and feelings and neural activity. In: Epistemology and emotions. Philosophy. Ashgate Publishing, Farnham, pp 167–184
30. Thagard P (2010a) EMPATHICA: a computer support system with visual representations for cognitive-affective mapping. In Proceedings of the workshop on visual reasoning and representation. Workshop on visual reasoning and representation. AAAI Press, Menlo Park, CA, pp 79–81
31. Findlay S, Thagard P (2011) Emotional change in international negotiation: analyzing the camp david accords using cognitive-affective maps. Group Decis Negot 1–20
32. McDoom OS (2012) The psychology of threat in intergroup conflict: emotions, rationality, and opportunity in the Rwandan genocide. Int Secur 37(2):119–155
33. Mercer J (2010) Emotional beliefs. Int Org 64(01):1–31
34. Sasley BE (2011) Theorizing states' emotions. Int Stud Rev 13(3):452–476
35. Thagard P (2010) Explaining economic crises: are there collective representations? Episteme 7(3):266–283
36. Ellemers N (2012) The group self. Science 336(6083):848–852
37. Tajfel H (1982) Social identity and intergroup relations, Reissue. Cambridge University Press, Cambridge

38. Haidt J (2013) The righteous mind: why good people are divided by politics and religion. Random House LLC, New York
39. Durkheim E (1971) The elementary forms of the religious life. Library of Alexandria, Alexandria
40. Geertz C (1973) The interpretation of cultures: selected essays. Basic Books, New York
41. Mitzen J (2006) Ontological security in world politics: state identity and the security dilemma. Eur J Int Relat 12(3):341–370
42. Van den Berghe PL (1978) Race and ethnicity: a sociobiological perspective. Ethn Racial Stud 1(4):401–411
43. Dawkins R (2006) Selfish gene 30 Anv. Oxford University Press, Oxford
44. Grosby S (2005) The primordial, kinship and nationality. When is the nation?: towards an understanding of theories of nationalism. Routledge, London and New York, pp 57–78
45. Smith AD (2009) Ethno-symbolism and nationalism: a cultural approach. Routledge, London; New York
46. Smith AD (1986) The ethnic origins of nations. Basil Blackwell, Oxford
47. Brass PR (1970) Elite groups, symbol manipulation and ethnic identity among the Muslims of South Asia. Political identity in South Asia. Curzon Press, London, pp 35–68
48. Laitin DD (2007) Nations, states, and violence. Oxford University Press, Oxford
49. Breuilly J (1993) Nationalism and the state, 1st edn. University of Chicago Press, Chicago
50. Hobsbawm E (1992) Nations and nationalism since 1780: programme, myth, reality, 2nd edn. Cambridge University Press, Cambridge [England], New York
51. Hobsbawm E, Ranger TO (1983) The invention of tradition. Cambridge University Press, Cambridge
52. Anderson B (2006) Imagined communities: reflections on the origin and spread of nationalism, Verso
53. Gellner E (1983) Nations and nationalism. Cornell University Press, New York
54. Fresque-Baxter JA, Armitage D (2012) Place identity and climate change adaptation: a synthesis and framework for understanding. Wiley Interdisc Rev: Clim Change 3(3):251–266

Conflict Cessation and the Emergence of Weapons Supermarkets

Gisela Bichler and Juan Franquez

Abstract It is commonly argued that the end of a conflict generates increased outflow of weaponry. A surplus of secondhand small arms (ranging from small caliber to military grade mediumrange motor tubes, and ammunition) make their way into the trade stream through makeshift weapons supermarkets. Anecdotally, case studies suggest four possible market structures: trade interchange markets, trade mediators, epicenters, and trade channels. Using dynamic actor-based simulation modeling (SIENA), this study captures changes in the gray market of gun trade following the end of armed conflict, testing the degree to which market activity evolves to reflect each structure. Information about small arms and ammunition transfers were obtained from UNcomtrade for 224 nations from 1997 to 2010. Suspicious and likely clandestine trade relations were extracted by pulling all transfers that were reported by only one party to the exchange. While interchange markets are predominantly associated with conflict cessation on a global level, significant regional variation exists. Policy implications are discussed.

Keywords Small arms · Supermarkets · Dynamic networks · Illicit trade · Conflict cessation

G. Bichler (✉)
Department of Criminal Justice, Center for Criminal Justice Research,
California State University, San Bernardino, USA
e-mail: gbichler@csusb.edu

J. Franquez
Research Associate, Center for Criminal Justice Research, California State University,
San Bernardino, USA

A. J. Masys (ed.), *Networks and Network Analysis for Defence and Security*,
Lecture Notes in Social Networks, DOI: 10.1007/978-3-319-04147-6_8,

1 Introduction

It is commonly argued that the end of a conflict generates increased outflow of weaponry. A surplus of secondhand small arms (ranging from small caliber to military grade medium-range motor tubes, and ammunition) make their way into the trade stream through makeshift weapons supermarkets [1, 2]. While much of this trade may involve legal transfers, it is more likely the case that full disclosure is not made by both parties to the trade. Though widely documented, little is known about the structure of these markets and how they fit into the overall illicit weapons trade. Anecdotally, case studies suggest four possible market structures.

1. When hostilities subside, do supplies reverse course as foreign military aide is refocused on emerging conflicts, i.e., *a trade interchange* market forms?
2. Or, do war-torn nations become trade brokers as individuals capitalize on stockpiles of surplus weaponry to gain financial solvency, i.e., post-conflict nations become *trade mediators*?
3. Alternatively, do post-conflict nations become central markets because they have established trade relations and there is little else to sell in the period immediately following the end of hostilities, i.e., the warring nation becomes a *trade epicenter*?
4. It is equally plausible that due to a weakened trade infrastructure weapons are shipped to a neighboring state that has the ability to access the international market, i.e., a *trade channel* forms funneling weapons out of the conflict zone.

To begin addressing these questions about market structure, this chapter uses dynamic actor-based simulation modeling software, called SIENA [3–5] to capture changes in the gray market of gun trade following the end of armed conflict.

Given the challenges associated with capturing accurate information about the volume and value of small arms transferred globally, this research relies on a relatively untapped source to estimate gray market activity.[1] Weapons transfers between nations, wherein only one party reported the trade to the United Nations Statistical Division (UNComtrade data), were used to construct annual trade networks (1997–2010) to represent clandestine trade activity[6, 7]. Though not necessarily illegal, failure to report shipments may be taken as an indication of an attempt to disguise weapons trade [8]. We chose to focus on the gray market because it is the focal point wherein legal and illicit activity intersect [2, 6, 9] and it is generally more visible than illicit trade.

---

[1] Commodities are traded in three ways: (1) legal trade occurs when all parties to the transfer comply with domestic and international regulations; (2) shady or clandestine trade activity involves some combination of legal and illicit activity, usually due to regulatory asymmetries between nations; and, (3) illegal trade typically involves underground economies where all aspects of the trade are illicit. The term gray market activity is generally used to refer to the second category of trade.

Using actor-based simulation models, it is possible to estimate the change in trade relations over time and test what type of trade structure is more likely to emerge post-conflict. The key advantage to using multivariate simulation models is that it permits the inclusion of competing explanations of structural change. Moreover, the model can isolate structural effects, within the context set by a range of covariates, including dynamic and dyadic (other networks) variables, and various interaction effects that may play a role in shaping trade patterns [4]. These interaction effects are critical to determining how conflict termination alters trade relations for specific nations. Thus, it is possible to test competing explanations of market structure, while controlling for the geo-political context of weapons trade—proximity based on shared land borders, existing military alliances, and network of insurgent activity. Going one step further, this study examines important regional differences in clandestine gun flow by comparing two case studies (Egypt and Angola) to global trends. Since conflict starts and stops repeatedly throughout the study period, these case studies offer an opportunity to capture repeated pulses, or outflows of suspicious arms, into the weapons trade.

Following a brief review of the nature of conflict-induced weapon supermarkets, this chapter introduces a market typology built from network metrics. Then, prior to describing the data and methodology, we provide a concise review of the assumptions underlying dynamic modeling with the intent to familiarize the layperson with this simulation technique. The findings are discussed in relation to two case studies, Angola and Egypt, so as to offer insight into how regional trade patterns fit within the context of global markets. The chapter concludes with some cautionary statements about the applicability of using dynamic network models to unravel how global trade structures evolve.

2 Conflict-Induced Weapons Supermarkets

2.1 Stochastic Trade Shocks

New information, such as military redeployments, generates a stochastic shock in the trade system [10–14]. This sudden, unanticipated change will be considered by individual nations as they adjust their pending weapons acquisitions [11, 15, 16]. Generally, the start and end of conflict are marked by dramatic change in gun trade activity [11, 15–18]. Figure 1 illustrates how nations may react throughout a conflict.

Leading up to the onset of hostilities, combatants begin to stockpile supplies. This may manifest as an increase in the volume of transfers from existing trade partners. Once conflict begins, an arms embargo may alter the course of trade flow. As depicted in Fig. 1b, suppliers may begin to use covert transfers, hidden through diversions, offsets, counter-trade, export-credits, military aide, etc., to continue supplying weapons to combatants. Over time more supply routes are established as other producers seek trade relations or foreign powers become embroiled in the

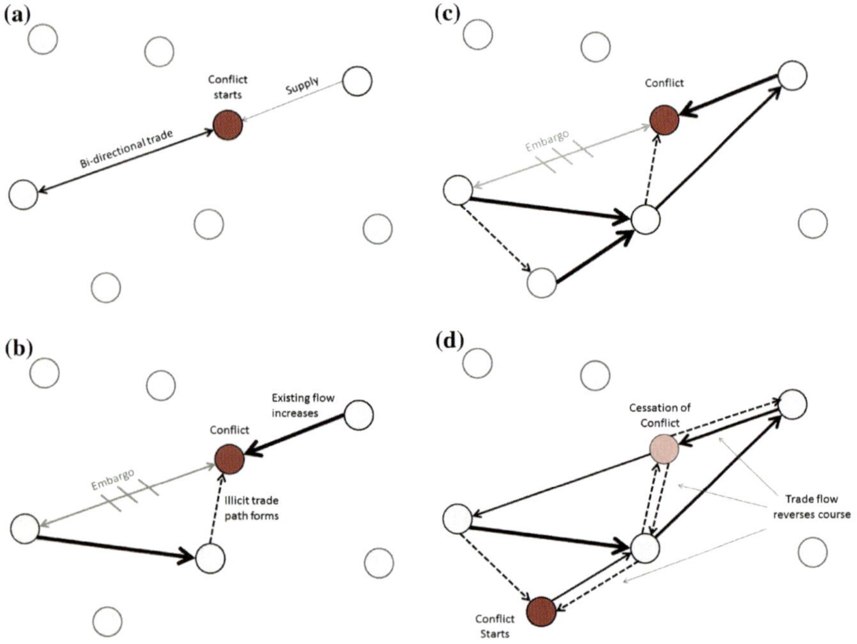

Fig. 1 Illustration of trade flow in response to conflict start and cessation. **a** Existing trade relations when conflict starts. **b** Undisclosed paths emerge to supply conflict. **c** Over time more trade relations form. **d** When conflict ends, trade flows reverse as demand increases elsewhere. *Note Dotted lines* represent undisclosed trade activity and *solid lines* are fully reported. *Circles* represent nations

conflict (Fig. 1c). Then, as the conflict subsides and other hostilities erupt elsewhere, the trade flow adjusts its course [12]. Weapons begin to flow out of the conflict zone, destined for other skirmishes. Whether intentionally or inadvertently (i.e., failure to disclose trade due to weakened infrastructure), this outflow is likely to be clandestine.

Arguably, one of the most critical periods to control the availability of weapons is *after* a conflict ends since weapons are recycled between conflict zones [17–19]. At the termination or significant de-escalation of conflict, a large pool of secondhand weaponry becomes available. Materials seized during combat are a valuable resource that can be traded for necessary materials and supplies when domestic production is still in disarray [10, 12, 20, 21]. Secondhand weapons face fewer international control mechanisms and are generally considered more likely to be transferred through illicit channels. The influx of these weapons is thought to facilitate a change in the pattern of weapons trade. Though not as long lasting as the effects of technological advances, many argue that these stochastic shocks temporarily flood illicit markets with used weapons ([20, 21]). Emerging from the literature are four explanations for how trade relations will be reshaped following conflict.

2.2 Market Typology

Interchange. Trade interchange markets may be the most likely structure to emerge following the cessation of hostilities. Exchange relations form naturally when nations supply different commodities, i.e., ammunition is provided by one country and weapons by the other [22]. These reciprocated transfers are likely to increase for several reasons following the cessation of conflict. Trade relations that were drawn upon to supply combatants become the ties used to distribute excess weaponry at the end of conflict; the trade flow reverses course. Moreover, war-related trade relationships are likely to be the most accessible at the time conflict subsides; thus, these existing trade paths that once supplied arms are likely to be used to supply other goods. Exchange patterns may also develop as an ally withdraws military assistance; with pending conflict emerging elsewhere the resources are extracted and redeployed [12, 18, 20]. Network analysis offers a statistic to capture this configuration. Reciprocity is a measure of the development of returned trade ties. As illustrated in Table 1, if nation i transfers weapons to country j, at a subsequent time period, country j will send small arms, accessories, or ammunition back.

Mediator. Historically, much evidence exists to support the argument that superpowers and major producers use indirect shipping channels, diverting small arms through intermediaries (e.g., [12, 15, 19–21, 23, 24]). Shipping weapons through intermediaries, sometimes with multiple transshipment points, permits supply nations to disassociate themselves from violations of international arms embargoes or for supporting rebel activity. For example, Naylor [20] argues that the Central Intelligence Agency (CIA) diverted weapons to support rebellions in Nicaragua and Angola, which ensued about the same time the USSR pulled out from Afghanistan. Conflict provides an opportunity to get away with surreptitiously transferring weapons to a country under an international embargo.

Conflict will increase this tendency: passing weapons through the conflict zone permits the illicit use of national supply routes to bypass international oversight [12, 16, 20]. When visualizing the trade network, the post-conflict nation will be positioned between many other pairs of countries if it is being used as a diversionary or transshipment point. The darkened circle in the second row of Table 1, labeled i, represents the post-conflict nation and it is situated between many other pairs of countries. Described in detail later in the chapter, the statistic able to identify this pattern in a trade network is called betweenness centrality.

Epicenter. During conflict the nation becomes a major consumer of weapons from various sources and upon cessation of hostilities, surplus secondhand weapons are sold widely ([20]; Naylor 1995). Buyers flock to the region as war resources are commonly available. With limited infrastructure and resources to rebuild the war-torn nation, small arms are one of the most likely commodities to be traded [18]. Weapons can be easily used as payment for other commodities, i.e., food supplies, general consumer goods, and construction materials [25]. When conflicts are supported by rival superpowers, the cache of leftover weaponry is greater and potentially

Table 1 Typology of weapons markets developing post-conflict

Trade type	Description	Trade network structure
Interchange	Exchange relations form naturally when nations supply different commodities, i.e., ammunition is provided by one country and weapons by the other. Following conflict, these trade relations are most apt to be used to unload excess weapons (e.g., [12, 18, 20, 22])	Nation *i* sends something to *j* and *j* sends something back (reciprocity). 
Mediator	Shipping weapons through intermediaries, sometimes with multiple transshipment points, permits supply nations to disassociate themselves from violating international arms sanctions or for involvement in supporting rebel activity (e.g., [1, 15, 20, 22]) Example: US → Pakistan → Afghanistan during the Soviet War in Afghanistan from 1979–1989	A nation (*i*) sits between pairs of other traders (betweenness centrality). 
Epicenter	During conflict the nation becomes a major consumer of weapons from various sources and upon cessation of hostilities, surplus secondhand weapons are sold widely (e.g., [1, 20]) Examples: Ethiopia, Eritrea and Somalia 1970s–1980s–1990s; Afghanistan 1980s	A nation buying from many nations will sells to many nations (indegree activity). 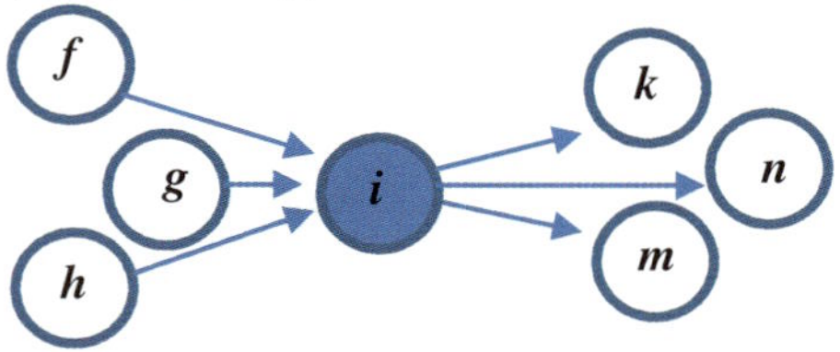
Channel	The nation in conflict becomes a central repository for weapons and upon conflict cessation does not have the infrastructure to sell off the cache. Consequently, weapons are transferred to an adjacent nation with the capacity to sell to a lot of nations [1, 19, 21] Examples: Vietnam → Thailand 1970s; Cambodia → Thailand 1980s; Ukraine → Siberia and Russia → Eastern Europe 1990s	A nation that receives from many others partners with a nation that sells to many (in–out degree assortativity). 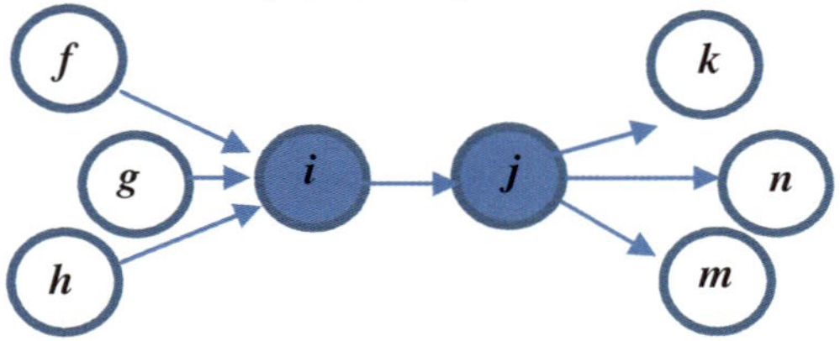

more diverse [26]. In network terms, this means that a country that received from many channels during a conflict is likely to export weapons through many channels following the conflict. Table 1 provides an illustration of what an epicenter would look like in a network: there would be many trade paths into and out of the nation pre-conflict and post-conflict, respectively.

Channels. The nation in conflict becomes a central repository for weapons and upon conflict cessation does not have the infrastructure to sell off the cache. Consequently, weapons are transferred to an adjacent nation with the capacity to sell to a lot of nations [19]. This activity forms a trade channel, wherein weapons are funneled en masse out of the post-conflict nation. Many examples of this trade structure exist. This type of network was widespread in Eastern Europe and Central Asia, where years of conflict have produced a dynamic flow of weapons from bordering countries creating markets that service many other nations outside of the region [19]. The network statistic that most closely resembles this tendency is in–out assortativity. As illustrated in the final row of Table 1, this pattern emerges when a nation that receives from many others (in degree centrality) is connected with another country that has high outflow to many other nations (out-degree centrality).

3 Modeling Dynamic Networks

Underlying dynamic models is the assumption that systems will continually evolve as actors (in this case we are talking about nations) seek to improve their position within the group. Even though decisions occur within a local context and are driven by individual preferences and motivations, when aggregated, they illustrate a dependence in the way the network changes. This dependence, and the situational context within which it develops, is visible in the network structure. As such, multiple forces are at work that will complicate how the trade structure develops.

3.1 General Trade Preferences

General trade preferences are taken to reflect trends in market activity. For instance, it is commonly argued that with the diffusion of weapons production capacity and decline of cold war politics, weapons trade is decentralizing [11, 27]. As more nations become involved in servicing or producing arms, accessories and ammunition, consumers have a greater number of suppliers to choose from. In network terminology, this means that the density of the market (cohesiveness of trade relations) should decline and that more nations should emerge as important players, producing an increase in the average number of trade partners each nation has. Moreover, regional suppliers will play a greater role in arming neighboring combatants [7, 9, 21]: while nations have more choices regarding where to obtain their weapons, the closest supplier may be the most efficient choice [15, 16, 19]. In addition to general trade preferences of actors, external influences will also shape what trade patterns emerge.

3.2 Facilitating Conditions

Evidence suggests that the geo-political context formed by shared land borders, military alliances, and insurgency networks should constrain illicit trade relations. Nations, and non-state actors, are apt to trade more frequently with others they share borders with [13, 16, 19]. Trade routes service all types of commodities and overland paths via truck or train are commonly used to move goods inland from seaports. As such, the geographic network of nations sharing borders sets a framework that shapes how illicit trade relations form.

Ties between nations formed by the politics of warfare will exhibit consistencies with weapons trade. This means that the web of weapons transfers should be entrained, or embedded, within military alliances and insurgent networks. Whether intended to support or suppress conflict, analysts argue that spikes in trade activity are produced by official and unofficial military assistance [10, 21]. For example, military expenditures directly correlate with fluctuating levels of arms transfers [14]. Russian weapons were sent to Arab states around the Six Day War and Syria after the break with Egypt [11]. More recently, cold-war aid from the U.S. went to Northern Iraqi groups opposing Saddam Hussein [12]. With growing coordination among insurgent groups (i.e., Al Qa'ida operated in 38 different countries from 1997–2010), it is likely that insurgent networks function similarly to military alliances between countries, channeling resources where needed [15, 19, 21, 22]. Thus, the evolving pattern of weapons transfers is constrained within geo-political networks.

4 Methodology

4.1 Source

Small arms[2] trade information was drawn from the United Nations Commodity Trade Statistics Database (UNCOMTRADE). This open-source database captures

[2] Small arms transfers included trade for the following customs codes: 930100 (military weapons), 930120 (rocket and grenade launchers, etc.), 930190 (military firearms), 930200 (revolvers and pistols), 930320 (sport and hunting shotguns), 930330 (sport and hunting rifles), 930510 (parts and accessories for revolvers and pistols), 930529 (parts and accessories for shotguns and rifles), 930521 (shotgun barrels), 930621 (shotgun cartridges), and 930630 (small arms ammunition).

international transfers of merchandise for over 170 reporter countries and their trade partners. Gathered and maintained by the United Nations Statistical Division (UNSD), the annual trade data are voluntarily reported. Total transfers, as measured in US dollars, were available for import, export, re-export, and re-import transfers[3] among recognized territories[4] [28]. Even though this is a widely used source of trade information, notable limitations exist, particularly for small arms trade. The data include only what nations are willing to disclose; for various reasons, i.e., political or military purposes, some trade is omitted or obscured. For example, transfers may be listed as shipping to a region and not to a specific statistical territory. Moreover, submitting territories are responsible for ensuring accuracy: due to problems with currency conversion or other technical issues, exports reported by one country often do not match values reported by trading partners. For these reasons, trade information was used to generate binary networks as described below.

4.2 Network Creation

Annual trade networks of small arms transfers were created for 14 years (1997–2010) between 224 countries.[5] To capture suspicious trade, transfers were examined as reported by both the origin and destination countries. When an import was claimed to be received but the exporting nation failed to report the transfer, the relationship between parties to the exchange was valued at "1", and when an export was reported but the import omitted, the tie was also valued at "1". In two situations relations were valued as "0": [15] if both nations reported the exchange, or [29] if no trade occurred. These directed trade networks captured trade flow from origin to destination. Of note, recursive activity wherein a nation re-exported and re-imported to itself was excluded.

---

[3] Re-import and re-export includes commodities that are considered to be *in-transit, passing* through a territory.

[4] A statistical territory is defined in the International Merchandise Trade Statistics: Concepts and Definitions, Revision 2 (United Nations Publications, Sales No. E.98.XVII.16) paragraph 64: as "the territory with respect to which data are being collected". Anything leaving this area is classed as an export even if the merchandise remains within the economic territory. An economic territory consists of the entire "geographic territory administered by a government within which persons, goods and capital circulate freely" and includes: airspace, territorial waters, continental shelf, territorial enclaves, and free zones, bonded warehouses or factories under customs control that are operated by offshore enterprise. When a statistical territory partially coincides with the economic territory of a nation, trade statistics fail to provide a complete record of inward and outward flows of goods. For the most part, the territories included in this study are identifiable and recognized economic territories.

[5] To maintain consistency throughout the study period, two pairs of nations were collapsed into a single entity: Belgium and Luxembourg became Belgium-Luxembourg, and Serbia and Montenegro were combined.

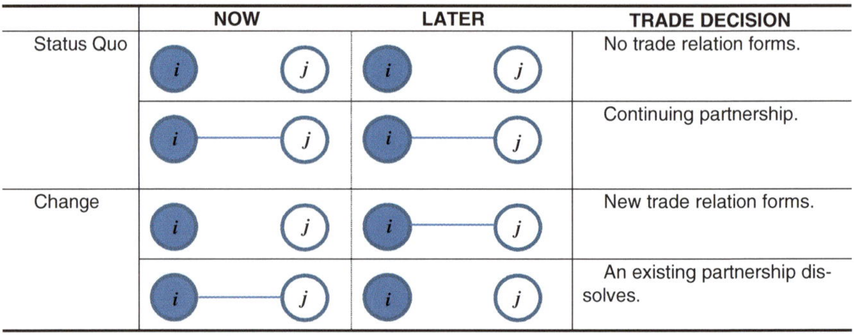

Fig. 2 Trade options available between the current time and a later observation

Dynamic models assume that trade relations evolve over time as nations form and terminate trade partnerships to fulfill individual needs. Change in these trade networks is actually the dependent variable of interest. This is a major departure from conventional analysis. As shown in Fig. 2, members of a network have four possible decisions to make between time periods. Maintaining the status quo involves keeping an existing trade relation or continuing to refrain from sending arms to a particular country. What we are most interested in are situations wherein a country decides to form a new trade relation or dissolves a previously existing partnership. A critical threshold of stability is needed to run the model.

Jaccard coefficients capture the structural stability of the network between each pair of successive years, i.e., 1997–1998, then 1998–1999 and so forth. This proportional statistic is calculated by taking the number of ties that exist divided by the number of ties that exist plus all change, i.e., ties that were formed and the ties that were dissolved [5]. Table 2 reports Jaccard coefficients for all observation periods.[6] On average, the observation periods exhibit sufficient stability. About 33 % of the trade relations are stable between successive years, this is sufficient to support running a dynamic model [5].[7]

[6] An observation period involves two networks from successive years.

[7] As a point of comparison, we examined the reported trade activity through the same set of observations, where both partners to the exchange acknowledge the transfer of weapons (this may be considered legal or fully disclosed trade). Among these partnerships, 69 % of reported trade relations with an annual value over $500,000 USD are stable and 48 % of low value trading partnerships continue across observations. Thus, the illicit network is much less stable than the fully documented trade activity.

Table 2 Changes in weapons transfer activity between years observed

Period	Status quo		Change		Jaccard coefficient[a]	Change rate (SE)[b]
	No trade	Stable trade	New partners	Trade dissolves		
1997–1998	47,670	799	733	750	0.350	18.22 (0.82)
1998–1999	47,722	751	698	781	0.337	16.85 (0.74)
1999–2000	47,549	707	954	742	0.294	23.20 (1.07)
2000–2001	47,451	830	840	831	0.332	19.63 (0.84)
2001–2002	47,334	874	948	796	0.334	21.05 (0.90)
2002–2003	47,242	880	888	942	0.325	22.02 (0.91)
2003–2004	47,170	845	1,014	923	0.304	25.10 (1.06)
2004–2005	47,078	912	1,015	947	0.317	23.83 (0.94)
2005–2006	47,003	983	1,022	944	0.333	23.74 (0.89)
2006–2007	46,990	989	957	1,016	0.334	23.32 (0.90)
2007–2008	46,991	994	1,015	952	0.336	24.20 (0.96)
2008–2009	46,914	1,009	1,029	1,000	0.332	24.97 (1.03)
2009–2010	46,981	1,058	933	980	0.356	22.24 (0.86)

[a] Multiplying Jaccard coefficients by 100 reveals the percent of the network ties that are that same at the second observation
[b] Final parameter estimates for the full model

4.3 Description of Networks

Table 3 reports descriptive statistics for these networks. Over time the density of clandestine trading becomes more integrated. In addition, as reflected in the increasing average degree scores, traders are selling to or buying from a greater number of nations over time. However, this value is not large enough to suggest that centralization is increasing. Increasing centralization would suggest that one or a few key players are beginning to monopolize trade. Instead, these scores indicate that many countries are trading among each other. This pattern is consistent with prior research that suggests a diffusion of weapons technology has occurred following the end of the Russo-American cold war [9].

Though change is found, there is a notable presence of dominant trading nations. The inset of Fig. 3 depicts all trade relations aggregated for the entire study period. The total aggregated network contains 7,208 unique trade partnerships among 224 countries. Countries, shown as rounded squares vary in size, with the larger symbols representing higher betweenness centrality—high scoring nations can be interpreted as being consistently positioned within the illicit flow of arms. The United States, Germany, United Kingdom, Canada and France are critical trade brokers. Selecting the illicit trade relations that are present throughout the entire study period of 14 years, the main image provides greater insight into dominant trade pathways. Only 48 trade relations are found every year among the 34 nations depicted.

Table 3 Description of clandestine small arms trade, 1997–2010

Networks	Int. transfers (ties)	Density (%)	Avg. degree centrality
1997	1,549	3.10	6.92
1998	1,532	3.10	6.84
1999	1,449	2.90	6.47
2000	1,661	3.30	7.42
2001	1,670	3.30	7.46
2002	1,822	3.60	8.13
2003	1,768	3.50	7.89
2004	1,859	3.70	8.30
2005	1,927	3.90	8.60
2006	2,005	4.00	8.95
2007	1,946	3.90	8.69
2008	2,009	4.00	8.97
2009	2,038	4.10	9.10
2010	1,991	4.00	8.89

Note Density reports the percent of existing trade relations found relative to the number of unique trade partnerships that *could* have occurred. For example, in 1997 only 3.1 % of all possible trade combinations between pairs of countries were observed. Average degree centrality reports the average number of illicit trading partners per country

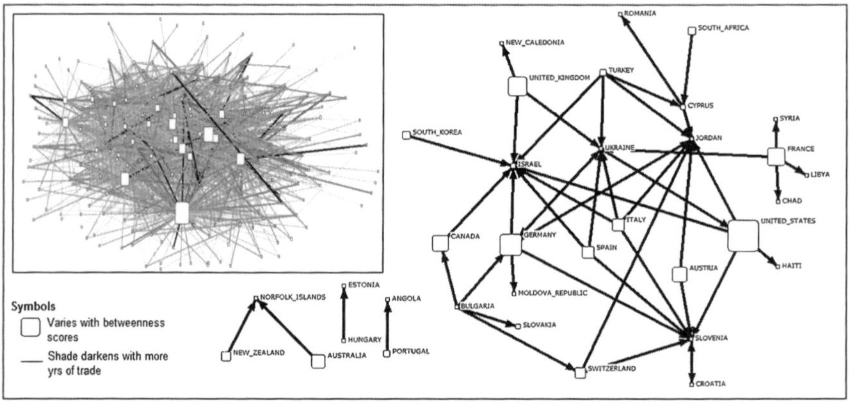

Fig. 3 Consistent illicit trade activity among 224 nations, 1997–2010

Examining degree centrality scores reveals that ten nations appear to consistently dominate misreported global trade in small arms (see Table 4). The nations listed in bold font are unique to only one type of degree centrality. It is notable that the Czech Republic narrowly misses inclusion in out-degree centrality; it was ranked 11th.

Table 4 Top trading nations involved in clandestine transfers, 1997–2010

	Normed out-degree		Normed in-degree	
Rank	Nation	Score	Nation	Score
1	United States	24.95	South Africa	14.83
2	Germany	24.47	United States	13.81
3	France	23.93	**Canada**	12.78
4	South Africa	23.61	France	12.42
5	**China**	23.22	Austria	12.42
6	United Kingdom	22.04	Germany	12.33
7	Italy	20.82	United Kingdom	11.31
8	**Russia**	19.67	Italy	11.18
9	**Switzerland**	19.19	**Israel**	10.63
10	Austria	19.09	**Czech Republic**	9.13
Centralization				
	21.5 %		**11.3 %**	

Note Bolded text indicates the named county has a top 10 ranking on only one of the statistics reported in the table

4.4 Variables

Constant Structural Effects. Six structural effects were used. All equations reported below were taken from the RSiena manual [3]. Unless specified, all effects were generated as evaluative functions. Evaluative functions estimate the tendency of an actor to change trade ties. Two structural measures are used to control for general changes in trade structures over time: out-degree density and transitivity. The out-degree density effect is a measure of outgoing interconnectivity as calculated by:

$$S^{net}_{out-density}(x) = x_{i+} = \sum_j x_{ij} \tag{1}$$

Many forms of transitivity may exist in a network. Transitivity is generally defined as the tendency for closure to emerge when the trade between three countries is considered. For example, if the United States sends weapons to the United Kingdom and to Canada, then it is expected that at some point, the United Kingdom and Canada will also form a trade relationship. Transitivity is considered a higher order structure, meaning that it is a structure that emerges among three countries rather than a pair. For this reason, any model looking into the existence of lower ordered patterns forming between pairs must include some measure of transitivity. Since no evidence of transitivity is suggested by the prior literature, a count of transitive triplets was used. Transitive triplets counts the number of times country *i* trades with two others (*j* and *h*) whom also trade with each other (a trade relation also exists between *j* and *h*).

Capturing Market Structure. Each type of market described previously can be identified with a different structural metric. When pairs of countries exchange weapons within a year, then a reciprocal trade relation forms. In this study, reciprocity is used to represent the emergence of an *interchange market*. For each nation, it is possible to sum the number of reciprocated ties that occur within the observation period with the following equation:

$$S^{net}_{reciprocity}(x) = \sum_{j} x_{ij}x_{ji} \tag{2}$$

Betweenness centrality is a count of the number of times a country sits between pairs of others in the illicit trading network [3, p. 100]. This statistic is used to determine whether weapons trade is likely to involve *mediators* or brokers. As operationalized here, this score is calculated with:

$$S^{net}_{betweenness}(x) = \sum_{j,h} x_{hi}x_{ij}\left(1 - x_{hj}\right) \tag{3}$$

Nations with high betweenness in a directed network are positioned to broker trade among others. When conflict ends, the remaining oligarchs within a war-torn nation may initiate trade of secondhand weaponry as an initial business venture to regain financial solvency. Capitalizing on the immediate surplus of arms, this trade activity will exhibit the brokerage structure within annual trade data as recent incoming weapons begin to be passed along.

Defined as the cross-product of a country's in- and out-degrees, the *in-degree related activity* effect indicates whether a nation holds a central position as a weapons supermarket. High scores would suggest that a nation receives weapons from many countries and then sends arms to many others. In other words, the nation is a trade *epicenter*. Ripley and colleagues [3, p. 102] define this measure as:

$$S^{net}_{in-activity}(x) = x_{i+}x_{+i} \tag{4}$$

Identifying a central trade *channel* between pairs of nations may suggest that rather than a single nation operating a supermarket, a pair of countries exhibit the tendency to be well positioned within the gun flow. *In–Out Degree assortativity* is taken to reflect the tendency for a nation with high in-degree centrality to transfer weapons to another country with high out-degrees. In RSiena [3, p. 104], this is calculated by the following formula where c is 2:

$$S^{net}_{inout-assortativity}(x) = \sum_{j} x_{ij}x^{1/c}_{+i}x^{1/c}_{j+} \tag{5}$$

Facilitating Conditions. Four covariates were included to estimate direct effects and interaction effects (with structural characteristics described above).

The shared borders covariate was a binary network of land borders.[8] Shared borders were identified through visual inspection of maps. This network includes two components comprised of 157 countries (70 % of the sample) joined through 585 shared land borders. All island nations were classed as isolates.

[8] Ties are coded 1 for a shared land border and 0 if the countries do not share a border.

Transnational Insurgent Ties. This is a derived 1-mode network linking countries that share active terrorists groups. Groups were considered active if they claimed responsibility for an attack during the year. Originally compiled by the National Consortium for the Study of Terrorism and Responses to Terrorism (START), terrorism incident information spanned from 1970 to 2011.[9] Ties are not directed.[10] In total, the derived country-to-country network includes 86 nations and 614 ties distributed among four components.

As illustrated in Fig. 4, ties are formed among nations because of the presence of active insurgent groups. Symbol size varies to illustrate degree centrality and the shade and width of ties reflect the number of shared groups over time. Thicker and darker lines indicate stronger transnational associations given the activity level of shared insurgent groups.

The *military alliance network* was formed by linking each country to the military alliances and colonial, or administrative, dependencies it has been involved with since World War II.[11] Information for this network was developed from membership lists for 26 different alliances, including the 5 Power Defence Arrangement, the Collective Security Threat Organization, NATO, and etc. This network does not illustrate directionality but only shows that a connection exists between two or more countries through an alliance. In total, 27 colonial or administrative dependency collectives were also included. The network is valued to indicate the number of alliances in common among pairs of nations. These strategic associations suggest possible avenues for which small arms and light weapons can be bought and sold internationally. It is expected that stronger ties between nations, more alliances, should promote the formation of gun trade relations, particularly reciprocated activity. The derived country-to-country network includes 213 countries linked through 12,808 ties. Betweenness centrality scores are highest for the United States, New Zealand, Australia, United Kingdom, France, South Africa, Pakistan, Mauritania and Algeria; normed scores range from a high of 7.0–2.7 respectively. Degree centrality scores rank the United Kingdom and France the highest; normed scores are 18.1 and 16.1 respectively. With a network average of 98 partners (SD 76.7), the centralization score of 13.1 % highlights the impact colonial ties have in generating relations.

---

[9] The three agencies that were used to supplement the database included: Pinkerton Global Intelligence Service (1970–1997), Center for Terrorism and Intelligence Studies (1998–2008), and the Institute for the Study of Violent Groups (2008–2011). START is an entity within the United States Department of Homeland Security Center of Excellence in Maryland.

[10] On average, 11 insurgent groups were active within each country (Med. = 1; Std Dev. = 27). India was a major outlier with about 295 active insurgent groups over the 14 year period examined.

[11] Since alliances and members are two different modes and the purpose of the research was to examine trade relations among countries, a one mode, country-to-country network was derived from this listing.

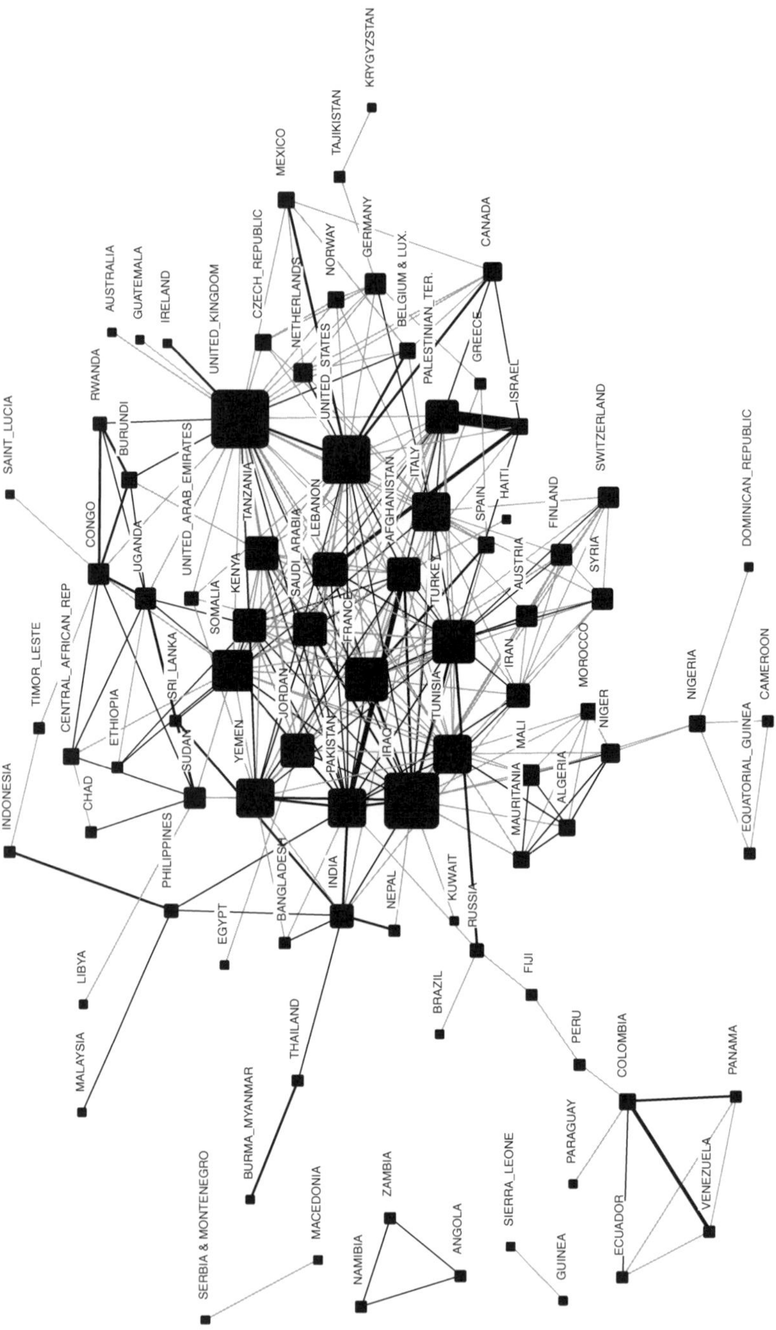

Fig. 4 Transnational insurgent ties, 1997–2010

Cessation of Armed Conflict. Conflict information was retrieved from a publicly available database—the Uppsala Conflict Data Program (UCDP)[12] compiled by Uppsala Universitet. This source records whether nations are involved in either intrastate[13] or interstate[14] conflict. A dichotomous variable was generated by assigning countries a value a "1" for the year conflict ended and one subsequent year. Scores of "0" were assigned if no conflict was observed that year or an existing conflict continued. Within the 14-year period, approximately 35 countries (min. = 29; max. = 42) were involved in conflict at any given time. On average, conflict ended for 7 countries (min. = 2; max. = 13) during each year observed.

5 Findings

The coefficients and odds ratios reported in Table 5 suggest that globally, from 1997 to 2010, there was a significant increase in the number of nations involved in reciprocating illicit weapons transfers. In fact, the odds of a country returning a trade relation are 342 %. And while there is also evidence that more nations positioned themselves between trade partners (Est. = 0.28; OR 1.33), this effect is considerably smaller and less likely to form than trade interchange (reciprocity). Two facilitating conditions were found to increase the likelihood of trade: sharing a land border and being involved in more military alliances (recall that the military alliance network is valued with higher scores reflecting more associations). Material to this investigation is the positive, significant effect of conflict cessation. The year a conflict ends and the year immediately following are associated with an 81 % increased likelihood of developing trade ties (Est. 0.60; OR = 1.81). When this effect is added to the structural effects associated with each type of weapons market, the results suggest that during the study period examined, the tendency to form weapons interchange markets was at least two times greater than all other trade market structures. This result suggests that the dominant tendency is for illicit weapons transfers to reverse course following the cessation of hostilities.

Turning to the two case studies, Egyptian centered trade activity tends to reflect the dominant global pattern and the Angolan egonetwork exhibits a substantially different market structure. Prior to discussing these results it should be noted that the Jaccard coefficients for both models suggest that there was a very low threshold of stability. For Egypt, trade relations were consistent between years about 19.4 % of the time and within the Angolan network, 23.4 % of trade ties were stable. This suggests a great deal of change was happening and this generates poorer T ratios

[12] This database provides information on countries around the world that have been engaged in conflict (intrastate or interstate) between the years of 1975–2011. Data is accessible at: http://www.ucdp.uu.se/gpdatabase/search.php.

[13] This is defined as conflict between two parties within a recognized state's boundaries.

[14] This type of conflict involves different nations.

Table 5 Impact of conflict cessation on clandestine weapons transfers, 1997–2010

	Full network				Egonets			
	Structural model		Conflict model		Egypt case study		Angola case study	
	Coefficient (SE)	Odds ratios	Coefficient (SE)	Odds ratios	Coefficient (SE)	Odds ratios	Coefficient (SE)	Odds ratios
Structural effects								
Control variables								
Outdegree density	−7.05 (5.93)	0.00+	−6.95 (1.64)*	0.00+	−3.50 (0.48)*	0.03	−5.11 (1.30)*	0.01
Transitive triplets	0.09 (0.01)*	1.09	0.00 (0.01)	1.00	0.12 (0.25)	1.13	1.21 (2.17)	3.35
Market structures								
Reciprocity (interchanges)	1.82 (0.86)*	6.20	1.23 (0.17)*	3.42	0.81 (0.46)	2.25	−6.02 (2.81)*	0.00
Betweenness centrality (mediators)	0.32 (0.12)*	1.37	0.28 (0.01)*	1.33	0.22 (0.05)*	1.25	−3.46 (0.58)*	0.03
In-degree activity (epicenters)	0.25 (0.73)	1.28	0.22 (0.16)	1.25	0.20 (0.22)	1.23	3.09 (0.88)*	22.04
In–out degree assortativity (channels)	−0.07 (0.08)	0.94	−0.02 (0.02)	0.98	0.03 (0.06)	1.03	0.71 (0.91)	2.03
Facilitating conditions								
Border network, *static dyadic*	–	–	1.21 (0.11)*	3.35	−0.21 (0.18)	0.81	2.47 (1.04)*	11.86
Alliance network, *valued, static dyadic*	–	–	0.22 (0.08)*	1.24	−0.01 (0.03)	0.99	−1.11 (0.69)	0.33
Insurgent network, *static dyadic*	–	–	0.00 (0.00)	1.00	−0.03 (0.06)	0.97	0.10 (0.06)	1.10
Conflict effects								
Conflict cessation	–	–	0.60 (0.29)*	1.81	0.35 (0.32)	1.41	−0.63 (1.93)	0.53
Conflict cessation * Rate of change	–	–	−0.08 (0.05)	–	0.21 (0.51)	–	0.80 (0.55)	–
Calculated effects								
Conflict cessation * reciprocity (interchanges)	–	–	1.79 (na)	6.01	1.14 (na)	3.11	−6.60 (na)	0.00
Conflict cessation * betweenness (mediators)	–	–	0.85 (na)	2.33	0.54 (na)	1.72	−4.05 (na)	0.02

(continued)

Table 5 (continued)

	Full network				Egonets			
	Structural model		Conflict model		Egypt case study		Angola case study	
	Coefficient (SE)	Odds ratios	Coefficient (SE)	Odds ratios	Coefficient (SE)	Odds ratios	Coefficient (SE)	Odds ratios
Conflict cessation * in-degree activity (epicenters)	–	–	0.78 (na)	2.19	0.53 (na)	1.43	2.51 (na)	12.25
Conflict cessation * in–out assortativity (channels)	–	–	0.54 (na)	1.71	0.36 (na)	1.43	0.12 (na)	1.13
MODEL stability								
% of Ties stable (average Jaccard coefficient)	32.95 %	–	32.95 %	–	19.40 %	–	23.35 %	–
% good T ratios	68.42 %	–	62.50 %	–	70.83 %	–	91.67 %	–
The grand mean of conflict was used for calculated effects	–	–	.938	–	.936	–	.926	–

$^*p < .05$

and consequently, unreliable standard errors. With this limitation in mind, the following discussion concerns large and significant effects.

Illicit trade involving Egypt and among trade partners shows a tendency toward interchange markets (reciprocity coefficient 0.81; OR = 2.25; n.s.) and mediation (betweenness coefficient = 0.22; OR = 1.25; $p < .05$). However, unlike the global market, this trade is not likely to occur within the border, alliance, or insurgent networks. Of these negative coefficients, the border effect is strongest. This means that changes in trade are likely to occur with non-bordering nations. Conflict cessation does materially increase the likelihood of a shift in trade structure in favor of reciprocated exchanges.

Within Angola's trade region significantly different patterns exist. First, over time it is more likely the case that illicit trade activity will concentrate in epicenters. This is evident from the positive, significant coefficient for in-degree activity and the negative, significant coefficients observed for reciprocity and betweenness. Among facilitating conditions, trade relations are significantly more likely to involve nations that share borders and much less likely to occur along military alliances. Conflict cessation serves to temporarily depress these effects; however, this evolving trade structure still favors weapons supermarkets centered on nations that are in post-conflict status.

The combined results found for outdegree density and transitive triples indicate that illicit trade networks show less cohesion over time. Outdegree density declines globally, and within both regional trade networks. This means that the tendency accruing from illicit weapons transfer decisions is a general preference for sparse, less complex trade patterns. With the exception of the Angolan egonet, transitivity is not found. Recall that transitive triplets count the number of times trade relations form between pairs of recipients. For example, a transitive triplet would exist if the United States sent weapons to Israel and Pakistan, and in the subsequent year, Israel sent weapons to Pakistan as well.

The results show some preference toward the development of epicenters where a few nations acts as illicit distributors. This tendency for heavy consumer nations to supply others provides an investigatory focal point. As Tihjuis suggests, some nations are best positioned to act as ''shipping locks'' that raise and lower weapons between the legal and illegal markets [2, 30]. Epicenters critical to both the legal and illicit domains are of particular concern.

5.1 Implications

> *The global diffusion of military capability is a process of enormous consequence, for it enables states to wage war and to engage in other forms of repression and violence. That states' leaders perceive a need to acquire and maintain military capability is […] a basic tenet of world politics […]. But because this perceived need is a near constant, other forces must be considered in order to account for shifting patterns in global arms production and transfers* [11; 226].

Global Trade

This study investigated how the cessation of conflict affected the illicit trade of small arms, 1997–2010. Underlying this analysis is the assumption that when armed conflict ends, small arms become an important trade commodity for nations facing the challenge of recovering from war. Taking advantage of available resources, temporary supermarkets emerge to unload weapons, ammunition, and other associated materials. While history is replete with many examples of these makeshift trade centers, the present study is the first to examine their development.

By controlling for the general trends in market structure, the dynamic model isolated conflict-oriented supermarket structure from general market trends. Post-conflict, the results suggest that on a global scale, small arms are most likely to reverse course along the trade ties that supported the initial acquisition of weapons. This effect is more than double the tendency to form trade epicenters and more than three times the tendency to form the trade channels described previously (e.g., [11, 20]). Thus, the "supermarkets" that formed following major conflicts such as Vietnam, are either the exception to the rule, an artifact of their era, or a regional trade structure that does not reflect global tendencies.

From a network perspective, this makes sense. Trade partnerships that supply hostilities are in social terms the closest associations post-conflict that are capable of converting surplus weaponry into cash. While prior scholarship in this field did not discuss this pattern, the tendency for a network to evolve in favor for greater reciprocity is well established within social network research (i.e., [4, 31]). In fact, the pattern is so prevalent that SIENA developers argue that reciprocity should be included in all dynamic models [3–5]. Since reciprocity is a lower order structure, this model included a measure of transitivity. No evidence of transitivity emerged at the global level, suggesting that the reciprocity effect is robust. Emerging trade structures (1997–2010) do not favor the expected supermarket structure anticipated in the extant literature.

Not surprising, two facilitating conditions exhibited positive effects; reciprocity is more likely to favor bordering nations or those with whom the post-conflict nation has strong military alliances. Though many nations dispute the location of borders, national neighbors are relatively constant. Extending this line of research, examining the porousness of border control would enhance our ability to predict which route weapons may take, i.e., considering the corruptibility of border officials and the impassibility of mountainous terrain [32]. Moreover, adding connectivity through key shipping routes may strengthen the precision of this model.

Considering that we used a valued measure of military alliances, wherein higher scores indicated that pairs of nations were party to more common alliances, increasing the number of shared alliances will greatly strengthen this trade effect. Moving forward, testing for structural equivalence among allies divided by rival

---

[15] The reader is reminded that this category also includes an extended array of light arms and military weapons, such as land mines and rocket grenade launchers, and all of the associated ammunition and accessories.

factions, i.e. NATO and the Warsaw Pact, would improve these models by controlling for long-term political influence associated with reactive asymmetry and technological innovations endemic to the ongoing arms race [11]. Kinsella [11] recommends including a co-integration correction factor for definable regions (Middle Eastern Security Complex) and subnetworks (Persian Gulf).

Policy aimed at controlling the flow of small arms post-conflict is best to target recent incoming supply routes.[15] Detailed information about who was initiating trade was not available from the data source used here. Thus, it is not possible to determine whether weapons flow out through the same dealers or companies that originally supplied the arms. This is a limitation of using aggregated data. Subsequent studies are needed to confirm whether this is the case. Such a study would contribute significantly to several related fields as the networks, mechanisms, and money laundering tactics used to smuggle weapons into conflict zones are also conduits for other goods, i.e., illicit drugs, precious gems, timber ([12, 13, 29]). Strong arguments suggest that this overlap is facilitated by two critical factors: (1) the presence of brokers—individuals, companies, and nations—that are able to operate in several nations [2, 12] and (2) use by insurgent groups of seized or illicit resources to fund arms purchases (i.e., [12, 20]).

6 Case Studies

The global system offers a baseline against which to investigate regional dynamics. The Egyptian egonet of weapons transfers provides a glimpse into the effects of conflict cessation on the Middle East security complex. The Middle East security complex is generally considered to include four regions: the Persian Gulf (Bahrain, Iran, Iraq, **Kuwait**, **Oman**, **Qatar**, **Saudi Arabia**, **United Arab Emirates**, and **Yemen**), Eastern Mediterranean (**<u>Egypt</u>**, **Israel**, **Jordan**, **Lebanon**, and **Syria**), Maghreb (Algeria, Chad, **Libya**, Mali, Mauritania, Morocco, **Niger**, and **Sudan**) and the Horn (Djibouti, Ethiopia, Somalia, and Tunisia). Bolded nations appear in Egypt's local trade network studied here. In total, 19.4 % of the trade partners present in the egonetwork were located within the region. This low level of trade with nearby states suggests that Egypt's weapons trade does not evolve to favor neighbors and this is not consistent with global trends. However, this case study mirrors the global network in terms of the effects of conflict cessation on illicit weapons trade. Post-conflict weapons are more likely to reverse course, flowing out of war-torn nations along supply ties. Six nations exhibit stable, illicit reciprocal ties with Egypt: Canada, France, German, Italy, Turkey, and the US. Other nations of interest because they receive small arms over the course of many years are: Spain, China, Austria, Switzerland, United Kingdom, Cyprus, and South Korea. In addition, Egypt appears to send weapons to: Sudan, Czech Republic, Chile, and Estonia. With the exception of Sudan, none of these critical trade partners are part of the Middle East security complex. This means that only ephemeral weapons trade partnerships are formed with others in the region.

The Angolan civil war (1975–2002) that directly followed its independence from Portugal was fought between the MPLA (People's Movement for the Liberation of Angola) and UNITA (National Union for the Total Independence of Angola) for control of the country and had ties with Soviet and United States powers, respectively. Cold-war era superpowers attempted to influence the political development of this newly independent country. The United States, although issuing an embargo against the country, continued to divert shipments into Angola to further assist UNITA [20]. Trade post-conflict within the Angolan ego-network exhibited a strong preference for an illicit epicenter. Closer inspection of the countries involved in this trade reveals little stability in partners. For example, while Angola routinely exchanges weapons with Namibia and Portugal, it receives from many more nations including: Brazil, China, France, Spain, and the United States. Moreover, the set of nations exporting to Angola change each year. With few stable, reciprocal weapons trade partnerships with neighboring countries, or even other nations in the region, Angola may operate more as a conduit into Africa. Additionally, within Angola's egonet, nations were more likely to alter trade habits with bordering nations. While this modeling technique appeared to capture evolving market structure and had sufficient sensitivity to estimate regional differences, it is not without its limitations.

6.1 Limitations

Several improvements to the modeling process are possible. To advance this line of research to the next level: (1) more work is required to better understand the temporal nature of system shocks, (2) successive periods of trade need to be a bit more stable, and (3) we need to conduct a systematic exploration of how to better measure important covariates.

Estimating Time Dependent Effects. Previous studies question the efficacy of using formal embargoes on nations during conflict, as nations are shown to ignore these mandates by trading directly or sidestepping trade restrictions through indirect deals with third parties [18, 19, 33]. Adding to prior work in this area, the present study suggests that trade controls may not work because they may not be in place long enough. One of the most critical periods to control the availability of weapons is *after* a conflict ends since weapons are recycled between conflict zones [17, 18]. This study used a two year window to capture the effects of conflict cessation. As discussed shortly, investigating the temporal lags associated with changes in international policy is a fruitful research direction.

Two types of stochastic or random shocks influence trade systems: market drift results from the lasting effects of change and temporary shocks arise from immediate and sudden change in local contexts. The present study did not include measures of market drift that may occur because of technological developments, diffusion of weapons technology and subsequent growth of new manufacturing centers, or the realignment of political blocks as nations continue to maneuver in the post 9/11 era. Controlling for market drift while estimating temporary stochastic shocks is likely to improve the utility of estimates generated. However, before such a model can be tested, additional research is needed to determine the duration of temporal effects.

Market drift is longer term and is likely to follow a measurable trajectory. This means that while the initial shock or change is stochastic, what follows is not a random long term effect. For instance, wealthy nations are quick to adopt technology and this tends to generate a domino effect as nations recoup some of the cost of rearmament by selling a portion of their existing supplies to less affluent nations. In turn, this generates resale to leading developing nations, and so on. Where strong service industries exist, existing supplies circulate longer, and consequently, the change effect associated with newly introduced weaponry is prolonged, as late adopters dally in the arms race. This extension of the transfer chain suggests two things. First, several overlapping supply chains exist, one for each major product class, and second, the time effect for each will vary. It follows that future inquiries should focus on a single commodity class (i.e., AK-47's or automatic assault rifles).

Regarding the stochastic shocks generated by the eruption of hostilities, two temporal effects are likely. Immediate change is inevitable for several reasons. For instance, as dealers race to supply combatants, arms transfers will flow into the region and upon conflict cessation, a temporary surge in outflow occurs. More lasting, but still temporary effects will also happen as superpowers, their allies, and neighboring countries develop contingencies and adjust their own weapons policies [11]. Each of these reactions must be modeled differently. With little prior work in this area, using a sensitivity procedure wherein a set of effects, i.e., 1 year, 2 years, 3 years, and etc., is systematically tested would offer a suitable start to determining the duration of temporal shocks.

Model Instability. A related issue is that any major development or deployment during conflict will generate trade pulses that are likely to be more visible at the regional level. This may account for degree of instability found with the two egonetworks examined (Egypt and Angola) relative to the level of consistency found in the global market. SIENA does not do well when excessive change exists. Addressing this limitation will improve model convergence and remove bias from the standard error estimates, thereby improving our ability to identify significant effects among the set of covariates. Perhaps the most logical way to add stability to the network is to incorporate other types of trade activity, such as legal trade. Aggregating legal and illicit trade would require generating a valued dependent variable, where one score, say a value of 1, is assigned for illicit trade activity and a value of 2 assigned for legal trade. Unfortunately, this capability is not currently available; SIENA can only model binary dependent variables. However, the concept of network entrainment may offer a solution to this problem.

Illicit trade is embedded within legal market activity. Returning to Fig. 1, the solid lines represent joint trade that is fully disclosed whereas the dotted lines illustrate the grey or illicit market. A more complete picture of the network may emerge from using legal trade as another covariate. A temporally-lagged, variable capturing legal trade activity can be introduced for each observation period. For example, legal trade for 1997 can be introduced to account for the shift in illicit trade patterns between 1997 and 1998.

Improving the Measurement of Covariates. One variable that did not perform as expected was the insurgent network. Political relations are widely acknowledged as being a critical factor influencing trade activity. Since armed conflict tends to involve recognized governing bodies and insurgent groups, two types of political alliances were included. One of the reasons why the insurgent network failed to add materially to the model is that the trade activity investigated was drawn from the UNCOMTRADE data warehouse. This source captures legal and marginal, or grey, market activity, but fails to include completely illegal trade. Since insurgent and rebel forces are likely to use marginal and illegal means to obtain arms, a substantial portion of their trade behavior is missing or obscured. Fortunately, there may be a partial solution to this issue.

Much of the illegal trade involves extended diversionary paths wherein a combination of falsified paper trails and transshipments are used to move weapons along indirect routes. Networks can be examined for these kinds of structures. Focusing on either the originator or the receiver, it is possible to determine whether stochastic shocks, like conflict cessation, are associated with an increase in the number of indirect paths a nation is party to.

A second explanation for the weakness of the insurgent variable is that it was aggregated. Ties between insurgent groups operating in different countries are likely to be more transitory than formal alliances between nations. Using annual networks offers two benefits: first, the resulting model will have greater temporal sensitivity and second, lag effects can be used to better account for possible resource sharing within specific groups, i.e., Al Qa'ida.

7 Conclusion

Conflict cessation is linked to the development of weapons supermarkets, dealing primarily in second hand guns [1, 15, 22]. Post-war recovery and rebuilding efforts must include strong weapons control policies. A critical aspect of successful arms control policy is to know where to focus attention. Since trade networks are dynamic, responding to stochastic shocks within the context of international and regional circumstances, a multivariate modeling technique is needed. Arguably, investigating the illicit flow of small arms may offer greater harm reduction for two reasons. First, these weapons are more prevalent amongst insurgent groups, kill more people, and are easier to conceal and transport because of their light weight [17–19, 34]. Second, the networks used to smuggle weapons into conflict zones are also conduits for other illicit goods (e.g., [11, 12]). Thus, the continued investigation into the structure of post-conflict weapons markets with dynamic network techniques is supported.

References

1. Naylor RT (1998) The rise of the modern arms black market and the fall of supply-side control. Transnational Organized Crime 22
2. Tijhuis E (2006) Transnational crime and the interface between legal and illegal actors. Doctoral dissertation, Universiteit Leiden
3. Ripley R, Snijders T, Lopez P (2011) Manual for RSiena. University of Oxford, Department of Statistics, Nuffield College, UK
4. Snijders T (2011) Network dynamics. In: Scott J, Carrington PJ (eds) The SAGE handbook of social network analysis. SAGE, Thousand Oaks, pp 501–513
5. Snijders T, Van De Bunt G, Steglich G (2010) Introduction to stochastic actor-based models for network dynamics. Soc Netw 32:44–60
6. Bichler G, Malm A (2013) Small arms, big guns: a dynamic model of illicit market opportunity. Global Crime 14(2–3):261–286
7. Kinsella D (2006) The black market in small arms: examining a social network. Contemp Secur Policy 27(1):100–117
8. Fisman R, Wei SJ (2009) The smuggling of art, and the art of smuggling: uncovering the illicit trade in cultural property and antiques. Am Econ J: Appl Econ 1:82–96
9. Kick EL, McKinney LA, McDonald S, Jorgenson A (2011) A multiple-network analysis of world system of nations, 1995–1999. In: Scott J, Carrington PJ (eds) The SAGE handbook of social network analysis. Sage, Thousand Oaks, pp 311–328
10. Craft C, Smaldone JP (2002) The arms trade and the incidence of political violence in Sub-Saharan Africa, 1967–1997. J Peace Res 39(6):693–710
11. Kinsella D (2002) Rivalry, reaction, and weapons proliferation: a time-series analysis of global arms transfers. Int Stud Quart 46(2):209–230
12. Marsh N (2002) Two sides of the same coin? The legal and illegal trade in small arms. Brown J World Aff IX(1):217–228
13. Mehalko L (2012) This is gun country: the international implications of US. Gun control policy. Boston Coll Int Comp Law Rev 35(1):297–330
14. Smith RP, Tasiran A (2005) The demand for arms imports. J Peace Res 42(2):167–181
15. Austin K (2002) Illicit arms brokers: aiding and abetting atrocities. Brown J World Aff 9(1):203–216
16. Vines A (2005) Combating light weapons proliferation in West Africa. Int Aff 81(2):341–360. doi:10.1111/j.1468-2346.2005.00454.x
17. Holdstock D, Jarquin A (2002) Commentary: conflict—from causes to prevention? Br Med J (Int Ed) 324(7333):345
18. Renner M, Peterson JA (1997) Small arms, big impact: the next challenge of disarmament. Worldwatch Institute, Washington, D.C
19. Joseph K, Susiluoto T (2002) Tackling small arms trafficking in the OSCE. Helsinki Monit 13(2):179–192. doi:10.1163/157181402401452825
20. Naylor RT (1993) The insurgent economy: Black market operations of guerrilla organizations. Crime, Law Soc Change 20:13–51
21. Neuman SG (1995) The arms trade, military assistance, and recent wars: change and continuity. Ann Am Acad Polit Soc Sci 541:47–74
22. Mandel R (1999) Deadly transfers, national hypocrisy, and global chaos. Armed Forces Soc (0095327X) 25(2):307–327

23. Hartung WD (2008) An unstoppable arms trade? World Policy J 25(3):137–140
24. Spapens T (2007) Trafficking in illicit firearms for criminal purposes within the European union. Eur J Crim Law Crim Justice 15(3/4):359–381
25. Krause K (2001) Norm-building in security spaces: the emergence of the light weapons problematic. Research Group in International Security, pp 247–263
26. Sanjian GS (2003) Arms transfers, military balances, and interstate relations modeling power balance versus power transition linkages. J Conflict Resolut 47(6):711–727
27. Grillot SR, Apostolova D (2003) Light weapons, long reach: bulgaria's role in the global spread and control of small arms. J South Eur Balkans 5(3):279–297
28. United Nations (2009) International recommendations for distributive trade statistics. Statistical papers: series M No. 89, New York, from http://unstats.un.org/unsd/trade/M89%20EnglishForWeb.pdf. Accessed 19 Nov 2012
29. Beittel JS, Library of Congress (2012) Mexico's drug trafficking organizations: source and scope of the rising violence. Congressional Research Service, Washington, D.C
30. Tijhuis E (2011) The trafficking problem: a criminological perspective. In: Manacorda S, Chappell D (eds) Crime in the art and antiquities world: illegal trafficking in cultural property. Springer, New York, pp 87–98
31. Wasserman S, Faust K (1994) Social network analysis: methods and applications. Cambridge University Press, Cambridge
32. Efrat A (2010) Toward internationally regulated goods: controlling the trade in small arms and light weapons. Int Org 64(01):97–131
33. Moore M (2010) Arming the embargoed: a supply-side understanding of arms embargo violation. J Conflict Resolut 54(4):593–615
34. Herron P, Marsh N, Schroeder M, Lazarevic J (2011) Larger but less known: authorized light weapons transfers. Small arms survey 2011: states of security. Cambridge University Press, Cambridge, pp 8–41

A Conspiracy of Bastards?

Simon Bennett

Abstract We assume that those charged with public safety and security act in the public interest. Generally this is the case. However, as the Holloway Road incident and Hillsborough cover-up demonstrate, sometimes public servants act in their own interests. Using actor-network and other sociological theories the author demonstrates how such deviant behaviour is organised and maintained. Several lessons are drawn. For example, that police statements should not be privileged over other types of witness statement. Also, that public servants, their representatives and Members of Parliament are not above using the gutter press to propagate false accounts. Finally, that the truth is sometimes distorted through the application of crude stereotypes. The Holloway Road incident and Hillsborough cover-up teach an important lesson—that effective checks-and-balances on the power of the State are essential for the safeguarding of liberty and justice. Such checks-and-balances include a free press and accessible and timely decision-review systems.

Keywords Police · Miscarriages of justice · Social theories of risk · Lessons · Checks-and-balances

1 Introduction

We assume the emergency services unfailingly work in the interests of citizens. Exceptionally, however, the emergency services may work not in citizens' interests, but in their own institutional (that is, selfish) interests.

S. Bennett (✉)
Civil Safety and Security Unit, University of Leicester, Leicester, UK
e-mail: sab22@le.ac.uk

A. J. Masys (ed.), *Networks and Network Analysis for Defence and Security*, Lecture Notes in Social Networks, DOI: 10.1007/978-3-319-04147-6_9,

In the United Kingdom recent events have led some to question whether the Police Service unfailingly acts in the public interest. Specifically, concerns have been raised that the orchestrated cover-up of policing failures at the 1989 Football Association Challenge Cup Semi-Final [1] between Liverpool FC and Nottingham Forest FC may have prevented lessons being learned, thereby endangering the safety of those attending subsequent sporting events (both in the UK and elsewhere). Other episodes, like the 1979 death of a special-needs teacher in Southall following a confrontation with police, the assaults visited on a group of teenagers in the Holloway Road by the Special Patrol Group in 1983, brutal confrontations during the Miners' Strike of 1984–1985, the adulteration of police notes following the bloody 18 June 1984 confrontation at Orgreave coking works,[1] the institutional racism that undermined the Metropolitan Police Service's (MPS's) investigation of the 1993 murder of student Stephen Lawrence,[2] the slaying by police of an innocent migrant worker, Jean Charles de Menezes, two weeks after the London bombings of 7 July 2005,[3] police officers supplying information to journalists for money,[4] the alleged character assassination of a senior Member of Parliament in September 2012[5] and the November 2012 arrest of five Kent Police detectives[6] on suspicion of persuading criminals to confess to crimes they did not commit have raised questions about police integrity and intentions. Writing in December 2012, Graeme Archer [2], winner of the United Kingdom's Orwell Prize for Political Blogging and self-confessed 'Tory Boy',[7] observed:

> What sickens most, I think, are the cover-ups, a sin of the police leadership as much as the rank and file. The lies about Jean Charles de Menezes in particular, coming so fast on the trauma of 7/7, will stay with me for life. These cover-ups can happen on an almost corporate level—think what we learned about Hillsborough. I never believed the

---

[1] Fearing some of its officers had committed perjury, misconduct in a public office and assault during the 18 June 1984 Orgreave confrontation (some of which was filmed by the news media) in 2012 South Yorkshire Police referred itself to the Independent Police Complaints Commission.

[2] In his 1999 report Lord Macpherson claimed the MPS's murder investigation had been hampered by what he termed 'institutional racism'.

[3] The 7 July 2005 London bombings (sometimes referred to as 7/7) killed fifty-two and injured over 700. The attacks were perpetrated by four home-grown Islamist terrorists. 'Rucksack bombs' were detonated on three tube (Metro) trains and a London bus.

[4] In 2013 Detective Chief Inspector (DCI) April Casburn of the Metropolitan Police Service received a sentence of 15 months' imprisonment for attempting to sell information to the *News of the World* (NOTW), a (now defunct) Sunday tabloid owned by Rupert Murdoch.

[5] The Metropolitan Police Service attributed comments to Andrew Mitchell, Conservative Party Chief Whip, that were either not made, or were made in a manner less threatening and less pompous than that claimed by officers on duty in Downing Street. Such was the initial furore over what the officers claimed Mitchell had said that he resigned his post as Chief Whip. The 'Plebgate' affair occurred against the background of a deteriorating relationship between the Police and Coalition Government over a 20 % cut to the Police budget. The release of CCTV images discredited the police account.

[6] Those arrested included a Detective Inspector.

[7] Traditionally British Conservatives ('Tories') have backed the Police.

'institutionalised racism' accusation, but were I a police officer, I'd be more concerned that my organisation could find itself accused of 'institutionalised lying'.

The chapter uses social theories of risk (for example Actor-Network Theory and theories pertaining to high-reliability, active learning and organisational culture)[8] to understand why and how those in positions of trust can act against what is commonly understood to be the public interest, and what costs—social and economic—are incurred.

2 Case Studies

2.1 Holloway Road

In January 2013 former London mayoral candidate and deputy chair of the police and crime committee Jenny Jones called for the disbandment of the Metropolitan Police Service's Territorial Support Group (TSG).[9] Jones claimed its hubristic attitude undermined community policing and public confidence:

> They think they are incredibly special. That generates a feeling that they can do things differently from other police officers. If you see police acting like some sort of paramilitary body I think it is bad for the police [3].

The TSG succeeded the Special Patrol Group (SPG). Established in 1961 the SPG provided "a centrally based mobile squad for combating particularly serious crime and other problems which could not be dealt with by local Divisions" [4]. As the decade progressed[10] the need for mobile units trained in crowd-control became more acute [5]. The London of the 1960s saw many anti-Vietnam War protests culminating in the violent Grosvenor Square riots of 1967–1968 [6]. *Life* correspondent Loudon Wainright offered this description of one:

> The crisp, late-afternoon light in Grosvenor Square was tinged with white from smoke bombs, and here and there out of the howling mob a bobby's blue-domed helmet sailed high into the air. It was a wild scene, weirdly anachronistic in the violent shuddering of huge banners and the surge of horses against a thicket of arms reaching upward to claw the

[8] There are two quite different approaches to measuring and mitigating risk: quantitative and qualitative. The latter recognises that risks are an emergent property of both technological *and* social arrangements.

[9] The TSG are charged with counter-terrorism, crime reduction and providing support to local units during public disorder (for example, during the English Riots of August 2011).

[10] The 1960s and 1970s saw significant unrest. If it coalesced around the Vietnam War in the 1960s, in the 1970s it coalesced around worker compensation and other employment issues. In 1972 nearly 24 million working-days were lost through stoppages. Britain was on its way to becoming 'The Sick Man of Europe'. Edward Heath's Conservative government put what it believed to be the obvious question: 'Who governs Britain?'.

The MPS deployed significant numbers of officers during the 2009 G20 protests at which newspaper vendor and bystander Ian Tomlinson died. The paramilitarism is noteworthy. Is this in the British tradition of 'policing by consent'?

> riders down. To see it was to watch a revolutionary mural come to life. Stones, ball bearings and showers of clods ripped from the moist London lawn flew against the lines of policemen, placard poles became spears, men shouted, fell and were trampled, young women with long hair and high boots fought with feet and fists and teeth to storm past the human barricades and blood ran down pale faces [7].

The SPG eventually fielded eight units. Each unit was made up of an inspector, three sergeants and thirty constables. The Special Patrol Group were distinguished by their mobility, esprit-de-corps, public-order training and distinctive transportation—the SPG used green Ford Transit vans at a time when most Metropolitan Police vans were dark blue. The cadre developed a reputation for toughness. A raid on SPG officers' lockers turned up "illegal truncheons, knives, two crowbars, a whip, a 3 ft wooden stave and a lead-weighted leather stick …. One officer was caught trying to hide a metal cosh …. Another officer was found with a collection of Nazi regalia" [8]. The SPG was disbanded in 1986.

According to McConville and Shepherd the "much-feared" SPG contributed to a 'crisis of legitimacy' vis-à-vis British policing in the 1970s and 1980s:

> That crisis centred on a system of policing that was increasingly seen to be: coercive in nature—most clearly dramatised in the accelerating deployment of riot-trained specialist squads such as the Special Patrol Group; lacking accountability—with the autonomy of chief constables a matter of concern, and a complaints system which lacked the confidence of both the public and police officers [9].

On the evening of 23 April 1979 at an anti-fascist demonstration in Southall, west London, a special-needs teacher named Blair Peach died from a single blow to the head. In 2010 three-thousand previously secret documents were made public. The documents "appeared to confirm the long-held suspicion that Peach was likely to have been killed by an officer from ... the Special Patrol Group" [10]. Commander John Cass, who conducted the investigation into Peach's death, concluded that the fatal blow was "almost certainly" dealt by one of six SPG officers. Some of these officers lied to protect the culprit's identity [10]. Addressing Cass's report in 2010 the Metropolitan Police Commissioner remarked: "I am sorry that officers behaved that way, according to Mr Cass" [10].

On 28 June 1979 the House of Commons discussed the performance and conduct of the Special Patrol Group [11]. Several MPs argued that the SPG did not enjoy the full confidence of the public. "Does the Minister accept that there is a tremendous gulf between the coloured community and the police generally, especially the Special Patrol Group?.... Unlike most Members of Parliament I personally witnessed the Special Patrol Group constantly in action at Grunwick.[11] I was appalled by what I saw. I abhor violence" remarked one MP. The debate continued: "Is it not important for the public to have confidence in all sections of the police force? Is it not a serious matter when the Special Patrol Group clearly does not enjoy that degree of confidence? Is the Home Secretary[12] aware that there is a strongly-held view that members of the Special Patrol Group have a hostile attitude towards the trade union and labour movement?" [11].

According to Rollo [12] the SPG were the *primary* cause of antipathy between ethnic minorities and the MPS. In the opinion of the National Council for Civil Liberties (NCCL)[13] police actions increased rather than reduced the chances of street-level confrontation [13]. The NCCL claimed elements of the police displayed racist behaviour. Following the death of Blair Peach at Southall the Home Secretary, William Whitelaw, promised to reform the SPG. Mooted changes

---

[11] The Grunwick dispute of the late 1970s saw the first deployment of the SPG to a confrontation between strikers and owners. There was violence on the picket-line. Five hundred and fifty arrests were made during the two-year strike. The strike failed, weakening the trades union movement. The course of the dispute set the tone for the Thatcher Years. For example, significant numbers of police were deployed during the Miners' Strike of 1984–1985. Prime Minister Thatcher, who referred to the strikers as "the enemy within", presented the strike as a trial of strength between parliamentary democracy and mob-rule. She encouraged the police to defend parliamentary democracy. Her death in April 2013 occasioned further assessment of her behaviour during the Strike. Speaking on the 13 September 2012 edition of BBC Radio Four's *Today* programme former Home Secretary Jack Straw commented: "The Thatcher government, because they needed the police to be a partisan force, particularly for the miners' strike and other industrial troubles, created a culture of impunity in the police service. They really were immune from outside influences and they thought they could rule the roost and that is what we absolutely saw in south Yorkshire [a reference to the South Yorkshire Police's behaviour after the Hillsborough disaster]".

[12] The British Cabinet Minister responsible for the police is called the Home Secretary.

[13] The NCCL was renamed Liberty.

included increasing the numbers of supervisory ranks, limiting the length of service to 4 years (presumably in an effort to prevent the development of an aberrant and potentially lethal SPG police sub-culture) and a more decentralised command and control structure to re-integrate the SPG with London's increasingly diverse communities [13]. Population mobility and globalisation ('transnationalism') was transforming London into a vibrant, multi-ethnic World City [14].

In the early evening of an August Saturday in 1983, patrolling SPG units were subjected to some shouted abuse by a group of teenagers.[14] Later in the evening four members of one of the SPG units (Carrier N33) assaulted a different group of teenagers[15] just off the less-than-salubrious Holloway Road in north London:

> The [five] boys were not arrested or accused of any offence [by the SPG]. One of them was kneed in the face and had his nose broken. Another was punched in the stomach and kicked in the face. A third was also punched and kicked in the face, and a 13-year-old boy was hit with a truncheon. All the boys were treated in hospital [15].

One boy testified he had seen police officers shouting and banging their truncheons when their Transit passed a group of black teenagers. Some minutes later a Transit pulled up alongside him and he was assaulted. Another boy claimed he had been held by the neck, punched in the eye and repeatedly kicked in the legs. A young girl said she had witnessed police officers using their truncheons on the youths.

Admitting the assaults had taken place the Metropolitan Police Service paid £5,000 damages to three of the youths. According to *Police Review* "concern over the Holloway case… provoked almost daily comment in newspapers and on the radio and TV" [16].

Scotland Yard's Complaints Investigation Bureau (CIB) probed the assaults for two-and-a-half years. Despite the investigation generating seventy-five statements (from officers, victims and witnesses) [17] in February 1986 the Police Complaints Authority (PCA) claimed it did not have enough evidence to charge anyone… or even to institute disciplinary proceedings. The PCA did, however, ask the MPS to parade the thirty officers on duty on the evening in question to warn them against unlawful behaviour. According to the PCA:

> They were told [by a Deputy Assistant Commissioner] in no uncertain terms of the anger and disquiet felt about the incident. They were told that, although the officers in only one of the vans were involved, all the officers in that van [Carrier N33] must have known what happened [15].

How did the officers of Carrier N33 behave *post* the assaults? In an episode reminiscent of the conspiracy scene in the movie *Serpico*[16] where corrupt officers

---

[14] There were three SPG Transits patrolling the area on the evening in question.

[15] Caucasians and Afro-Caribbeans.

[16] Frank Serpico worked undercover for the New York Police Department (NYPD) at a time when corruption was rife. In 1967 he provided credible evidence of police corruption. No action was taken. Serpico sensed his life might be in danger. With only one ally, another police officer, he went to the *New York Times*. The Mayor of New York established the Knapp Commission to

congregate in a park to persuade the protagonist to participate in a police racket, the officers of Carrier N33 went to a park "to agree to say nothing about the assaults to anyone else" [15]. Carrier N33's sergeant "... said the purpose of the meeting in the park, which all the officers under his command had attended, was to discuss the investigation. He had advised his officers that they had a choice between saying nothing at all to the investigators or telling all that they knew" [15]. Carrier N33's driver lied about the incident until February 1986 when he confessed to senior officers.

On 7 February 1986 in a now-famous editorial titled 'A conspiracy of bastards' *Police Review's* Brian Hilliard[17] claimed the conspiracy of silence was damaging the MPS. On 21 February 1986 *Police Review* ran the headline 'Holloway assaults: immunity likely for police witness'. At a Scotland Yard Press Conference on the same day the MPS announced the setting up of a 'hot-line'. Four calls were received from officers willing to give evidence. The trial of constables Edward Main, Michael Gavin, Nicholas Wise and Michael Parr commenced on June 16 1986 at London's Central Criminal Court. The constables entered pleas of not guilty to charges of actual bodily harm (ABH) and conspiracy (with their sergeant) to pervert the course of justice. In his summing-up the judge observed: "You find yourself almost knee-deep in lies being told by police officers" [15]. Wise, Gavin and Main were found guilty of assault. The sergeant was found guilty of misconduct. The prosecutions vindicated the tough stance taken by *Police Review's* editor, Brian Hilliard. Hilliard had determined that honest police officers should not be dragged into the gutter by a small minority. In his 7 February 1986 editorial 'A conspiracy of bastards' Hilliard wondered whether SPG officers' conduct had been shaped by MPS culture:

> One may well speculate on the quality of supervision which permits constables to work in a climate in which they feel entitled to administer summary injustice.[18]

The officers' behaviour is all the more remarkable when considered against the furore over an earlier police assault case. A few months before the Holloway Road incident an Islington-based constable named Brian Renton had knocked out the

(Footnote 16 continued)
investigate corruption. Serpico testified to the Knapp Commission in October and December 1971. He was the first NYPD officer to go public about police corruption. In 1973 De Laurentiis released a Sidney Lumet-directed movie about Serpico's NYPD career. Before testifying, Serpico was shot in the face in a drugs raid. Fragments entered his brain. He believed he was set up by the two NYPD officers who accompanied him on the raid. It was not Serpico's colleagues who called for an ambulance, but a resident.

[17] Brian Hilliard, editor of *Police Review*, died in 2010. A former police officer, Hilliard campaigned tirelessly for justice. For the sake of honest police officers and the public Hilliard drew attention to the dark underbelly of the British constabulary—the minority of officers whose behaviour undermined the good work done by the majority. Through his integrity and courage Hilliard showed that journalism can be a force for good.

[18] Employees' perceptions and behaviour are subject to numerous influences, including the perceptions and behaviour of those at the top of the organisation.

eye of a man he had arrested. The officer was sent down for 2 years. The MPS agreed damages of *circa* £200,000 with the victim. *Police Review* reported:

> Senior officers are known to be particularly concerned that the Holloway assault, in which one of the suspects was serving at Islington police station, took place within a few hundred yards of the restaurant in the Renton case and within 3 months of that incident. The officers concerned in the Holloway assaults, therefore, were certainly aware of the restaurant attack and of Renton having been suspended from duty and charged [16].

The Miners' Strike of 1984–1985 saw ugly confrontations between police officers and strikers. Miners were charged by police on horseback. Police officers were assaulted by miners. Determined to counter mass-picketing the Police interdicted flying pickets. One National Union of Mineworkers (NUM) official (who later became a Member of Parliament) claimed the SPG had been deployed to the coalfields. Established to deal with public disorder but mired in allegations of ill-discipline and racism the SPG disappeared shortly after the strike ended.

2.2 Hillsborough

On April 15 1989 a Football Association Challenge Cup Semi-Final between Nottingham Forest and Liverpool was staged at the Hillsborough football stadium in Sheffield, South Yorkshire, England. Like many English football grounds, Hillsborough was an old stadium that had been remodelled and refurbished over the years. Built in 1899, it was situated in a densely populated suburb that made crowd-management difficult. Further, by 1989 the layout of the ground reflected an aggressive crowd-control design paradigm[19] that included high fences (2.5 m at Hillsborough) to separate rival fans and prevent pitch invasions (the top-sections of the pitch-side fences were angled to discourage climbing) and two CCTV-informed control rooms, one for the ground staff and another for South Yorkshire Police (SYP). It is reasonable to conclude that the embedded crowd-control paradigm influenced the perceptions, thinking, tactics and behaviour of both Sheffield Wednesday's stewards and SYP Officers. It may also have influenced the perceptions of those not directly involved, including journalists: "[Hooliganism] was such an easy, lazy narrative, swallowed because of the preoccupations of that era … and because Heysel was not such a distant memory" [18].[20] The paradigm's inception can be traced to the 1977 McElhone investigation *Report of the Working Group on Football Crowd Behaviour*. In the 1970s and 1980s there was a

[19] Paradigms shape perceptions and influence behaviour. They suggest what should be noticed and what discounted.

[20] Thirty-nine Juventus fans were killed in the 1985 Heysel Stadium disaster when a retaining wall collapsed. The fans were running from a group of Liverpool supporters. The Liverpool supporters had scaled a fence that separated them from their rivals. Six hundred fans were injured. English football clubs were banned from European competition until 1990. The ban on Liverpool FC's participation ran for another year. Fourteen Liverpool fans were convicted of involuntary manslaughter.

The Miners' Strike 1984–1985 ranged police against miners in an often violent dispute. Theoretically the National Coal Board (which ran the mines) was owned by the British public

perception that crowd behaviour was poor and that remedial measures (like membership schemes and perimeter fences) were needed. The Prime Minister of the day, Margaret Thatcher, was determined to eradicate football hooliganism.

The ground's topography added to its latent risks. Specifically, supporters accessed the West Stand's pens 3 and 4 (the loci of the disaster) via a downward-sloping tunnel. Although there was room in the pens to either side (2 and 5) supporters congregated in 3 and 4. Egress was via a small gate onto the pitch (all pitch-side gates were locked) or back through the tunnel. The tunnel's downward slope might have added impetus to the crowd. The initial surge happened when SYP Officers opened some exit gates to relieve crushing on the concourse and at the turnstiles. A SYP Officer remarked that if the gates had not been opened "[S]omeone will get killed". Crowd-crush incidents were not unknown at Hillsborough. At the 1981 FA Cup Semi-Final between Wolverhampton Wanderers and Tottenham Hotspur there was crushing on the concourse, at the turnstiles and on the terracing. Lessons appear not to have been adequately learned. The match kicked-off. The crushing worsened. Perceived initially as crowd disorder, nothing was done. Eventually both the stewards and SYP Officers realised they were watching a disaster unfold. By that time it was too late for 96 supporters. In the immediate aftermath of the disaster SYP constructed a biased account of the events of 15 April, 1989.

First, the force altered officers' written accounts: "Some 116 of the 164 [police] statements identified for substantive amendment were amended to remove or alter comments unfavourable to SYP" [1]. The 2012 Hillsborough Independent Panel Report noted: "[C]orrespondence between SYP and the Force solicitors

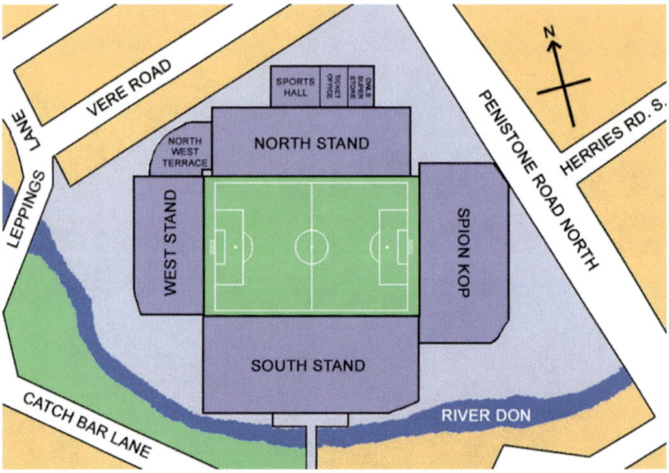

Hillsborough football ground. Stadia footprints are constrained by local features like housing, roads, railways and rivers. Narrow streets and access points induce congestion

[Hammond Suddards] reveals that comments within officers' statements 'unhelpful to the Force's case' were altered, deleted or qualified .... [O]fficers were discouraged from making criticisms of senior officers' responses ... 'key' words and descriptions such as 'chaotic' were counselled against and, if included, were deleted" [1]. Redacted statements included:

- Basically it was chaos
- [Police numbers] appeared to be a bit thin on the ground for the numbers of people involved
- It seemed very bad that only one in our serial—the sergeant—should have a personal radio. We had great difficulty in finding out what ... was happening and for too long a time we were basically working in the dark [19].

According to the Hillsborough Independent Panel, senior officers' efforts to construct a biased account of the disaster did not receive the unanimous support of street-level officers:

> A significant number of SYP officers were uncomfortable with the methodology adopted in reviewing and altering their initial accounts and with the role of the SYP solicitors .... One officer stated he had accepted the changes only because he was suffering from depression and post-traumatic stress. He considered it an 'injustice for statements to have been "doctored" to suit the management of South Yorkshire Police'. Another officer had accepted the process, but had not realised how much of his statement had been removed [1].

The constabulary were not the only agents of the State to engage in revisionism and construction. The South Yorkshire Metropolitan Ambulance Service (SYMAS) used the same tactic when reviewing employee statements. While Lord Justice Taylor, chair of the public inquiry into Hillsborough, declared there was no reason to remove opinion, he nevertheless accepted SYP's final versions of officers' statements.

Secondly, SYP co-ordinated efforts to impugn the reputation of Liverpool FC supporters. A Sheffield-based press agency called Whites acted as the conduit for misinformation. In 2012 one Liverpool MP suggested SYP may have operated a 'black propaganda' unit [20]. The 19 April 1989 edition of *The Sun*[21] daily newspaper (a tabloid with a right-of-centre editorial bias) carried the front-page banner headline 'The Truth'. There were three sub-headlines: "Some fans picked pockets of victims"; "Some fans urinated on the brave cops"; "Some fans beat up PCs giving the kiss of life". The story ran:

> Drunken Liverpool fans viciously attacked rescue workers as they tried to revive victims of the Hillsborough soccer disaster, it was revealed last night. Police officers, firemen and ambulance crew were punched, kicked and urinated upon by a hooligan element in the crowd. Some thugs rifled the pockets of injured fans as they were stretchered out unconscious on the pitch. Sheffield MP Irvine Patnick [Conservative] revealed that in one shameful episode a gang of Liverpool fans noticed that the blouse of a girl trampled to death had risen above her breasts. As a policeman struggled in vain to revive her, the mob jeered: 'Throw her up here and we will **** her' [fuck her] .... One furious policeman who witnessed Saturday's carnage stormed: 'As we struggled in appalling conditions to save lives, fans standing further up the terrace were openly urinating on us and the bodies of the dead' [21].

The article quoted one 'high-ranking' police officer as saying: "The fans were just acting like animals. My men faced a double hell—the disaster and the fury of the fans who attacked us".

THE Sun

THE TRUTH

Tycoon Adnan thrown in jail

GATES OF HELL

•Some fans picked pockets of victims

•Some fans urinated on the brave cops

•Some fans beat up PC giving kiss of life

DI GRIEVES FOR LEE, AGED 14: Pages 2 and 3

---

[21] The newspaper's editor at this time was Kelvin MacKenzie.

Reacting to what they perceived to be biased reporting, some Liverpudlians agitated for a boycott on *The Sun* newspaper (see weather-beaten poster below). To this day some Merseyside newsagents refuse to sell *The Sun*.

Many newspapers regurgitated what they were fed. On the same day that *The Sun* printed its 'The Truth' banner the *Daily Telegraph* went with 'Police tell MP of attacks on them as they helped injured'; the *Sheffield Star* with 'Fans "made sex jibes at body"'; the *Daily Express* with 'Police Accuse Drunken Fans: Police saw "sick spectacle of pilfering from the dying"'; the *Daily Mail* with 'They were drunk and violent and their actions were vile'; and the *Daily Star* with 'Dead Fans Robbed by Drunk Fans'.[22] The *Daily Mirror* (a tabloid with a left-of-centre editorial bias) bucked the trend and went with: 'Nightmare that will never leave me: The moment fans told me: "Steve, they're being crushed"' [1].

[22] To mark the ninth anniversary of the disaster Liverpool University's Football Industry Group surveyed the attitudes of 1,350 members of the public. Nearly 60 % of those polled in Sheffield blamed Liverpool supporters for the disaster. Nearly 90 % of those polled in Liverpool blamed SYP. In both Liverpool and Sheffield the vast majority had never changed their view of culpability.

Daily Express

WIN A FABULOUS FORD FIESTA PAGE THREE

Officers 'saw sick spectacle of pilfering from the dying'

POLICE ACCUSE DRUNKEN FANS

THE HORROR OF HILLSBOROUGH

Decision

No way! Mr Alway's a loser in love

Lights

DAILY Mirror

An amazing story of miracles, mercy and courage FROM THE INSIDE

NIGHTMARE THAT WILL NEVER LEAVE ME

Leigh, 14, is soccer horror victim No. 95

The moment fans told me: Steve, they're being crushed

STEVE McMAHON'S AGONY

See Centre Pages

The Hillsborough Independent Panel Report noted: "[F]rom the mass of documents, television and CCTV coverage disclosed to the Panel there is no evidence to support these allegations other than a few isolated examples of aggressive or verbally abusive behaviour clearly reflecting frustration and desperation" [1]. The panel concluded that the falsehoods originated with "... a local Sheffield press agency [Whites Press Agency] informed by several SYP Officers, an SYP Police Federation spokesperson and a local MP" and that the Federation was given "informal support" by the SYP Chief Constable (the highest-ranking SYP officer) [1]. According to Herbert [22] the SYP Officers' "concocted stories to the agency [Whites] were so severe that some were 'watered down' before they were filed".

In a 1998 letter to the Hillsborough Family Support Group football fan John Barry claimed one of his fellow-students on the Sheffield Business School MBA programme, a "middle-ranking" police officer, had told him over a drink that he was engaged in black propaganda:

> He told me that he had been asked by his senior officers to put together the South Yorkshire Police evidence for the forthcoming inquiry. He said that 'we are trying to concoct a story that all the Liverpool fans were drunk and we were afraid that they were going to break down the gates so we decided to open them' [23].

In October 2012 that police officer, by now Chief Constable of West Yorkshire, resigned. In his resignation statement Sir Norman Bettison denied any wrongdoing: "The suggestion that I would say to a passing acquaintance that I was deployed as part of a team tasked to 'concoct a false story of what happened', is ... wrong" [24]. When the Hillsborough Independent Panel published its report, Margaret Aspinall, chairperson of the Hillsborough Family Support Group, said this about the behaviour of South Yorkshire Police *post* Britain's worst sporting disaster: "Without the truth you cannot grieve and where there is deceit you get no justice" [25]. *The Independent's* Ian Herbert observed of the Hillsborough saga: "[L]ies can be halfway around the world before the truth has its trousers on" [18]. If we take just one lesson from Hillsborough it is that we would be wise to treat official accounts with scepticism.[23] Triangulate and the truth will emerge.

3 Analysis: Theoretical Underpinnings

3.1 Actor-Network Theory

Actor-network theory (ANT) provides a way of understanding complex systems (like health care, power generation, the military-industrial complex or commercial aviation) or complex activities (for example, the development and maintenance of a narrative (story), world-view or belief system). ANT makes no value judgments. The methodology is amoral, Machiavellian even. It generates an unembellished factual record. ANT posits that systems—physical, bureaucratic or ideological—emerge from a process of 'heterogeneous engineering' [26].

[23] The *modus operandi* of academic inquiry is methodical scepticism. However authoritative its origin, no one account is privileged over any other.

The Hillsborough Memorial at Liverpool FC's Anfield stadium

A system consists of numerous, more-or-less-aligned human and non-human (tangible and intangible) actants [27, 28]. A system is a *hybrid collectif*. The closer actants' alignment (the more in-step the actants) the more resilient and effective the system (physical, bureaucratic or ideational) [29, 30]. ANT acknowledges reality's contiguousness (connectedness). It acknowledges enrolled ('translated') actants' interdependence and mutual shaping (affective interaction). It recognises humanity's restlessness—its tendency to *purposeful action*:

> The theory's aim is to describe a society of humans and non-humans as equal actors tied together into networks built… to achieve a particular goal, for example the development of a product [like an investment bond, air service or story] [29].

Viewed through the lens of ANT human enterprise involves the purposeful assembly ('translation') of human and non-human elements[24] (core employees, consultants, auditors, lawyers, landlords, educators, research laboratories, hardware, real estate, finance markets, professional associations, rules, laws, strategic plans, organisational cultures, histories, accounts, beliefs, values, etc.) to achieve some preconceived goal, like the manufacture of motor cars, extraction of oil from under the sea, government of a country, establishment of a dominant narrative/story, reputation defencc or maintenance of privilege. Actor-networks require maintenance. They can weaken as actants resist translation and/or migrate to other networks.

---

[24] Heterogeneous engineering.

3.2 Isomorphic Learning[25]

Disasters are quite rare events. Hyperbolic media reporting creates a false picture of their frequency: "The Asian tsunami, the flooding of New Orleans and now the Kashmir earthquake—it often seems like the world is about to end.... It is important to keep things in perspective, however.... Relatively few of us will experience the trauma of a natural or man-made disaster. Most of us will die peacefully in our own bed (or one loaned to us by a hospital or hospice)" [31]. Because disasters are so rare it is important that all useful information is made immediately available.

Isomorphic learning—the capacity to learn from incidents and accidents in companies or industries other than one's own—is only possible when information is shared: "[F]or an industry to be able to learn from the experience of managing… risks, individual organisations need to be able to learn from the experiences of each other" [32]. Further, for benefits to accrue, that information must be accurate. Mitigative measures based on inaccurate information may not work and are wasteful of scarce resources (money, personnel, hardware, time, etc.). There are two types of learning: passive and active. In the former, awareness does not lead to action. In the latter it does.

Japan's coastline has been threatened by tsunamis throughout history. An earthquake in 1896 generated a tsunami 25 m high. An earthquake in 1993 sparked a tsunami that ranged in height from 10 to 20 m. The Japanese and Russian coastlines were impacted. Over 200 people died—the largest tsunami-related death-toll in 50 years [33]. On 11 March 2011 a 14 metre-high tsunami inundated the Fukushima Daiichi nuclear power plant on the north-eastern coast of Japan. The seawater overwhelmed the plant's generators. With backup power off-line the reactors overheated. Meltdowns occurred in three of the plant's six boiling-water reactors. A large-scale civilian evacuation was ordered. According to media reports the 40-year-old plant's design assumed a worst-case 5.7 m-high tsunami. A 2008 internal report claimed the plant could face a tsunami of up to 10.2 m. Managers sidelined the report [34]. There were no additional mitigative measures. There was no active learning. The Fukushima Daiichi cleanup will cost *circa* $100 billion and take 40 years.

3.3 Highly-Reliable Organisations

The question of whether or not accidents can be prevented divides theorists. Those like Professor Charles Perrow argue that accidents in complex, tightly-coupled

---

[25] Britain's Professor Brian Toft developed the theory of isomorphic learning in journal papers like 'The failure of hindsight' (*Disaster Prevention and Management*, 1992) and books like *Learning From Disasters: A Management Approach* (Perpetuity Press, 1997).

systems are all but inevitable. Those like the University of California at Berkeley's Todd LaPorte, Gene Rochlin and Karlene Roberts argue that accidents can be prevented with the right organisational structures, practices and culture. Normal accident theory (NAT) articulates the view that accidents are inevitable; High-reliability organisation (HRO) theory the view that they are preventable [35, 36]. Mason offers this definition of the high-reliability organisation:

> Successful cultures become susceptible to hubris and carelessness. One antidote for organisational hubris is the highly reliable organisation (HRO) model, based on the concept of mindfulness. These organisations are constantly aware of the possibility of failure, appreciate the complexity of the world they face, concentrate on day-to-day operations and the little things, respond quickly to incipient problems and accord deep respect to the expertise of their members. They value knowledge and expertise highly, *communicate openly and transparently, and avoid concentrations of power or corruption by setting up independent units with countervailing powers* [my emphasis] [37].

The concept of mindfulness—effectively a heightened state of awareness and constructive self-critique—draws on the work of Karl Weick [38, 39].

3.4 Organisational Culture

Used in the context of social theories of risk, the term culture means 'the way we do things around here'.[26] Organisational culture is multi-layered [40], hard to map and difficult to manipulate:

> Organisational culture is seldom monolithic. Organisations often consist of numerous sub-cultures, constituted in part through workers' shared interests, beliefs, skills and profession. Other sources of cultural cleavage within organisations include class, gender and work location .... Occupational groups with a strong self-identity [for example police officers] may develop an organisationally transcendent supra-culture [41].

Subcultures "complicate culture management" and may generate "inconsistencies or conflicts" [42]. Sub-cultures spawn local norms that may be considered aberrant. Bruce [43] described one mental institution where "patients were managed within two quite different frameworks. The consultants classified patients according to formal diagnostic schema ... [b]ut the nurses ... had a much simpler system that reflected their working concerns. They labelled patients as 'wetters' or 'wanderers'". While the crude labelling may have made the nurses' work easier, the resulting depersonalization hardly served to dignify their charges or profession.

There are two schools of thought on the possibility of engineering cultural change. Functionalists claim culture can be re-engineered through, for example, exhortation, sloganeering and training. Because they believe culture to be an emergent property of social groupings and organisational forms, Interpretavists

[26] This is a slight variation on the McKinsey organisation's original definition of corporate culture: "[Corporate culture is] how we do things around here".

doubt that culture can be re-engineered. In *All Together Now* British entrepreneur Sir John Harvey-Jones described how organisations frustrate reform:

> Large organisations have been built on rigid hierarchical models which make team working difficult. People are difficult to involve if they are submerged by layers of supervisors, each of whom acts as a filter, adjusting the message to what they think their superiors will listen to. Large organisations have... many buffers between decisions and results [44].

4 Analysis: A Systems Perspective

4.1 Actor-Network Theory

The Hillsborough Injustice went unpunished for over two decades. Actor-network Theory explains why: South Yorkshire Police created a heterogeneous and extensive actor-network to support its version of events. The *hybrid collectif's* heterogeneity gave it stability:

> Heterogeneity is [a] central aspect of a stable network. The more the diverse elements are interrelated, the more... stable a network becomes, because each element is kept in place by a number of elements, each one concerned with a different aspect of the element which is kept in place. In order to disconnect an actor from a network, many connections have to be untied... [29].

South Yorkshire Police translated actants to produce and sustain a false account of the events of 15 April, 1989. It cultivated this network-of-deception regardless of the considerable emotional costs borne by victims' relatives and friends. Actants included:

- A Member of Parliament
- The Police Federation
- A press agency
- The local and national press
- A deviant sub-culture amongst some South Yorkshire Police officers that sanctioned the adulteration of police statements to meet some predetermined, selfish goal. This sub-culture manifested in the days following the Orgreave confrontation when police statements were adulterated to incriminate picketing miners. Ninety-five miners were prosecuted. All walked free: "[T]he prosecution withdrew after the police's oral and written evidence... had been discredited. Each prosecution was supported by two police officers making near-identical statements. One [officer] admitted... that sections of his statement had been dictated by a plainclothes Officer.... One officer's signature was analysed and found not to have been in his handwriting. Michael Mansfield QC, who represented three acquitted miners, described South Yorkshire police's evidence then as 'the biggest frame-up ever'" [45]

- Negative perceptions of Liverpool FC's fans *post* Heysel
- Negative perceptions of football fans generally. Fans were stereotyped as drunken, ill-disciplined louts more interested in casual violence than association football
- Perceptions of the City of Liverpool as a locus for civil unrest. The July 1981 Toxteth riots lasted for 8 days. Sparked by a police arrest, the riots saw seventy buildings burned to the ground or demolished and 460 police officers injured
- Perceptions of the City of Liverpool as a repository of failed, old-style 'windbag' socialism and rust-belt industries. The right-wing press framed the city as an anachronistic drag on a booming Thatcher's Britain. At least one Cabinet minister shared this view. Following Toxteth Mrs Thatcher's Chancellor, Sir Geoffrey Howe, suggested Liverpool be subject to a policy of 'managed decline'. Howe wrote to his Prime Minister: "I fear that Merseyside is going to be much the hardest nut to crack.... We do not want to find ourselves concentrating all the limited cash that may have to be made available into Liverpool and having nothing left for possibly more promising areas such as the West Midlands or, even, the North East.... It would be even more regrettable if some of the brighter ideas for renewing economic activity were to be sown only on relatively stony ground on the banks of the Mersey.... I cannot help feeling that the option of managed decline is one which we should not forget altogether. We must not expend all our limited resources in trying to make water flow uphill". Conscious of the incendiary nature of his comments he cautioned: "This is not a term for use, even privately" [46].
- The authoritarian character of the government of the day (Mrs Thatcher's visceral dislike of hooliganism and determination to eradicate it was well-known)
- The pro-police stance of the government of the day: "Within days of taking office in May 1979 the Tories honoured their election promises to 'uphold the rule of law' by granting substantial pay rises to the police... [this helped prepare] the ground for controlling working-class action" [12]. Ex-Home Secretary Jack Straw has remarked: "The Thatcher government... created a culture of impunity in the police service. They really were immune from outside influences and they thought they could rule the roost".
- The location of the Hillsborough stadium that made it difficult to monitor and control approaching crowds (see illustration)
- The latent errors/resident pathogens inherent in the design of the stadium (for example, the downward slope of the tunnel that led to the terraces; crush barriers that could buckle; pitch-side gates that were not only locked but were too small to allow a rate of egress onto the pitch sufficient to relieve crushing in the pens; a pitch-side fence that was angled in such a way that even the most agile could not climb it to escape).

These actants were translated by South Yorkshire Police and the local branch of the Police Federation to produce a *hybrid collectif* that supported their preferred narrative (that ill-disciplined and drunken fans contributed to the disaster and that SYP was a victim of circumstance). Significant and sustained press reporting strengthened the network and embedded SYP's narrative in the public consciousness.

In 2009 Gordon Brown's Labour government excepted official documents pertaining to the 1989 disaster from Britain's Thirty Year Rule.[27] Officers' doctored statements were key actants in the SYP/Police Federation actor-network. Exposed as partial fabrications these statements could no longer play their assigned role in the actor-network. As more and more actants resisted translation and deserted (sometimes to the rival Hillsborough Independent Panel actor-network) so the SYP/Police Federation actor-network crumbled. Kelvin MacKenzie's fulsome apology for his 19 April 1989 'The Truth' front-page was but one example of a SYP/Police Federation actor-network actant resisting translation. Publication of the HIP Report triggered an avalanche of desertions from the SYP/ Police Federation actor-network. Apologies came thick and fast. Even the South Yorkshire Police Federation apologised.

Of all the countervailing voices, the Hillsborough Independent Panel—with unrestricted access to official records—posed the most serious challenge to the SYP/Police Federation narrative. Those who knew the truth about Hillsborough probably realised the writing was on the wall the moment the Thirty Year Rule was waived and the Panel established. These two actions prevented the dominant translation of the events of 15 April 1989 from, to borrow a phrase from Donnelly [47], 'shaping and determining future translations'.

As actants resisted translation there was a 'rush to apology'. South Yorkshire's Chief Constable apologised to both victims' relatives and Liverpool fans generally. South Yorkshire Police Federation issued a statement:

> The South Yorkshire Police Federation fully endorses the apology provided by the Chief Constable, David Crompton, to the families of those who sadly lost their lives in the Hillsborough disaster [48].

South Yorkshire Police Federation claimed: "South Yorkshire Police is a very different organisation to that of 1989, as is the police service generally" [48].[28] The Prime Minister apologised. "I am profoundly sorry for the way the force failed on 15th April 1989 and I am doubly sorry for the injustice that followed and I apologise to the families of the ninety-six and Liverpool fans" said Cameron. Mayor of London Boris Johnson apologised for a 2004 *Spectator* article that repeated some of the allegations made against Liverpool supporters at Hillsborough. "I'm glad that this independent report has finally nailed the myth that drunken fans were in any way responsible for the deaths of 96 people. That was a lie that unfortunately and very, very regrettably got picked up in a leader in the

---

[27] Under the rule State documents must remain secret for three decades.

[28] Consider this Police Federation claim in relation to recent lapses in standards of conduct, including the alleged character assassination by Downing Street police officers of a senior politician, the imprisonment of a Detective Chief Inspector for attempting to sell information to a low-brow right-wing tabloid and the 27 March 2013 imprisonment of two ex-police officers for selling information to journalists. One received ten months the other 2 years.

Spectator in 2004, which I was then editing" said Johnson. The Chairman of the Football Association apologised. Even Kelvin MacKenzie, the *Sun* editor responsible for the 19 April 1989 banner headline 'The Truth' that so angered Liverpudlians, apologised to victims' relatives. MacKenzie said he had been "totally misled" by authority figures:

> Twenty-three years ago I was handed a piece of copy from a reputable news agency in Sheffield in which a senior police officer and a senior local MP were making serious allegations against fans in the stadium. I had absolutely no reason to believe that these authority figures would lie and deceive over such a disaster. As the Prime Minister has made clear these allegations were wholly untrue and were part of a concerted plot by police officers to discredit the supporters thereby shifting the blame for the tragedy from themselves. It has taken more than two decades… and a 2-year inquiry to discover to my horror that it would have been far more accurate had I written the headline The Lies rather than The Truth. I published in good faith and I am sorry that it was so wrong [49].

Of course, in publishing the original story MacKenzie forgot one of the rules of good journalism—where possible, corroborate. Running a story on the strength of a single press release is risky. The Whites press release was an effective actant in the SYP/Police Federation actor-network. It translated the *Sun* newspaper thereby expanding the *hybrid collectif*. To borrow a phrase from Callon and Latour [27] for over two decades the Whites press release 'bent space around itself'. Texts act.

Viewed through the lens of actor-network theory the sergeant whose officers perpetrated the Holloway Road assaults was the prime mover in promoting an actor-network whose output was, effectively, a conspiracy of silence. The thirty or so SPG officers directly or indirectly associated with the incident were translated in the hope that they would not implicate colleagues. Key moments in the life of the SPG's actor-network included the meeting in the park. Actants included the thirty officers and the SPG's distinctive sub-culture that emphasised collegiality and mutual support. Other actants included the Conservative government's pro-police stance, perceptions of the inner-city (and especially such neighbourhoods as Holloway Road) as lawless, negative perceptions of teenagers, racist attitudes and the willingness of some SPG officers to use unreasonable force (perhaps involving the use of unauthorised weapons like crowbars or staves).

Actor-networks are subject to destabilising environmental forces. Hilliard's February 1986 publication of the 'Conspiracy of bastards' editorial, the MPS offer of immunity, the setting up of the hot-line and telephone calls from four officers acted to destabilise the network. In making the calls the officers resisted translation. In time they were translated by the prosecution. Defections significantly reduced the capacity of the SPG's network to 'shape and determine future translations'.

4.2 Isomorphic learning

Isomorphic learning is possible only if certain preconditions are met. Amongst the most important are that information on incidents and accidents must be:

Accurate
Complete
Timely

For over two decades the information pertaining to the onset and management of the Hillsborough disaster was neither accurate nor complete. For example, according to HIP, drunkenness played little or no part in the aetiology of the disaster:

> There was no evidence to support the proposition that alcohol played any part in the genesis of the disaster and it is regrettable that those in positions of responsibility created and promoted a portrayal of drunkenness as contributing to the occurrence of the disaster and the ensuing loss of life without substantiating evidence [1].

On the other hand, something that was largely overlooked—the inability of South Yorkshire Police officers to correctly interpret the crowd's comments (pleas for help and action) and behaviour and deploy resources in the most effective manner possible—did play a part in the aetiology of the disaster. References to the confused state of the police response were either removed or edited to show SYP in a better light. Initially SYP officers, perhaps conscious of previous incidents like Heysel, used a hooliganism paradigm to interpret the crowd's comments and behaviour. With the unfolding disaster framed in this way it is unsurprising that officers were slow to act. The fact that it took over two decades for the truth to be placed on public record meant that lessons that could have been learned about crowd-control, organisational cultures and sub-cultures, intelligence-gathering, interpretive paradigms, intra and inter-service communication and co-ordination and post-incident investigation were not learned. Because of this failure it is possible that lives were needlessly lost both in the United Kingdom and elsewhere. That is a major indictment of South Yorkshire Constabulary and South Yorkshire Metropolitan Ambulance Service. When considering statements produced by the emergency services, including the police, one would be well-advised to action the ABC mantra used by crime-scene investigators: Assume nothing; Believe nobody; Check everything.

It is worth noting there are barriers to isomorphic/active learning other than the deliberate construction of a distorted narrative account by those with a vested interest. "[I]n the real world our capacity to learn and apply the lessons of the past may be compromised by interceding social, economic and political dynamics" writes Bennett [50]. Interceding dynamics include:

- Lack of political will
- Lack of funds
- Lack of expertise
- De-sensitisation to the consequences of incident and accident
- A feeling that to do something would be to admit culpability
- The media spotlight shifting to other issues
- Fatalism ('what will be will be')
- Complacency

4.3 Highly-Reliable Organisations

Highly-reliable organisations (HROs) "… value knowledge and expertise highly, communicate openly and transparently, and avoid concentrations of power or corruption by setting up independent units with countervailing powers" [37]. It is informative to analyse SYP's behaviour in relation to each of these HRO requirements:

HROs 'value knowledge'

The South Yorkshire Police Force's self-interested construction of a misleading account of the events of 15 April 1989 proves that the SYP of the 1980s did not appreciate the importance of knowledge.

HROs 'communicate openly and transparently'

The South Yorkshire Police Force's communication of information to actants may have been open, but much of it was misinformation. The SYP actor-network was in part built on lies.

HROs 'avoid… corruption by setting up independent units with countervailing powers'

The deception perpetrated by SYP appears to have been sanctioned by the Chief Constable. That is, it was sustained from the very top of the organisation: "[T]he SYP Police Federation, supported informally by the SYP Chief Constable, sought to develop and publicise a version of events that focused on several police officers' allegations of drunkenness, ticketlessness and violence among a large number of Liverpool fans. This extended beyond the media to Parliament" [1]. It is unlikely that independent units could or would achieve much in an organisation that was rotting from the top. This was Frank Serpico's experience at NYPD. Because of NYPD's seeming indifference to police corruption, Serpico sought allies and redress outside the force. He took his story to the *New York Times* (a key actant in Serpico's anti-corruption actor-network). It worked, but at a considerable cost to the under-cover officer [51].

Given the above it is reasonable to conclude that the SYP of the 1980s could not be described as a highly-reliable organisation. Any organisation that ignores employees' systematic lying can only be described as a highly unreliable organisation (HUO). A HUO is an organisation undeserving of public trust, funding and respect.

4.4 Organisational Culture

Organisational culture may be defined as 'the way we do things around here'. Given that officers' statements pertaining to both Orgreave and Hillsborough were doctored, it is clear the South Yorkshire force of the 1980s harboured a sub-culture that considered the falsification of evidence an acceptable expedient. As mentioned earlier, organisations consist of numerous sub-cultures [40]. In this case certain officers elevated the securing of prosecutions and cultivation of a reputation for competence above all other considerations—including the need to tell the truth. Most worrying, perhaps, was that in the case of the Hillsborough cover-up the Chief Constable appeared complicit [1].

The behaviour of the South Yorkshire force in regard to Orgreave and Hillsborough raises several questions. First, if officers could lie about Orgreave and Hillsborough, how many other false accounts have been produced? How many people who should have walked free were put behind bars on the strength of adulterated police statements? Secondly, how did this deviant sub-culture come to life? Thirdly, what sustained it? Fourthly, how many other forces harboured similar sub-cultures? Fifthly, how can such organisational pathogens be countered?

Regarding the second and third questions, it is worth reconsidering what ex-Labour Home Secretary Jack Straw observed about the police service during the Thatcher Years, namely that the Conservative Government's generally supportive attitude towards the police fostered a culture of impunity. Feelings of impunity can produce aberrant behaviour. If one feels favoured it is tempting to believe the rules no longer apply. It is tempting to believe that the rules apply only to those who are not favoured—like the miners who struck in 1984–1985. (Mrs Thatcher described striking miners as 'The Enemy Within' and launched 'the dash for gas' [52]). Hubris can lead one into dark places [53].

Processes of enculturation ensure a sub-culture's survival. In *Watching police, watching communities* McConville and Shepherd discuss how long-serving officers use inducements and sanctions to ensure new recruits conform to the preferred outlook and *modus operandi*:

> [B]ehind assertions of the value of group solidarity lies a less acceptable reality…. The initiate is made fully aware of the lowly status of the probationer and of the importance of acquiescence in group norms…. In one form or another the process involves undermining the self-esteem of recruits, increasing their sense of dependency, and throwing them back upon the group…. Talking about the strategies, many officers told us, could not capture the insidious and powerful nature of the group's will imposed upon the recruit [9].

Given the Special Patrol Group's distinct command-and-control structure, training regime and *modus-operandi* it was almost inevitable that it should develop a unique culture—one that sometimes expressed itself in prejudice, intimidation, victimisation, violence and, in the case of teacher Blair Peach, killing. The SPG had its own central command-and-control structure—A.9 at Scotland Yard. Officers were trained in riot-control. They had equipment commensurate with that role. They were on call 24 h a day and could spend extended periods cruising the streets in their distinctively liveried carriers. Despite certain privations there were three applicants for every position—testament to officers' perception of the SPG as an elite force. The organisational culture of the SPG of the 1970s and 1980s was unlike that of the MPS generally. Here was an elite force whose members normalised what some MPS officers (especially community beat officers) regarded as socially unacceptable attitudes and behaviours. The SPG with its 'hot-house' atmosphere, engineered elitism, can-do attitude and operational latitude spawned a deviant sub-culture. Rollo claims the SPG saw itself as Defender of the Status-Quo: "The Special Patrol Group could not stop protestors, whether they were black, unemployed or dockworkers. But it could dole out violence as a deterrent" [12]. To the extent that the SPG defended the status-quo it performed a *political* role.[29]

The behaviour of SPG officers following the death of Blair Peach and Holloway Road assaults confirms McConville and Shepherd's observations about the role of police subculture in determining the tenor of inner-city policing. Immersed in the SPG's informal code of silence, officers kept their counsel regardless of the resulting injustice and impact on the Metropolitan Police Service's reputation:

> The key features which account for the temper of inner-city policing... result from... police subculture rather than from the pathologies of individual officers. This police subculture is held in place by networks of understanding... transmitted within... work-group settings.... This occupational culture is... held in place by a sacred and overarching code of secrecy... [9].

The behaviour of quite large numbers of South Yorkshire police officers following Orgreave and Hillsborough shows that deviant behaviour can spread beyond individual police units. Even the Chief Constable was implicated in the Hillsborough cover-up. Mason suggests the solution to unreliable behaviour may lie in "setting up independent units with countervailing powers" [37]. But how independent could an internal audit team be? As discussed, the indoctrination of police officers begins at training school and is continued in situ by time-served officers [9]. At NYPD officer Serpico felt he had no choice but to go public. It has been claimed the United Kingdom's Independent Police Complaints Commission has neither the resources nor power to effectively police the police. The Chair of the Home Affairs Select Committee has remarked:

> Nearly a quarter of officers were subject to a complaint last year [2012]. Many were trivial, but some were extremely serious, involving deaths in custody or corruption—it is an insult to all concerned to do no more than scratch the surface of these alleged abuses. The IPCC investigated just a handful and often arrived at the scene late, when the trail had gone cold. The Commission is on the brink of letting grave misconduct go uninvestigated [54].

Given that the Metropolitan Police Service's Territorial Support Group (TSG) can display some of the same characteristics as the SPG now is not the time to diminish oversight. It was a TSG officer's forceful shove that contributed to the death of newspaper seller Ian Tomlinson: "Tomlinson, 47, died shortly after being shoved to the ground by a riot policeman.... An inquest last year ruled that [PC] Harwood unlawfully killed him" [55]. Commander Julian Bennett, who chaired

---

[29] Was the alleged smear campaign against Andrew Mitchell MP politically-motivated—a means of intimidating a government that was cutting police budgets? Was the intended message: 'Look what we've done to your Chief Whip. Would you like to be next?'.

the disciplinary panel, commented: "PC Harwood's use of force in this case cannot be justified. His actions have discredited the police service and undermined public confidence in it" [55]. TSG operations seem to possess the same potential for lethality as those of the SPG.

Without effective checks and balances on officers' behaviour abuses may proliferate. Why conform if you believe the chances of being held to account are negligible? In the UK, police strategy and tactics are not subject to democratic control. Yes, laws are made by Parliament. But the way in which those laws are policed are decided by senior officers. Operational freedom allows officers to emphasise and de-emphasise. Should they wish, they can articulate political agendas.

5 Conclusions

The case studies suggest first, that those charged with ensuring the safety and security of the public can elevate their own interests above those of the people they are meant to serve. They suggest secondly that social theories of risk can help explain (a) deviant behaviour amongst police officers and other emergency services personnel and (b) the persistence of distorted narratives. They suggest thirdly that testimony provided by emergency services personnel should be assessed for accuracy and triangulated. In the event of an incident or accident *all* testimony should be viewed through the prism of methodical scepticism. No source should be privileged over any other.

In the Autumn of 2012 Nick Herbert resigned as Police and Criminal Justice Minister. Disturbed by the Mitchell affair he said: "[Police] corruption may not be endemic, neither is it an aberration… the cancer must be cut out before it spreads" [56]. The Prime Minister was said to have been shocked [56] by revelations that flowed in the wake of his Chief Whip's resignation. In March 2013 Andrew Mitchell queried the integrity of the MPS's investigation into the 'plebgate' affair. The Chair of the Home Affairs Select Committee suggested that to avoid any conflict of interest the Metropolitan Police Commissioner should have asked the IPCC to investigate [57]. [30] In December 2012 the *Daily Mail* newspaper reported that the number of British police officers being investigated for corruption had risen by 62 % [58].

Everyone is fallible—even those women and men who serve in the blue-light services. Think otherwise and you create a latent error/resident pathogen.

[30] These concerns about the integrity of MPS investigations bring to mind those of Detective Frank Serpico whose doubts about the integrity of NYPD investigations were validated. Serpico proved that NYPD was incapable of investigating itself. Was Commissioner Hogan-Howe's decision to allow the MPS to investigate itself an attempt to create an actor-network whose outputs could be guaranteed to support his officers' version of the Downing Street encounter with Mitchell?

Methodical scepticism is society's safety net. No actor or organisation has a monopoly on the truth—not even the police. Society forgets that at its peril.

In his influential *Police Review* editorial 'A conspiracy of bastards' ex-policeman and campaigning journalist Brian Hilliard observed:

> Police officers use 'bastard' as a term of hatred for the worst type of criminal. They apply it to the cruel, the vicious, the ruthless and the sneaking pervert [59].

It is disconcerting to note that some officers' behaviour in regard to the Holloway Road incident and Hillsborough disaster was cruel, vicious and ruthless. In each case officers put their own interests and those of the police service before the public interest.

References

1. Hillsborough Independent Panel (2012) Hillsborough. The report of the Hillsborough independent panel. The Stationery Office, London, Sept 2012
2. Archer G (2012) Say hello to the police federation. Wave goodbye to automatic respect. http://blogs.telegraph.co.uk/news/graemearcher/100194982. Accessed 22 Dec 2012
3. British Broadcasting Corporation (2013) Call to disband Met riot police over mistrust. http://bbc.co.uk. Accessed 5 March 2013
4. Metropolitan Police Service (2013) History of the metropolitan police: special patrol group. http://www.met.police.uk. Accessed 10 March 2013
5. Punch M (2011) Shoot to kill: police accountability, firearms and fatal force. The Policy Press, Bristol
6. Glancey J (2001) London: bread and circuses. Verso, London
7. Wainright L (1968) A look from afar at old glory. Life, 29 March, p 25
8. Lewis P (2009) Prosecutors to review Blair Peach death after 30 years. http://www.guardian.co.uk/uk/2009/dec/14/blair-peach-case-review. Accessed 26 March 2013
9. McConville M, Shepherd D (1992) Watching police, watching communities. Routledge, London
10. Lewis P (2010) Blair peach: after 31 years met police say 'sorry' for their role in his killing. http://guardian.co.uk. Accessed 10 March 2013
11. Hansard (1979) HC Deb 28 June 1979, vol 969, cc 638–641. House of Commons, London
12. Rollo J (1980) The special patrol group. In: Hain P (ed) Policing the police, vol 2. John Calder, London, pp 153–208
13. Watts-Pope D, Weiner NL (eds) (1981) Modern policing. Croom Helm, Beckenham
14. Bennett SA (2009) Londonland: an ethnography of labour in a world city. Middlesex University Press, London
15. Hilliard B (1987) The Holloway incident. Police Rev, 17 July
16. Police Review (1986) Holloway assaults: immunity likely for police witness, 21 Feb 1986
17. Police Review (1986) Met officers are reprimanded after inquiry into assaults. Police Rev, 7 Feb 1986
18. Herbert I (2012) Hillsborough: everyone in football has waited too long for the lies to be exposed, 13 Sept. http://independent.co.uk. Accessed 9 March 2013
19. Alleyne R (2012) New Hillsborough inquiry shows police were in 'chaos' as the tragedy unfolded. http://telegraph.co.uk. Accessed 11 Feb 2013
20. Prince R (2012) Hillsborough documents may show existence of 'black propaganda' police unit, says MP. Daily Telegraph, 12 Sept
21. The Sun (1989) The truth. The Sun, 19 April 1989

22. Herbert I (2012) The book that foretold truth of Hillsborough. The Independent, 22 Sept 2012
23. The Telegraph (2012) Hillsborough police chief 'boasted about smearing fans', Parliament told. http://telegraph.co.uk. Accessed 8 March 2013
24. West Yorkshire Police Authority (2012) West Yorkshire Police Authority announces Chief Constable's Resignation, 24 Oct 2012. http://www.westyorkshire.police.uk. Accessed 8 March 2013
25. British Broadcasting Corporation (2012) Hillsborough files: reaction to release of government papers. http://www.bbc.co.uk/news/uk-19569749. Accessed 27 March 2013
26. Law J (1987) Technology and heterogeneous engineering: the case of the Portuguese expansion. In: Bijker WE, Hughes TP, Pinch TJ (eds) The social construction of technical systems: new directions in the sociology and history of technology. MIT Press, Cambridge, pp 111–134
27. Callon M, Latour B (1981) Unscrewing the big leviathan: how Actors Macro-structure Reality and How Sociologists Help them to Do So. In: Knorr-Cetina K, Cicourel AV (eds) Advances in social theory and methodology: towards an Integration of micro and macro sociologies. Routledge and Kegan Paul, Boston, pp 277–303
28. Latour B (2005) Reassembling the social: an introduction to actor-network-theory. Oxford University Press, Oxford
29. Stalder F (1997) Actor-network-theory and communication networks: toward convergence. http://felix.openflows.com/html/Network_Theory.html. Accessed 14 May 2009
30. Miettinen R (1999) The riddle of things: activity theory and actor-network theory as approaches to studying innovations. Mind, Cult Act 6(3):170–195
31. Bennett SA (2005) The importance of perspective. http://www.bbc.co.uk/leicester/content/articles/2005/10/13/earthquake_dr_simon_bennett_feature.shtml. Accessed 16 March 2013
32. Borodzicz EP (2005) Risk, crisis and security management. Wiley, Chichester
33. Titov VV, Synolakis CE (1997) Extreme inundation flows during the Hokkaido-Nansei-Oki tsunami. Geophys Res Lett 24(11):1315–1318
34. McCurry J (2011) Fukushima Daiichi nuclear power plant operator 'ignored tsunami warning'. http://www.guardian.co.uk/world/2011/nov/29/fukushima-daiichi-operator-tsunami-warning. Accessed 16 March 2013
35. McIntyre GR (2000) Patterns in safety thinking. Ashgate, Aldershot
36. Health and Safety Laboratory (2011) High reliability organisations: a review of the literature. Health and Safety Executive, Bootle
37. Mason RO (2004) Lessons in Organizational Ethics from the Columbia Disaster: Can a Culture be Lethal? Org Dyn 33(2):128–142
38. Weick KE (1987) Organisational culture as a source of high reliability. Calif Manag Rev 29:112–127
39. Weick KE, Sutcliffe KM, Obstfield D (1999) Organising for high reliability: processes of collective mindfulness. Res Organisational Behav 21:81–123
40. Eyre EC (1984) Mastering basic management. Macmillan Education, London
41. Bennett SA, Stewart N (2007) Employees' experience of, and attitudes towards team working at a National Health Service (NHS) District General Hospital. Risk. Manag: Int J 9:145–166
42. Armstrong M (1996) A handbook of personnel management practice. Kogan Page, London
43. Bruce S (1999) Sociology. A very short introduction. Oxford University Press, Oxford
44. Harvey-Jones J (1994) All together now. Heinemann, London
45. Conn D (2012) Miners' strike: police to be investigated over 'battle of Orgreave'. South Yorkshire force refers itself to the IPCC after claims of fabricated and co-ordinated evidence. http://guardian.co.uk. Accessed 18 March 2013
46. British Broadcasting Corporation (2011) Toxteth riots: Howe proposed 'managed decline' for city. http://bbc.co.uk. Accessed 18 March 2013
47. Donnelly PF (2007) Organizational forming in (a)modern times: path dependence, actor-network theory and Ireland's Industrial Development Authority. Dissertation, University of Massachusetts Amherst. Electronic Doctoral Dissertations for UMass Amherst. Paper AAI3347800

48. South Yorkshire Police Federation (2012) Hillsborough. http://www.southyorks.polfed.org/news_info.asp?id=119. Accessed 19 March 2013
49. Dex R (2012) Kelvin MacKenzie, the Sun's editor behind Hillsborough 'Truth' headline, offers profuse apologies. http://www.independent.co.uk/news/media/press/kelvin-mackenzie-the-suns-editor-behind-hillsborough-truth-headline-offers-profuse-apologies-8130665.html. Accessed 27 March 2013
50. Bennett SA (2001) Case studies in architectural surety. Institute of Civil Defence and Disaster Studies, Worcester
51. Maas P, Serpico F (2005) Serpico: the classic story of the cop who couldn't be bought. Perennial, New York
52. Macalister T (2013) King coal nears the end of the line. The Guardian, 8 March
53. Bennett SA (2006) After Hubris, Nemesis: Why flag carriers fail. Vaughan College, Leicester
54. The Huffington Post (2013) Independent Police Complaints Commission 'Under-Equipped and Hamstrung'. http://www.huffingtonpost.co.uk/2013/01/31/ipcc-criticised-by-home-affairs-select-committee_n_2591313.html. Accessed 23 March 2013
55. Walker P (2012) Ian Tomlinson case: PC Simon Harwood sacked for gross misconduct. http://www.guardian.co.uk/uk/2012/sep/17/simon-harwood-sacked-gross-misconduct. Accessed 23 March 2013
56. Helm T, Doward J, Boffey D (2012) Plebgate: senior Tory slams 'cancer' of corruption in UK police service. http://www.guardian.co.uk/uk/2012/dec/22. Accessed 23 Dec 2012
57. Legge J (2013) Andrew Mitchell complains to police watchdog over Scotland Yard handling of 'Plebgate' row report. http://www.independent.co.uk/news/uk/crime/andrew-mitchell-complains-to-police-watchdog-over-scotland-yard-handling-of-plebgate-row-report-8555427.html. Accessed 31 March 2013
58. Verkaik R (2012) Shocking 62 % rise in police officers being investigated for corruption with eight out of ten accused of illegally disclosing information. Anti-corruption units are facing a workload of 245 cases every month. http://www.dailymail.co.uk/news/article-2252395. Accessed 3 Jan 2013
59. Hilliard B (1986) A conspiracy of bastards. Police Rev, 7 Feb 1986

Decision Support Through Strongest Path Method Risk Analysis

Philip O'Neill

Abstract Decision support is required in situations that entail risk in order to mitigate potential loss or harm. Risk managers prepare contingency plans for this purpose. Moreover, emergency managers need to understand the consequences of their decisions during actual emergencies in order to minimize the consequences of the emergency on the population that it affects. The Strongest Path Method provides an analytical framework and tools for decision support in situations that entail risk resulting from complex functional dependency relationships. Infrastructure risk analysis will be used to illustrate the method.

Keywords Risk · Risk analysis · Infrastructure risk · Strongest path method · Mathematical models · Risk models

1 Introduction

The complex interconnectedness of infrastructure, processes, commodities and services in our society gives rise to risk. Failure of any particular system or service can cause far-reaching harm transmitted through other systems, infrastructures, processes, commodities and services. In order to safeguard ourselves, we need to be able to identify pathways of potential harm in order to mitigate potential loss.

This chapter will describe a modeling approach which was originated by the author for decision support and contingency planning in anticipation of the "Y2K bug" and which continues to be developed through critical infrastructure modeling projects being carried out by Public Safety and Emergency Preparedness Canada (PSEPC) and Emergency Management Ontario (EMO). As of this writing, the

P. O'Neill (✉)
Risk Logik, 80 Little Bridge Street, Almonte, ON K0A 1A0, Canada
e-mail: phil.oneill@risklogik.com

A. J. Masys (ed.), *Networks and Network Analysis for Defence and Security*, Lecture Notes in Social Networks, DOI: 10.1007/978-3-319-04147-6_10,

provinces of New Brunswick and Saskatchewan have initiated projects to build similar infrastructure models.

Since November 1998, I have developed a technique, known as the "***strongest path method***" (**SPM**), for performing risk analysis of multiple systems of systems that are highly interconnected. It was developed in response to the recognition that infrastructure components are strongly linked by functional relationships and that the functional relationships are more important than the components themselves.

The *connectedness* entails complex dependency relationships throughout the system of systems and these propagate risk. Users, managers and regulators of such systems who want to understand the impact and vulnerability of its component parts require a risk analysis method that deals explicitly with chains of dependency relationships. SPM provides such capability.

Not only are the strongest path or paths taken into consideration when estimating the potential impact of one entity on another but the compound effects of all pathways are included. The results of the path analysis can be used to prioritize risk and to prepare risk mitigation plans and contingency plans.

The underlying framework of the decision support method is risk analysis. The method and tools described in this chapter are intended to support decision making in situations where risk mitigation is the primary objective. In other words, decision makers who wish to minimize the possibility of loss in their planning domain might find SPM valuable.

2 Terminology

SPM is a constructive hierarchical method that models entities and dependency relationships from the fundamental level of individual entities and direct dependency relationships up to the level of aggregated systems and systems of systems. It is important at the outset to put forward well defined terminology in order to develop SPM and its tools.

Degree of Impact—a measure of the consequences of an event.

Impact can be positive of negative. For example, the consequences of buying a $10 lottery ticket might be winning $1,000 or losing $10.

Likelihood of Occurrence—a measure of the possibility or the relative level of belief in the possibility that an event will occur.

It is important to note that risk analysis is often carried out for systems in which not all of the potential events that might affect it are random. For example, deliberate human actions are not random events. For this reason, "likelihood of occurrence" is used instead of "probability of occurrence".

Risk—the possibility of loss (reference to Rowe [1]).
Under this paradigm there can be **2 *kinds of potential loss***:

1. loss due to injury, damage or destruction; and
2. loss due to missed opportunities.

In the context of infrastructure risk analysis it is the first type of loss that is of primary concern.

There are other definitions of risk currently in use. For example, ISO 310000 defines risk as "the effect of uncertainty on objectives". However they are all founded on the principle that risk has **2 *dimensions***:

1. degree of impact; and
2. likelihood of occurrence.

In SPM each entity is assessed with a degree of impact and a likelihood of failure; in other words, each entity is given a standalone risk assessment. This assessment can be made by expert judgment, statistical analyses, other modeling techniques such as simulation, or a combination of any of these.

Risk Index—the degree of impact multiplied by the likelihood of failure.

The risk index is of particular use in that it provides a single value which establishes a prioritization of multiple risks.

Entity—something that has a distinct existence of its own.

Entities can be non-physical. For example, a command, a piece of information or a body of knowledge can be an entity. As well, entities can be activities or processes. For example, a project can be an entity and, the activities that make up a project can be entities.

System—a collection of entities that act together to produce an output or outputs.

In the previous example, if activities in a project are modeled as entities, then the project itself is a system. From this point of view, a portfolio of projects could be regarded as a system of systems.

The term "system of systems" is often used to draw attention to the fact that collections of systems can also act together. For our purposes, because a system of systems satisfies the definition of *system*, let it be understood that a system can in fact be a system of systems.

The level of detail that is used to model a system is equivalent to the resolution of its component parts. Consequently, the end use of the model for decision support will determine the level of detail that should be represented by its entities.

Output—something that is produced by an entity or a system.

An output can be modeled explicitly as an entity in its own right. Otherwise, it can be taken into consideration implicitly in the scoring of dependency relationships that are involved with it.

Input—something that is acquired or received by an entity or a system.

An input can be modeled explicitly as an entity in its own right. Otherwise, it can be taken into consideration implicitly in the scoring of dependency relationships that are involved with it.

Dependency relationship—the transfer of output from a source entity to a dependent entity.

3 The Modeling Paradigm

Modern society can be viewed as a collection of networks that overlap and interact with each other. There are transportation networks, communications networks, energy networks, supply chains, distribution networks, social networks, cyber networks and so on.

SPM makes use of a mathematical object known as a *directed graph*. A directed graph is essentially a mathematical representation of a network in which the connections in the network have a direction. Any pair of distinct nodes A and B in the directed graph can have 0, 1 or 2 connections between them. If there is 1 connection then it either goes from A to B or from B to A. If there are 2 connections then 1 of them goes from A to B and the other goes from B to A.

Many decision making problems can be represented using a directed graph. These include:

- project management
- portfolio project management
- supply chain management
- business continuity planning
- military campaign planning
- analysis of influence diagrams
- infrastructure risk analysis

In order to motivate the development of SPM, infrastructure risk analysis will be used to focus the discussion and illustrate tools and techniques.

Typically planners, both in private enterprise and public service, from the national level down to the local community level, divide their planning domain into coherent subsets called sectors. For example, for emergency planning in the province of Ontario, infrastructure is divided into:

- Food
- Water
- Health Care
- Electricity
- Telecommunications
- Public Safety and Security
- Finance
- Natural Gas
- Oil
- Transportation
- Government

Each sector takes inputs from other sectors and by means of its own enablers and activities, produces outputs that are in turn taken as inputs by other sectors. As well, there are controls and regulations that govern the activities of any sector along with monitoring and verification agents who oversee the activities and processes.

Planners describe their domains in terms of the enablers, actions, controls, agents, inputs and outputs that exist in their sector. That which exists as distinct and individual in a sector will be referred to as an ***entity***. An interaction between two entities will be referred to as a ***relationship***.

For modeling purposes, a ***sector*** can be regarded as a particular categorization of entities. For any given sector, it is possible to identify the networks that it contains, overlaps and interacts with. By using a ***directed graph***, we can model networks of entities and relationships.

The risks in our society that result from *interconnectedness* can be characterized as stemming from *dependency relationships* which exist at the level of the individual *connections*. In our modeling paradigm, the connections are relationships and hence dependency relationships will be represented as edges and the strength of the relationship will be represented by a weight called the *degree of dependence*.

A dependency relationship is a special kind of relationship in which something is passed from one entity to another. The *something* need not be material (such as electricity or water) but can be immaterial (such as data or instructions). Node B depends on node A if and only if node A provides node B with something. So if edge (A, B) exists in the graph then B depends on A.

Direct dependency relationships are well understood by domain experts; indirect dependencies are not. While experts have much insight and intuition about direct dependencies, analysis is needed to verify or correct intuition and to synthesize expert knowledge into a comprehensive view of indirect dependencies. There are two main challenges in this (1) capturing expert knowledge (2) constructing a complete strategic picture with prioritization of indirect dependences for purposes of contingency planning.

4 Risk Analysis

In his landmark book of 1977 which was revised and augmented in 1988, William D. Rowe described an overarching approach for risk analysis [1]. Although not ground breaking in terms of new methods, it consolidates the fundamentals of risk analysis into a unified body of knowledge coupled with a general purpose method for undertaking any kind of risk analysis, even qualitative risk analysis in cases where probability estimates and other quantitative measurements are not practical or may even be impossible.

The body of his work primarily concerns the development of a general purpose method for carrying out "risk assessment". In his framework, risk is the possibility of loss in which loss is assessed according to 2 dimensions: (1) degree of impact and (2) likelihood of occurrence. His general method is sufficient for our purposes. We will proceed from the degree of impact and likelihood of failure of individual entities to aggregate assessments of global impact and global vulnerability of each entity as the result of multiple pathways of impact on other entities and multiple pathways of vulnerability from other entities.

Likelihood estimates can be made in terms of "degree of belief" or "expert judgment" if necessary. The degree to which an analysis is quantitative versus qualitative depends upon the degree to which objective measurements can be made.

The risks in our society that result from *interconnectedness* can be characterized as stemming from *dependency relationships.* A dependency relationship is a relationship in which *something* is passed from one entity to another. The "something" need not be material (such as gasoline or water) but can be immaterial (such as data or instructions). Node y depends on node x if and only if x provides y with something. So if edge (x, y) exists in the graph then y depends on x. A pair of entities may have a mutual dependence, referred to as ***interdependency***, in which case both edges (x, y) and (y, x) will exist in the graph. Such relationships need not be symmetrical; that is, the entities do not necessarily have an equal degree of dependence on each other.

In our *interconnected* society, the dependency relationships themselves exist at the level of the *connections.* In our modeling paradigm, the connections are relationships and hence dependency relationships will be represented as edges in the graph and the strength of the relationship will be measured by a weight called the *degree of dependence*.

Direct dependency relationships are well understood by domain experts; indirect dependencies are not. While experts have much insight and intuition about direct dependencies, analysis is needed to verify or correct intuition and to synthesize expert knowledge into a comprehensive view of indirect dependencies.

5 Directed Graphs

A directed graph is defined by a set of nodes and a set of ordered pairs of nodes called edges. More formally, a *directed graph* G is a set of *nodes* $N = \{x_1, x_2 \ldots x_n\}$ and a set of *directed edges* $E = \{(x_i, x_j), \forall\, i \in N_O, j \in N_D\}$, where N_O is the subset of N that are origin nodes of edges and N_D is the subset of N that are destination nodes of edges.

We will use nodes to represent entities and directed edges to represent relationships in our modeling paradigm. An edge that is directed from node x to node y is defined by the ordered pair (x, y). For our purposes loops, *i.e.* edges of the form (x, x), are not needed and will not be allowed. Each of the edges can be assigned a *weight*. For our purposes the weight will represent the *strength* of the associated relationship. In order to develop the analytical techniques that follow, it is sufficient to use *high*, *medium* and *low* as weights which we will later associate with a numerical scale.

Lower case italic letters such as x, y and z will be used to represent *nodes*. Following the usual convention, (x, z) will represent the *directed edge* from node x to node z. In our modeling construct, directed edges of the form (x, x), (sometimes called loops), are not allowed; in other words, x and z must be *distinct* nodes.

A ***directed path*** in G is a set of nodes $\{x_1, x_2 \ldots x_k\} \subseteq N \wedge (x_i, x_{i+1}) \in E$, $\forall$ $i = 1, 2 \ldots k - 1$. In plain language, a directed path is a sequence of nodes with the property that each node is connected to its respective successor by an edge in G. For example, in Fig. 1, $\{s, w, x, y\}$ is a directed path; while $\{s, w, x, v\}$ is not, because (x, v) does not exist.

Note, that according to this definition, any edge $(x, z) \in G$ is also a path of length 1. Furthermore, if $(x, z) \in G$, we say that z is ***adjacent*** to x. And in the context of modeling dependency relationships, we also say that z is a ***direct dependent*** of x.

Because we will only consider directed paths, we will simply use ***path***, to mean *directed path*. The set of all paths from x to z in G will be represented by the notation $[x, z]$. Bearing in mind that $(x. x)$ is not allowed, all paths of the form $\{x = x_1, x_2 \ldots x_k = x: k \geq 2\}$ are called *cycles*; therefore, we use the notation $[x, x]$ to refer to all of the cycles passing through x. The notation $[x, G]$ represents the set of all paths from x to all nodes in G (including x itself). Similarly, $[G, x]$ represents the set of all paths from all nodes in G, (including x itself), to x.

In order to derive a method for estimating the impact of every node in G on all nodes in G, we will use path analysis. In particular, the analysis will be based on *paths of strongest impact*, from any node x to any node z (including $z = x$). Next, we will define what we mean by "paths of strongest impact", we will refer to them as ***strongest paths***, and we will use the notation $[[x, z]]$ to mean the set of strongest paths connecting x to z.

Each of the edges of G can be assigned a ***weight***. For our purposes the weight will represent the ***strength*** of the associated relationship. In order to develop the analytical techniques that follow, it is sufficient to use ***high***, ***medium*** and ***low*** as weights.

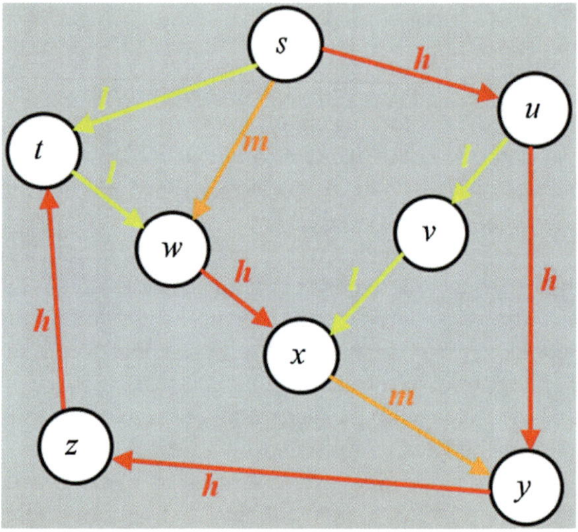

Fig. 1 A directed graph with weighted edges that are color coded

A directed graph with weighted edges can be depicted using a diagram such as Fig. 1. For this graph, $N = \{s, t, u, v, w, x, y, z\}$ and $E = \{(s, t), (s, w), (s, u), (t, w), (w. x), (u, v), (u, y), (v, x), (x, y), (y, z), (z, t), (t, w)\}$. The weights l, m and h stand for low, medium and high as the strength of each relationship. Color coding of the edges according to weight (red for high, orange for medium and yellow for low) allows visual distinction in the diagram.

For our purposes, a sub-graph of G, call it G', is a graph determined by a subset E' of E such that only end-points of edges in E' are in N', the node set of G'. It will be useful to identify particular sub-graphs of G that are determined by the weights of the edges in G. For example, we might want to consider the sub-graph G' in Fig. 1 determined by the edges of degree h. In this case, $E' = \{(s, u), (u, y), (w, x), (y, z), (z, t)\}$ and $N' = \{s, t, u, w, x, y, z\}$.

Depending upon the context of the model, it is necessary to define criteria for the evaluation of scores. Because the modeling paradigm is based upon the notion of infrastructure "*failure*", it is important to establish what failure means.

A pragmatic and widely-used benchmark for defining "failure" is the ***minimum acceptable level of service*** (**MASL**). For a given infrastructure, the MASL is established based on the outputs (services, products, plans, directives, communications etc.) that it should supply to other infrastructures and the rate at which the outputs should be supplied. MASL is used by Emergency Management Ontario in its risk analysis framework.

By using MASL we can establish criteria for the direct degree of dependence of one infrastructure on another. For in as much as an infrastructure "*fails*" if it falls below its defined MASL, we define the following criteria:

- if failure of entity x inevitably leads to failure of entity y, then y has a ***high direct dependency*** on x, and conversely x has a ***high direct impact*** on y.

- otherwise, if failure of entity x leads to degradation of entity y to the extent that y must enact a contingency plan or resort to alternate operating procedures in order to stay above MASL, then y has a ***medium direct dependency*** on x and conversely x has a ***medium direct impact*** on y;
- otherwise, if failure of entity x leads to significant degradation of entity y, but y can stay above MASL without significantly changing its operating procedures, then y has a ***low direct dependency*** on x and conversely x has a ***low direct impact*** on y.
- otherwise, if failure of entity x does not lead to significant degradation of entity y, then y has 0 ***direct dependency*** on x, and conversely x has 0 ***direct impact*** on y.

In the worst case, a medium direct dependency relationship could result in marginal acceptable level of service, under a contingency plan or alternate operating procedures. Similarly, in the worst case, a low direct dependency could result in marginal acceptable level of service under normal operating procedures.

If x is adjacent to y, then these criteria provide relatively straightforward rules for measuring the degree of direct dependence of y on x and conversely the degree of direct impact of x on y. The challenge now, is to enlarge them in order to generalize from *direct* to *indirect* dependencies. The definition of degree of direct dependence is given with respect to a path of length 1, so by making a few more considerations we can use the same criteria for the degree of indirect dependence by assessing impact propagated over paths of any length.

The criteria describe the effect of a *high* impact event (*i.e.* less than minimum acceptable level of service) on a direct dependent. However, we also need to estimate the effect of a *medium* impact event (*i.e.* possibly marginal acceptable level of service under alternate operating procedures) and a *low* impact event (*i.e.* possibly marginal acceptable level of service under normal operating procedures) on a direct dependent. There are two dimensions for this estimate: (1) the degree of the triggering impact event; and, (2) the degree of the direct dependency relationship. It is reasonable to expect that a strong triggering event will have little impact if the degree of dependence is low; while even a relatively weak triggering event will be felt if the degree of direct dependence is high.

Thus, we estimate that the propagated impact can be no higher than the lesser of the triggering degree of impact and the degree of dependence. For example, a medium degree triggering impact acting over a low degree of direct dependence will cause a low impact because the degree of direct dependence is low; whereas, a medium impact trigger acting over a high degree of direct dependence may cause a medium impact because of the high degree of direct dependence.

In summary, we will use the following rule to estimate the propagation of impact along each edge in a path. For edge (x, y) with triggering impact at x of degree $= I(x)$ and degree of direct dependence of y on $x = D(x, y)$ then the degree of the triggering impact at y is given by:

$$I(y) = min\{I(x), D(x, y)\}$$

Just as the direct degree of dependence of y on x is based on the presumption of failure of x, the estimate of the degree of indirect dependence of y on x resulting from a particular path $\{x = x_1, x_2 \ldots x_k = y\}$ will be based on the presumption of failure of x.

Assume, therefore, that x has failed and that $I(x) = h$. Then by using the propagation rule, we find $I(x_2) = min\ \{h,\ D(x,\ x_2)\} = D(x,\ x_2)$ because $D(x, x_2) \leq h$. Subsequently, $I(x_3) = min\ \{D(x, x_2), D(x_2, x_3)\} = D^*$; $I(x_4) = min\ \{D^*, D(x_3, x_4)\}$, and so on As we proceed to y beyond the first edge in the path, each subsequent application of the propagation rule compares the lowest degree edge yet encountered with the degree of the next edge on the path and sets the triggering degree of impact to the lower value. Therefore, the indirect dependence of y on x resulting from the selected path is the degree of the lowest degree edge along that path.

Of all paths from x to y, we want to find the path or paths that propagate the greatest degree of impact from x to y. Equivalently, we want to determine the path or paths that determine y's greatest dependence on x. By the conclusion of the previous paragraph, we know that the greatest degree of impact is carried by the path or paths whose lowest degree edge is the highest among all paths connecting x to y. Any such path of shortest length will be referred to as a ***strongest path***.

For example, consider the graph in Fig. 1 and all paths connecting s to t. These are shown in Fig. 2. There are 4: Path 1 = $\{(s, t)\}$, Path 2 = $\{(s, w),(w, x),(x, y),(y, z),(z, t)\}$, Path 3 = $\{(s, u),(u, v),(v, x),(x, y),(y, z),(z, t)\}$, Path 4 = $\{(s, u),(u, y),(y, z),(z, t)\}$.

Using the propagation rule, we find that the impact of s on t from Path 1 is low by virtue of (s, t), the impact of s on t from path 2 is medium by virtue of (s, w) and (x, y), the impact of s on t from Path 3 is low by virtue of (u, v) and (v, x), and the impact of s on t is high from Path 4 by virtue of all of its edges being high degree. Therefore, the indirect dependence of t on s is high and the strongest path is Path 4.

In addition to the degree of dependence that is assigned to the edges of the graph, the nodes of the graph are assigned two values: (1) ***direct degree of impact***, and (2) ***direct likelihood of failure***. Respectively, these values are intended to measure the relative importance of a node in and of itself and the relative likelihood that it will fail independently.

At this point suffice it to say that the degree of impact assigned to a node is a weight that reflects its importance relative to the others and that the likelihood of failure is a weight that reflects the relative likelihood that a node will fail because of random internal causes or non-random external threats that target it.

In order to accomplish the risk analysis according to Rowe's framework, we need to estimate the cumulative degree of impact between every pair of nodes in the graph. While the strongest paths constitute the most significant influence of one node on another, we need to take into account all of the paths that connect a pair of nodes in order to assess the relative risk of every node.

To do this, we will define several utility functions. In what follows, we will use naming conventions, based on letters, for these functions:

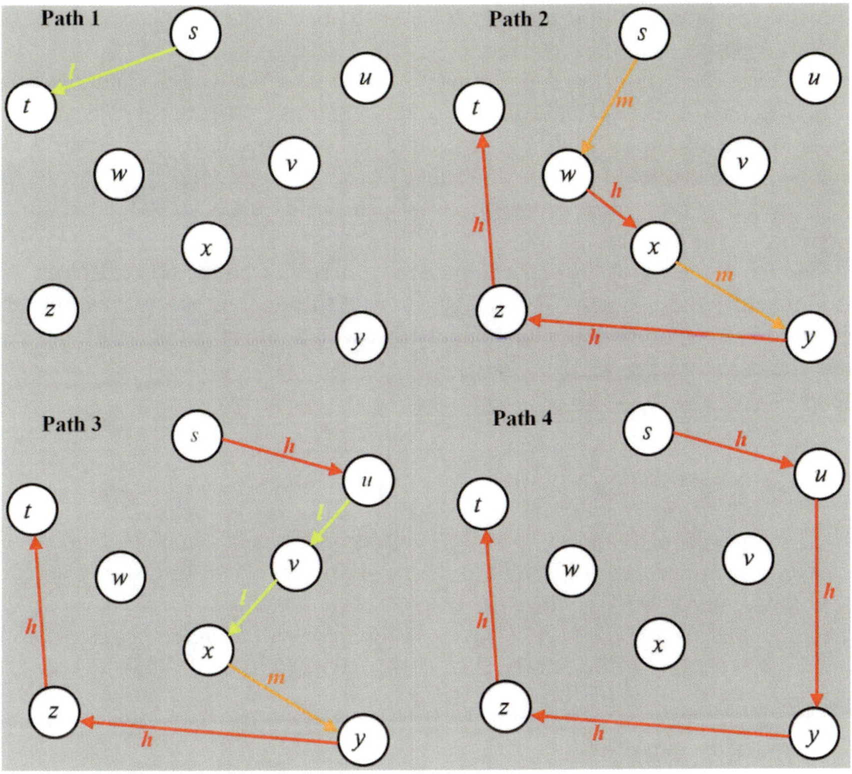

Fig. 2 All paths connecting *s* to *t* in the graph of Fig. 1

- *D*, degree of ***Dependence***;
- *I*, degree of ***Impact***;
- *F*, likelihood of ***Failure***; and,
- *L*, path ***Length***.

There are four possible cases for the argument of each function:

1. the argument is a single node, *x*;
2. the argument is a single edge, (*x*, *z*);
3. the argument is a single path, $\{x_1, x_2 \ldots x_k\}$;
4. the argument is a set of paths, such as [*x*, *y*], [*x*, *G*], [*G*, *x*], bearing in mind that any of these sets may be the null set, ∅.

If the argument is a single node, *x*, then *I*(*x*) is the direct impact of *x* on the general public without regard for any of its explicit dependency relationships in the model; *F*(*x*) is the deemed likelihood of failure of *x*, without regard for any of its explicit dependency relationships in the model; *D*(*x*) and *L*(*x*) have no meaning (possibly except trivially in terms of loops) and are consequently undefined (if for no other reason than loops are forbidden in our paradigm).

For example, $I(x)$ is the direct impact of node x on the general public without regard to any of its explicit dependency relationships in the model; $I(x, z)$ is the direct degree of impact of x on z from the direct relationship (x, z); $I([x, z])$ is the cumulative degree of impact of x on z taken over all paths; $I([[x, z]])$ is the degree of impact of x on z taken over a strongest path from x to z; $I(x, G)$ is the global impact of x on the entire graph G.

We have noted that all paths from x to y in G will be represented by $[x, y]$. Additionally, the notation $[[x, y]]$ will be used to represent any strongest paths from x to y in G, bearing in mind not only that such a path might not exist, but also that there might be more than one.

1. Strongest Path Impact of x on z

$$I([[x,z]]) = D([[x,z]]) \times I(x)^{L([[x,z]])}$$

The strongest path impact of x on z is the strongest path degree of dependence of z on x multiplied by $I(x)$ raised to the power of the length of a strongest path. Recalling that $I(x) \in [0, 1]$, the latter term acts as a damping factor. In other words, we allow that the ripple effect of propagated impact will decrease in severity as it moves away from x. Note also, that if $[[x, z]] = \varnothing$, then $D([[x, z]]) = 0$ and therefore $I([[x, z]]) = 0$.

2. Cumulative Impact of x on z:

$$\text{If } [x,z] = \varnothing, \text{ then } I([x,z]) = 0; \text{ otherwise,}$$
$$I([x,z]) = 1 - \prod_{(y,z)\in E} (1-\min\{D([[x,y]]), D(y,z)\}) \times I(x)^{L([[x,y]])+1}$$

The cumulative impact of x on z includes a term for every pathway that exists from x to z. Let $y \in N$ be the set of nodes on which z is directly dependent. For each of z's direct dependencies, the strongest path impact of z on x through y, is factored into the function. The contribution of each of these paths is damped by the factor $I(x)$ raised to the power of the path length. Similar to the binomial probability function, it compounds the effects of the terms. Note as well, that if $[x, z] = \varnothing$, then $I([x, z]) = 0$.

3. Global Impact of x on G:

$$I([x,G]) = \sum_{z\in N} (I([x,z]) \times I([z])) / \sum_{z\in N} I([z])$$

By "global" we mean the impact of any node x on the entire model. Consider the impact of a node, x, on every individual node, z, in the graph (including x itself). The cumulative impact of x on z is given by $I([x, z])$ and the relative importance of z itself is given by $I([z])$. The terms $I([x, z]) \times I([z])$ give the global significance of the impact of x on z. The sum of these products gives the impact of x on the entire model. This can be regarded as a sum of cumulative impact values weighted by direct impact values. In order to make a better relative comparison of different versions of a model, we normalize the weighted sum by dividing by the sum of the weights.

4. Cumulative Vulnerability of z from x:

$$F([x,z]) = I([x,z]) \times F(x)$$

There are two factors about x that influence the possible failure of z: the cumulative impact of x on z and the likelihood of failure of x. Therefore, the cumulative vulnerability of z from x is the product of the cumulative impact of x on z and the likelihood of failure of x.

5. Global vulnerability of z from G:

$$F([G,z]) = 1 - \prod_{x \in N}(1 - F([x,z)]))$$

The cumulative vulnerability of z from all nodes in the model is the binomial probability that a failure event of any node will cause z to fail.

6. Risk Index of x:

$$R(x) = I([x,G]) \times F([G,z])$$

The risk index of an entity is the product of the global impact of an entity times its global vulnerability. This provides a single score for comparing risk among all of the entities in a model.

The analytical tools we need for risk analysis are now complete, except for a description of how to find the strongest path degree of dependence between all ordered pairs of nodes. An application of any "*shortest path*" algorithm suffices; we will call the new algorithm, ***Strongest Path Degree.***

In graph G let G^* be the subgraph of G generated by edges of degree $\geq$ *. Assume that G has a scale for degree of dependence = $\{1,2, \ldots, k\}$. If $[x, z] \neq \emptyset$, let $S^*(x, z)$ be the length of the shortest path from x to z in G^*; if $[x, z] = \emptyset$ then $S^*(x, z) \equiv 0$.

The following algorithm is intended to show that the degree of the strongest path between every pair of nodes can be found in polynomial time, $O(n^2m)$, where n is the number of entities and m is the number of dependencies. This algorithm can be implemented using any shortest path algorithm. Efficiencies with the choice of shortest path algorithm and the organization of the search can be exploited to reduce the execution time.

Algorithm: Strongest Path Degree

Set $\boldsymbol{D([[x, z]])} \leftarrow \boldsymbol{0}$ for all ordered pairs of nodes $\{x, z\}, x \in N, z \in N$
For * = $\boldsymbol{k}$ to $\boldsymbol{1}$, step = $\boldsymbol{-1}$
For all ordered pairs $\{x, z\}$, calculate $\boldsymbol{S^*([x, z])}$
Set $\boldsymbol{D([[x, z]])} \leftarrow \boldsymbol{max\ \{D([[x, z]]), S^*([x, z])\}}$
Next *

End: Strongest Path Degree

The strongest path degree algorithm initializes all $D([[x, z]])$ to 0. It then searches for a shortest path connecting x to z in a sequence of sub-graphs starting with G^k, then G^{k-1} and so on to G^1. For each ordered pair {x, z} it saves the first non-zero shortest path length that it finds as $D([[x, z]])$.Subsequently it does not change that value of $D([[x, z]])$ throughout the remainder of the search, because it can only encounter ever decreasing values of $S^*([x, z])$. If a shortest path is not found, then there is no path from x to z and consequently, $D([[x, z]]) = 0$, its initial value.

5.1 Example

We will use an example model to illustrate the use of the tools that have been developed. The example model, shown in Fig. 3, is a small infrastructure model. There are 10 entities: Drinking Water, Local Electrical Distribution, Natural Gas Storage and Transport, Ambulance Services, Local Food Outlets, Local Food Distribution, Farm Food Production, Health Canada/food inspection, Hospitals & Clinics and Cyber Networks.

Each of these has been assessed with a degree of impact (the number labeled on the left side of each node in Fig. 3), high (score = 7, colored dark orange), medium (score = 5, colored orange) or low (score = 3 colored yellow).

As well, each has been assessed with a likelihood of failure (the number labeled on the right side of each node in Fig. 3), high (score = 7, border colored dark orange), medium (score = 5, border colored orange) or low (score = 3 border colored yellow).

There are 30 direct dependency relationships that have been scored high (score = 9, edge colored red), medium (score = 5, edge colored orange) and low (score = 3, edge colored yellow). All entities except local electrical distribution have been assessed as having medium dependence on cyber networks. Local electrical distribution, however, has been scored as having high dependence on cyber networks.

After path analysis is carried out and scores for global impact and global vulnerability are computed, a histogram of the risk indices is shown in Fig. 4. The highest risk entity in the model is local electrical distribution. The second highest is local food distribution. At medium risk are three entities: Health Canada/food inspection, Local Food Outlets and Cyber Networks. The remaining entities are of relatively low risk.

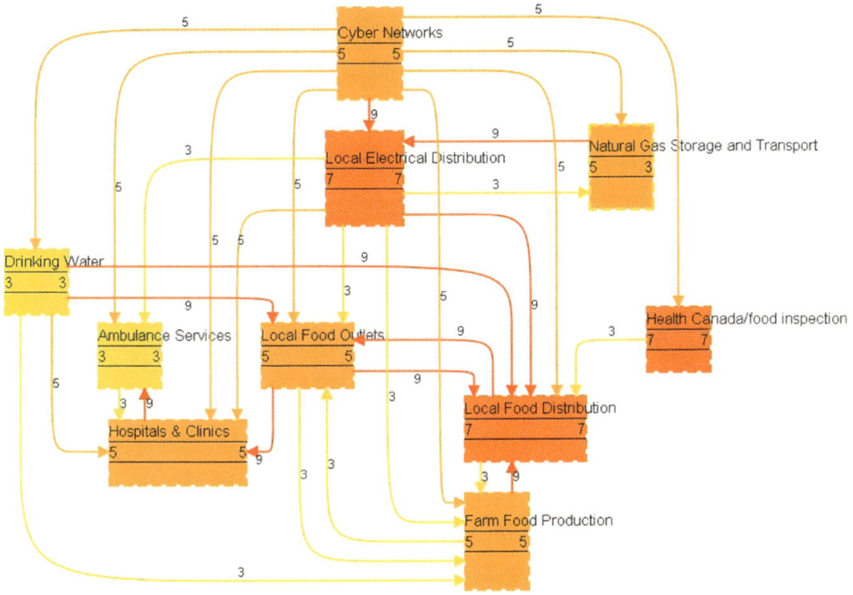

Fig. 3 A small infrastructure model

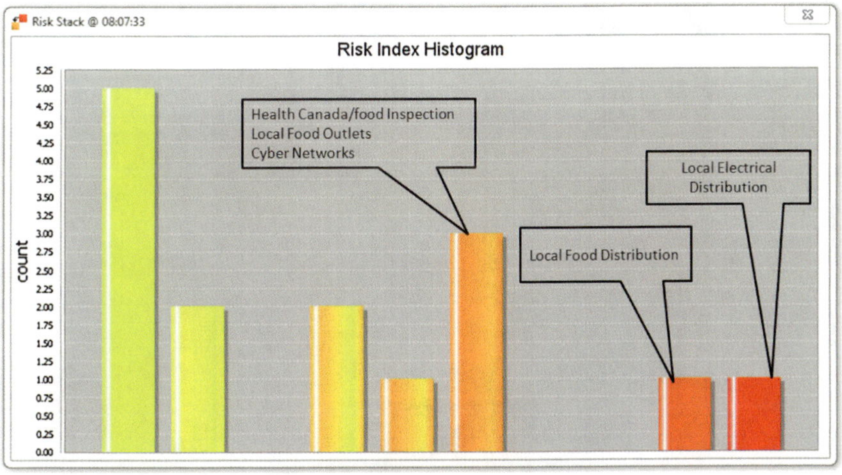

Fig. 4 Histogram of the risk indices for the model shown in Fig. 3

Even though Cyber Networks is assessed as medium risk in the model, we can still assess all consequent impact were it to fail. Figure 5 shows the sub-network of high dependency relationships in the model.

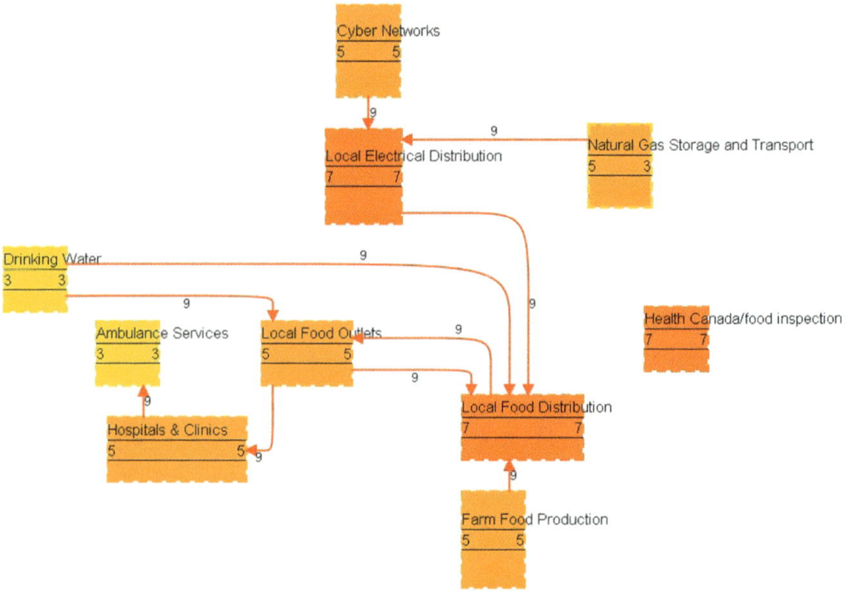

Fig. 5 High dependency relationships for the model in Fig. 3

Fig. 6 Paths of high impact from cyber networks for the model in Fig. 3

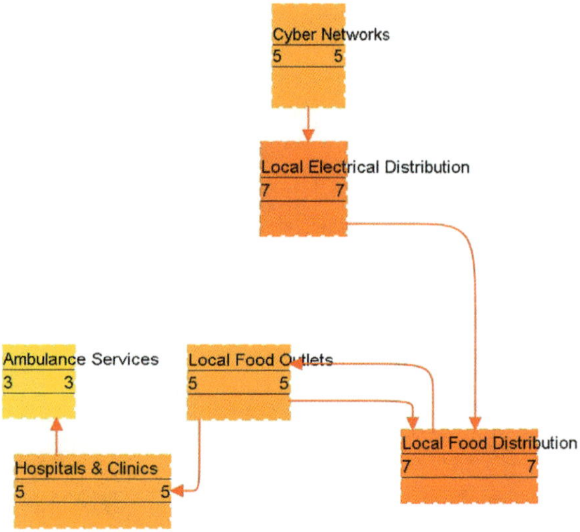

From this sub-network, we can isolate the paths of high impact emanating out of the cyber networks entity. These are depicted in Fig. 6.

While only local electrical distribution has a high direct dependence on cyber networks, Fig. 6 shows that local food distribution, local food outlets, hospitals & clinics and ambulance services would all fail if cyber networks fail.

6 Summary and Conclusions

A paradigm for decision support based on risk mitigation has been presented. Risk models in the form of a directed graph are constructed from entities which are assessed with a degree of impact and a likelihood of failure and dependency relationships between the entities which are scored for degree of dependence according to well defined criteria.

The paradigm allows the knowledge of experts to be used for infrastructure risk analysis. Results from other analytical models, such as simulations, can also be included in a model.

The paradigm includes utility functions for evaluating global impact and global vulnerability for every entity in the model. The calculations are based on the notion of the "strongest path" connecting every pair of entities in a model. An algorithm for calculating all strongest paths in a model was given.

As a result of performing the path analysis, such models reveal the potential consequences of the failure of any entity on all of the others. This enables contingency planners to anticipate all outcomes in any infrastructure damage scenario.

The paradigm together with calculation tools and graphical output features has been implemented as RiskOutLook®, java-based software available exclusively from Risk Logik.

Reference

1. Rowe WD (1988) An anatomy of risk. Krieger Publishing Company, Malabar, Florida

Critical Infrastructure and Vulnerability: A Relational Analysis Through Actor Network Theory

Anthony J. Masys

Abstract Threats to national security, such as that against critical infrastructures not only stem from man-made acts but also from natural hazards. Hurricane Katrina (2005), Blackout Canada-US (2003), Fukushima (2011), Hurricane Sandy (2012), and Alberta floods (2013) are examples that highlight the vulnerability of critical infrastructures to natural hazards and the crippling effect that failures can have on the social and economic well-being of a community and a nation. Focusing on the initiating event that precipitated the critical infrastructure failure does not capture the root vulnerabilities or 'resident pathogens' that are 'hard-wired' into the greater networked system. Through the complexity/systems lens of Actor Network Theory (ANT), this chapter explores how key 'actors' within a network can align other actors creating 'unseen' vulnerabilities.

Keywords Critical infrastructure · Actor network theory · Complexity · Systems thinking

1 Introduction

Hurricane Katrina (2005) devastated New Orleans thereby revealing inherent vulnerabilities that resided in the socio/political/ecological/technical infrastructure (system) of the city and the nation. These 'unseen' vulnerabilities that emerged at the 'seams' of interconnection and interdependencies can be characterized as 'resident pathogens', in that the hurricane as a 'triggering mechanism' interacted with the '…city's fragile physical environment, aging infrastructure, and declining economic and social structure' [12] as well as policies, regulations and politics. Comfort [12] asks the question '…Was the damage in New Orleans due to

A. J. Masys (✉)
University of Leicester, Leicester, UK
e-mail: Anthony.masys@gmail.com

A. J. Masys (ed.), *Networks and Network Analysis for Defence and Security*,
Lecture Notes in Social Networks, DOI: 10.1007/978-3-319-04147-6_11,

Hurricane Katrina, or was it some combination of human and technical factors that failed under the stress of the hurricane?'. Similarly, this question resonates with other events such as: Fukushima (2011) whereby an earthquake and resulting tsunami had a devastating effect on the Fukushima Diiachi nuclear power plant [23]; Hurricane Sandy (2012) that resulted in significant disruption to New Jersey and New York [19], Ash Cloud (2010) stemming from the eruption of Eyjafjallajökull and resulting in significant disruptions to air travel and trade in Europe [21]. All these cases highlight the interconnectedness and interdependencies that characterize how we live and what we depend upon.

As noted in Weick and Sutcliffe [50, p. 1], 'Unexpected events often audit our resilience' and thereby can highlight 'resident pathogens' that are 'hardwired' in policy, politics, regulations, procedures, practices, shaping decision making and actions. The case study of Hurricane Katrina as described by Comfort [12, p. 7] highlights '...serious failures in policy, planning, and practice at all four levels of government—municipal, parish, state, and federal—in reference to a city exposed to known hazards'.

Critical infrastructure (CI) can be broadly defined as '... the assets, systems, and networks, whether physical or virtual, so vital ... that their incapacitation or destruction would have a debilitating effect on security, national economic security, national public health or safety' [15]. The affect of such events as the Ash Cloud, Fukushima, Hurricane Katrina and Hurricane Sandy had on security and safety and in particular critical infrastructure was significant. It is thereby essential that the network implications and vulnerabilities that reside (embedded) within the CI be understood.

Modern infrastructure is characterized as complex coupled networks with inherent interdependencies and interconnectivity. Public Safety Canada [39] identifies ten critical infrastructure sectors:

- Energy and utilities
- Finance
- Food
- Transportation
- Government
- Information and communication technology
- Health
- Water
- Safety
- Manufacturing.

Understanding this complex coupled system requires models that embrace the inherent interdependency and relationality. Because of these interdependencies within and across critical infrastructures, failures within one network may cascade through dependent nodes in other networks. As noted in Vespignani [48, p. 428], '...in some cases the understanding of complex networks provides counterintuitive and surprising approaches to the engineering and management of complex socio-technical systems'.

Masys [32] describes in his study of the Uberlingen Mid-air collision, how inherent vulnerabilities emerged from the 'hardwired politics' and the interdependencies resident within the socio-technical system. In both [32] and [33], actors displaced in both space and time had a significant effect on the accident aetiology. With respect to critical infrastructure, Vespignani [48] describes how '...in power grids and other flow-carrying networks, the failure of a single node or line can trigger a domino effect (cascading failure) in which the overload induced by the flow redistribution may generate a global failure of the network'.

Network thinking, as described in Barabasi [3] opens novel perspectives to understanding complex systems such as markets and economic system, socio-technical systems, criminal and terrorist networks. This network thinking mindset leverages a topological analysis which is based upon classical graph theory through which interesting properties of the structure of a network system can be revealed. The properties of interdependency and interconnectivity resident within these networks can also be examined from a temporal perspective thereby revealing interesting dynamic properties of an evolving network. Such analysis can be instrumental in assessing vulnerabilities of critical infrastructures that can shape operating and design decisions. Drawing upon scenario planning as described in Masys [35] and network analysis [33–35], such vulnerability analysis addresses such questions as:

- How do we define the boundaries of the system
- What are the relevant threats and hazards associated with the system (an examination of the space of possibilities)
- How is resilience realized within the system
- How do the interdependencies shape the structure and dynamics of the system
- What are the uncertainties
- Do resident pathogens [such as normalization of deviance [47]] exist.

Such questions can only be answered through a holistic system of system approach that explores the space of possibilities that describe the structure and dynamics of the network. A vulnerability analysis examines the robustness and resilience of the network against random failures and targeted attacks. This is of particular importance considering the impact that critical infrastructure failures can have on national security, economy, health, safety and social well-being. For example the ash cloud resulting from the 2010 eruption of Eyjafjallajökull led to the disruption of some 100,000 flights and 10 million passenger journeys [18]. In the UK alone, Gross aviation sector losses in the UK tallied £375 million in April and May 2010 alone [37].

From a vulnerability analysis point of view, it is commonly agreed that infrastructures have become more complex and their behavior is hard to understand or predict. Complex network research [3–5] highlight how some elements (nodes) can evolve to figure prominently in a network and how these structures (topologies) can be susceptible to random failures and/or targeted attacks. Critical

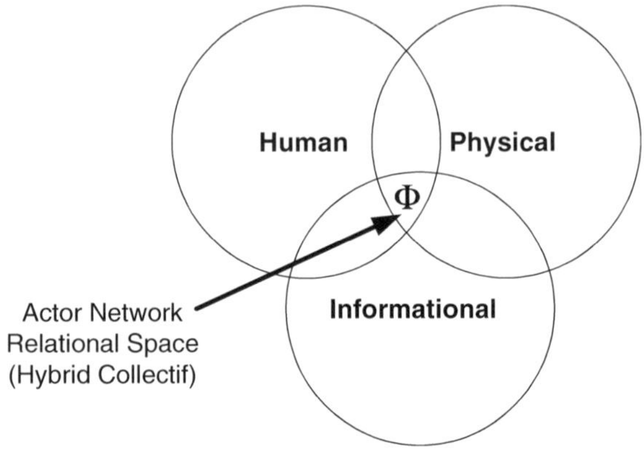

Fig. 1 Hybrid collectif [33, 34]

Infrastructure has spatial and temporal qualities making them complex and subject to a set of multiple hazards and potentially asymmetrical threats thereby revealing complex vulnerabilities. Kroger and Zio [25, p. 5] define vulnerability '…as a flaw or weakness (inherent characteristics, including resilience capacity) in the design, implementation, operation, and/or management of an infrastructure system, or its elements, that renders it susceptible to destruction or incapacitation when exposed to a hazard or threat, or reduces its capacity to resume new stable conditions'. This vulnerability quality of CI resonates with such cases as Hurricane Katrina that revealed hidden vulnerabilities in social, economic and management domains. Reduction of technological and social vulnerabilities calls for better system understanding and preventive analyses [25, p. 5].

Considering the interdependencies between and within CI sectors, it is apparent how a cascade across boundaries has the potential for multi-infrastructural collapse and unprecedented consequences. Interdependencies and the resultant infrastructure topologies can create subtle interactions and feedback/feed forward mechanisms that often lead to unintended behaviours and consequences. Analysis through a socio-technical system thinking lens [33, 34], recognizes the entangled state space described by these systems which can be conceptualized as the hybrid collectif [7, 33, 34], the intersection of the physical, human and informational domains (Fig. 1).

2 Actor Network Theory

Kroger and Zio [25, p. 47] describe the application of complex network theory methods to the analysis of CIs to support vulnerability analysis through a topological-driven and dynamical analyses. They describe how the interconnected structure of a CI can be represented by an unweighted network, where the edges between nodes are either present or not.

As described in Masys [33–35], sociology offers an interesting approach for looking at the socio-technical elements of complex systems through the application of Actor Network Theory (ANT). The systems perspective of ANT looks at the inter-connectedness of the heterogeneous elements characterized by the technological and non-technological (human, social, organizational) elements. Yeung [51] notes that much of the work that draws on actor network theory places its analytical focus on unearthing the complex web of relations between humans and non-humans. The socio-technical lens is highlighted in ANT, but not as separate social and technological analysis silos. It is well recognized that the interaction of non-human actors with the human actors (such as with CI operations) gives shape and definition to identity and action. Latour [28, p. 806] argues that '...it is impossible even to conceive of an artifact that does not incorporate social relations, or to define a social structure without the integration of non-humans into it. Every human interaction is socio-technical'. The 'social' is thereby described as 'materially heterogeneous' [7: 166]. As a unique worldview, ANT treats both human and machine (non-human) elements in a symmetrical manner, thereby facilitating the examination of socio-technical systems, where Callon [8: 183] argues, '...it is difficult to separate humans and non-humans, and in which the actors have variable forms and competencies'.

The 'systems' view facilitated by the actor network lens does not separate the traditional notions of the social from the technical. It recognizes as described by Coakes [11: 2], 'Socio-technical thinking is holistic in its essence; it is not the dichotomy implied by the name; it is an intertwining of human, organizational, technical and other facets'. Senge [42] argues that since the world exhibits qualities of wholeness, the relevance of systemic thinking is captured within its paradigm of interdependency, complexity and wholeness. Discrete occurrences in time and space are seen as entangled. Flood [20, p. 13] argues that '...they are all interconnected. Events can be understood only by contemplating the whole'. The holistic perspective of ANT makes it well suited to facilitate an examination of the complex socio-technical systems and matters pertaining to CI vulnerability analysis.

2.1 Methodology

As described in Powell and Owen [38, p. 141], 'ANT is concerned with tracing the transformation of heterogeneous networks (Law 1991) that are made up of people, organisations, agents, machines and many other 'objects'. ANT explores the ways that the networks of relations are composed, how they emerge and come into being, how they are constructed and maintained, how they compete with other networks, and how they are made more durable over time. It examines how actors enlist other actors into their world and how they bestow qualities, desires, visions and motivations on these actors [29].

This is enabled through a relational analysis described as 'following the actors' [29] and detailed in Masys [33, 34].

2.2 *Translation/Inscription*

Fundamental processes within ANT are inscription and translation. Inscription refers to the way technical artifacts embody patterns of use: Technical objects thus simultaneously embody and measure a set of relations between heterogeneous elements [2: 205]. Described in detail in Masys [34], technical objects and systems are inscripted with concepts of how it will be used. These inscriptions enable action at a distance by creating 'technical artefacts' that ensure the establishment of an actor's interests such that it can travel across space and time and thereby influence other work [26].

The process of translation is a key element in understanding how different elements in an actor network interact [43]. Translation rests on the idea that actors within a network will try to enroll (manipulate or force) the other actors into positions that suit their purposes. When an actor's strategy is successful and it has organized other actors for its own benefit, it can be said to have translated them. As articulated by Yeung [51: 6], 'Actors in these relational geometries are not static "things" fixed in time and space, but rather agencies whose relational practices unleash power inscribed in relational geometries and whose identities, subjectivities, and experiences are always (re)constituted by such practices'.

Within the context of the case study, an examination of actors such as those characterized from technologies to policies and the relations inherent within the actor network [34], facilitates an exploration of how these "actors" mediate action and how they are entangled in local socio-technical/political configurations. The lens of ANT facilitates the view of the world in terms of heterogeneous elements, thereby employing a "systems thinking" perspective of the problem space.

Chapman [9: 350] argues that accidents involving socio-technical systems are difficult to mitigate '...because the nature of complexity in these systems is not well understood by those who design, manage and operate them'. Chapman [9: 350] further argues for the requirement to progress better conceptual models and frameworks that reveal the inherent complexity and thereby make these complex socio-technical systems more transparent.

From an actor network perspective, when engineers work, they are typically involved in designing and building projects that have both technical and social content and implications [31]. Design can be construed as a process where various interests (from various parties within the process) are translated into technological solutions. In addition, the design encompasses organizational arrangements and procedures that must be followed to make the technology work properly (or as envisioned by the design team). Within this process, existing and legacy technology will be reinterpreted and translated into new ways of using it. To make the technology work, all these elements must be aligned, i.e. cooperating toward a common goal [1]. The inscribed patterns of use may not succeed because the actual use deviates from it. Rather than following its assigned program of action, a user may use the system in an unanticipated way; he/she may follow an anti-program [27, 36], perhaps influenced through cultural norms and standard operating procedures.

3 Case Study

As described in the US/Canada Power Outage Task Force Final Report (2004: 1), 'on August 14, 2003, large portions of the Midwest and Northeast United States and Ontario, Canada, experienced an electric power blackout. The outage affected an area with an estimated 50 million people and 61,800 megawatts (MW) of electric load in the states of Ohio, Michigan, Pennsylvania, New York, Vermont, Massachusetts, Connecticut, New Jersey and the Canadian province of Ontario. Power was not restored for 4 days in some parts of the United States. Parts of Ontario suffered rolling blackouts for more than a week before full power was restored'. The Ohio phase of the August 14, 2003, blackout was caused by deficiencies in specific practices, equipment, and human decisions by various organizations that highlight deficiencies in corporate policies, lack of adherence to industry policies, and inadequate management of reactive power and voltage as key actors in the disaster aetiology.

The impact of such an event is significant. As described by Public Safety Canada [39], the CI sectors (Energy and utilities, Finance, Food, Transportation, Government, Information and communication technology, Health, Water, Safety, Manufacturing) are interdependent. Society has come to depend on reliable electricity to service these CI sectors. Although detailed analysis highlights the failure in the system, applying an ANT perspective brings into the analysis a unique worldview and perspective that reveals hidden resident pathogens that are 'hardwired' into the system through inscription and shape the system structure and dynamics through translation processes.

4 Discussion

What we are seeing today is the emergence of highly interdependent systems (socio-technical, market economies, and critical infrastructures). Examples of these systems include the internet, communication technologies, power distribution grids, and transportation networks. These networks are characterized by their physical, spatial and temporal heterogeneity. Because of the inherent complexity in these systems, they challenge our understanding pertaining to their structure and dynamics and thereby our understanding regarding vulnerabilities and their impact on society [48, p. 425]. What arises from these characteristics is the necessity to revisit ones assumptions, challenge ones worldview and question ones notion of uncertainty. As described in Kroger and Zio [25, p. 9] uncertainty pervades these network structures affecting system behavior. Given that CI relies on intricate, often nonlinear interactions among a large number of interconnected and geographically distributed components of different types, including both technical and non-technical elements [25, p. 34], a holistic view of the system is required. It is not only important to understand the actors themselves but as well their connectivity, interdependencies and from an ANT perspective their influence on the system and how this influence (inherent relationality) shapes the system behavior and topology.

Vespignani [49: 984] argues that because of the interdependencies of the CI such as those demonstrated by power generation grids, telecommunication, transportation, 'The failure or damage ...would cause huge social disruption, probably out of all proportion to the actual physical damage'. What emerges is the realization that 'recent disasters ranging from hurricanes to large-scale power outages and terrorist attacks have shown that the most dangerous vulnerability is hiding in the many interdependencies across different infrastructures'. As reported in Brummitt et al. [6: 12159], 'when a tenth of humanity lost power over 2 days in India in July 2012, technical failure was not the only culprit. Like many recent blackouts, this outage resulted from couplings among systems, including extreme weather exacerbated by climate change, human operator errors, suboptimal policies, and market forces'. These interdependencies can be viewed in terms of the relationality inherent within the system. This relational (ANT view) reveals how localized damage can result in cascading system level failures.

Urry [44: 59], in his discussion of complexity and systems, remarks that there exists a '...profound disproportionality of 'causes and effects'. Such systems possess a history that irreversibly evolves and in which past events are never 'forgotten'. His statement resonates with the CI. Through the lens of ANT what emerges from the analysis is a network characterized by actors that are neither purely technical nor purely social, but rather what [7] terms 'a hybrid collectif' (Fig. 1). This actor network comprised of 'heterogeneous' elements/relations erases the dichotomy that traditionally exists between the human and nonhuman. Technology and systems do not operate within a vacuum. CI have inherent organizational goals that involve designs on who will use it, how they will use it and the processes to facilitate its use. Hence complex vulnerabilities emerges from the actor network.

The Blackout (2003) provides a rich example of how vulnerabilities emerge from the spatially, temporally and characteristically heterogeneous actors. In a sense, these vulnerabilities can be likened to 'resident pathogens' [40: 198] that reside 'inscripted' within the system and through the process of translation can precipitate cascading failures.

As described in the final report [46], August 14, 2003, blackout was caused by deficiencies in specific practices, equipment, and human decisions by various organizations that affected conditions and outcomes that afternoon. Deficiencies in corporate policies, lack of adherence to industry policies and regulatory framework inscripted into the system are symptoms of translation process that facilitated inadequate management of reactive power and voltage which at the sharp end caused the blackout. This realization shows the spatial, temporal and physical heterogeneity that characterizes the actor network. Following the actors described in the literature reveals interdependencies that shape perception, actions and decisions. For example, failure to conduct long-term planning system planning studies, failure to conduct contingency and extreme conditions assessments, and independent reviews reflect an organizational mindset that does not embrace foresight and relies on 'unchallenged' assumptions. This was coupled and supported by 'planning and operational requirements and standards that were sufficiently ambiguous' (U.S.-Canada Power System Outage Task Force Final Report). This 'resident pathogen' thematically can be captured under the idea of 'illusions of invulnerability' and certainty, a mindset that accepts the status quo.

It is well documented in the literature that maintaining situation awareness (SA) is one of the most critical and challenging features for those operating complex socio-technical systems such as that within aviation, medicine and the nuclear industry [17]. Masys [33] describes how SA emerges from a network construct shaped by the translation and inscription process within a complex socio-technical system. In this case study, situation awareness was adversely affected by the inherent lack of procedures to ensure that its operators were continually aware of the functional state of their critical monitoring tools. This coupled with ineffective internal communications procedures and the lack of additional or back-up monitoring tools to understand or visualize the status of their transmission system shaped, through the translation process, a situation awareness [33] that did not embrace foresight but rather can be traced to an illusion of certainty and invulnerability

Diane Vaughan [47] coins the term 'normalization of deviance' to describe how organizations may erode safety by essentially accepting assumptions and mindsets that deviate from sound practice. What can be argued is the opposite of Murphy's law: what can go wrong goes right and then we draw the wrong conclusions. This can result in a 'drift into failure' [14] through the acceptance of practices, procedures and decisions that are characterized as a normalization of deviance shaped (translated) by policy, regulations, environment. This acts as 'resident pathogens' waiting to be unleashed. It is not the fact that they exist but rather that the network is interdependent thereby connecting actors displaced in space and time. Decision taken in the past under certain conditions become entangled in new conditions. This is the legacy systems and policies meeting the realities of the day that as described in Masys [33, 34] reify as policy violations, inadequate tools and systems, inadequate training of personnel to handle emergency conditions, lack of joint procedures or guidelines on when and how to coordinate a security limit violation.

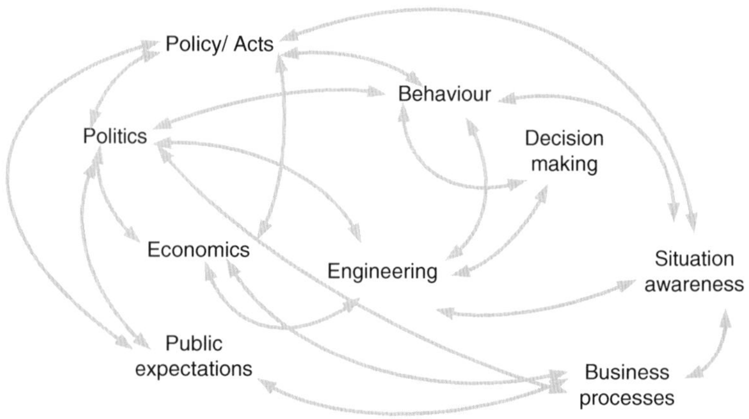

Fig. 2 Actor network- inscription and translation diagram

Cause 2

Of particular note was the failure to adequately manage tree growth in its transmission rights-of-way. The industry failed to maintain equipment ratings through a vegetation management program. A vegetation management program is necessary to fulfill NERC Policy 2, Section A, Requirement 1 (Control areas shall develop, maintain, and implement formal policies and procedures to provide for transmission security… including equipment ratings). However the vegetation management requirements were not defined in NERC Standards and Policies. The lack of defined standards and polices inscripted into the system facilitated a translation process that shaped regional and local management practices and mindsets. Following the actor, as described in ANT, reveals these inherent interdependencies and vulnerabilities. The relations between the actors emerge as key drivers in understanding the dynamics, structure of the aetiology of the disaster.

Take for example how some of the policies or guidelines were inexact, non-specific, or lacking in detail, thereby 'translating' divergent interpretations among reliability councils, control areas, and reliability coordinators. This shaped perceptions and actions with regards to implementing reliability standards. What can be construed as an illusion of invulnerability, reified as a lack of urgency in rectifying the deficiencies. Following the actors (Fig. 2) reveals how this 'mindset' shaped auditing processes regarding compliance and reliability requirements. Translation becomes apparent whereby as reported '…if those policies are ambiguous and do not make entities' roles and responsibilities clear and certain, they allow companies to perform at varying levels and system reliability is likely to be compromised' (U.S.-Canada Power System Outage Task Force Final Report).

The network does not exist in a vacuum. Context and environment are integrated into the actor network. As noted in the report (U.S.-Canada Power System Outage Task Force Final Report):

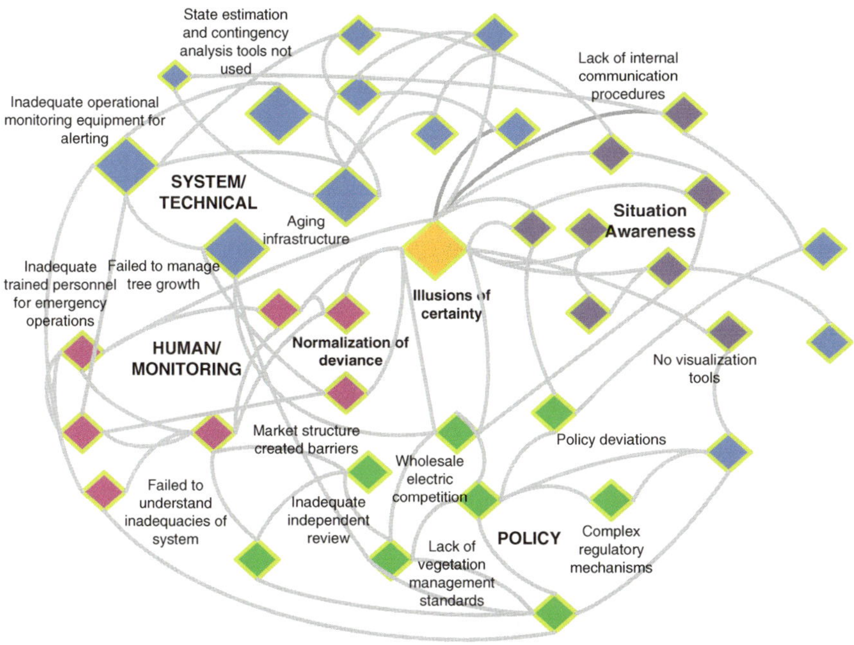

Fig. 3 Actor network (resident pathogens)

> Besides blaming high inter-regional power flows for causing the blackout, some blame the existence of those power flows upon wholesale electric competition. Before 1978, most power plants were owned by vertically-integrated utilities; purchases between utilities occurred when a neighbor had excess power at a price lower than other options. A notable increase in inter-regional power transfers occurred in the mid-1970s after the oil embargo, when eastern utilities with a predominance of high-cost oil-fired generation purchased coal-fired energy from Midwestern generators. The 1970s and 1980s also saw the development of strong north-to-south trade between British Columbia and California in the west, and Ontario, Québec, and New York-New England in the east. Americans benefited from Canada's competitively priced hydroelectricity and nuclear power while both sides gained from seasonal and daily banking and load balancing.

Johnson [24] argues that 'Deregulation has created new market pressures for innovation across many national infrastructures. The inherent complexity of the market structure creates barriers lack of complex regulatory mechanisms and the stress created by large transfers of energy across aging infrastructures and unprecedented distances were implicated in pan-regional blackouts across areas of North America' [24].

Policy, politics become hardwired into the system and through that affecting behaviour, decision making, situation awareness (Fig. 3). Small changes can have extraordinary and unpredictable consequences. This arises from the inherent complexity within the CI characterized as a 'hybrid collectif' [7], the intersection of the human, informational and physical domains (Fig. 1).

Johnson [24] notes, that many of the reports that were published after the European and North American blackouts of 2003 refer to a 'failure of imagination'. We were unprepared for the infrastructure vulnerabilities that have been created through the development of national and international energy markets. This failure of imagination reflects a mindset of illusions of certainty and invulnerability that emerged through the inscripted policies, regulation and market pressures and translated to design and operational decisions Fig. 3.

The performance of critical infrastructure relies on intricate, often nonlinear interactions among large number of interconnected and geographically distributed components of different types, including both technical and nontechnical elements. [25, p. 34]. An ANT analysis reveals how these heterogeneous elements interact thereby revealing hidden resident pathogens (Fig. 3). ANT thereby provides a valuable approach to foresight and possibility space exploration [35].

Emerging from the hybrid collectif (Fig. 1) are mental models '... deeply ingrained assumptions, generalizations, or even pictures or images that influence how we understand the world and how we take action' [42]. As described by Masys [35] with reference to the oil and gas industry safety culture, several processes through which mental models become flawed in industrial settings, resulted in the misreading of situations [10] which resonates with this case study. These processes include '... retaining outdated knowledge that no longer applies, accepting unreliable sources of information at face value, and missing out on critical data because of poor communication within the work organization' [9]. Illusions of certainty have everything to do with expectations. As [50] argue within the context of organizations:

... Expectations are built into organizational roles, routines, and strategies. These expectations create the orderliness and predictability.... Expectations, however, are a mixed blessing because they create blind spots. Blind spots sometimes take the form of belated recognition of unexpected, threatening events. And frequently blind spots get larger simply because we do a biased search for evidence that confirms the accuracy of our original expectations.

The analysis reveals that the vulnerabilities exist not in and of themselves but as a function of the relationality of the network, their interconnectivity and interdependencies (Fig. 3) that transcend what is social and what is technical thereby situating the accident aetiology within a decentered network construct (hybrid collectif). As argued by Helbing [22, p. 51], '...systemic failures and extreme events are consequences of the highly interconnected systems and networked risks humans have created. When networks are interdependent this makes them even more vulnerable to abrupt failures'. The actor network emerges as a ''hyper-connected world'' characterized by ''hyper-risks'' [22, p. 51] (Fig. 3).

As described in this actor network analysis and supported by Helbing [22, p. 51], 'Many disasters in anthropogenic systems should not be seen as 'bad luck', but as the results of inappropriate interactions and institutional settings. Even worse, they are often the consequences of a wrong understanding due to the counter-intuitive nature of the underlying system behaviour'. This requires a paradigm of network thinking and becomes apparent through the lens of actor network theory.

The Nonlinear interactions that transcend physical, spatial and temporal domains characterize the case study and actor network analysis. Vulnerability analysis reveals a complex, coupled network thereby suggesting that design and operation of such systems requires a requisite complex perspective to understand structure and dynamics. Systems thinking and network thinking become key e-nablers in CI management. Helbing [22, p. 53] argues that '...Individual risks may rightly have been viewed as small, but the risk to the system as a whole was vast'. Coping with networked 'hyper-risks' requires a humble recognition of the complexity of CI systems. ANT explores the ways that the networks of relations are composed, how they emerge and come into being, how they are constructed and maintained, how they compete with other networks, and how they are made more durable over time. It examines how actors enlist other actors into their world and how they bestow qualities, desires, visions and motivations on these actors [29].

Zolli et al. [52, p. 53] argues that 'the best way to determine the appropriate levels of inoculation for such super spreaders is to model crises before they occur'. Critical infrastructure vulnerability analysis requires the same level of modeling. As described in Schoemaker and Day [41: 84], '...the key to safety and reliability is to spot problems early and share them among well-trained personnel. ...accessing distributed intelligence takes a culture of alertness and information sharing across multiple social networks'. The technical, organizational, social, economic and political disconnects described above underscore why it is important to embrace multi-vocality and multiple perspectives in complex system design and operation [35].

5 Conclusion

Comfort, Hauskrecht and Lin [13] argue that the measurement of risk in metropolitan regions exposed to a range of hazards represents a complex set of the interactions between social and technical systems that requires an integrated research approach This highlights the requirement to evaluate the vulnerabilities inherent in the interconnected system (actor network) characterized by the hybrid collectif. Drawing upon ANT, Dolwick [16] argues that: '...if one were to try to draw a map of all of the actors present in any interaction, at any particular moment in time, instead of a well-demarcated frame, one would produce a highly convoluted network with a multiplicity of diverse dates, places and people'. The interdependencies and interconnectivity within the system contributes to its vulnerability (Perrow 2005). Comfort [12, p. 9] argues that '...the vulnerability of technical systems that support basic operations in a city cannot be calculated separately, but rather must be based upon careful estimates of the degree of interdependence or dependence across the entire socio-technical system that provides services to an urban region'.

Understanding, modeling and analyzing these systems with regards to vulnerability continues to be a challenge. A dynamic heterogeneous network comprised of evolving social needs, aging CI, new and legacy policies, regulations and procedures impact and interact with CI design and operations. Seeded within this actor network are potential 'resident pathogens'. Disconnects between policy, regulations, procedures, equipment, training became apparent through the critical infrastructure incident. It can be likened to a 'normalization of deviance' that permeated the network space through inscripted policies, technical capabilities/limitations, training. Through the process of translation the drifting into failure became a network dynamic characterized by the resident pathogens within a hybrid collectif (Fig. 1).

What ANT reveals is that vulnerabilities emerge within the hybrid collectif and thereby reflect the interdependencies and inherent relationality that crosses the human, physical and informational domains. By not delineating between the social and the technological (socio-technical) and employing the complexity/systems lens of ANT, a network mindset is introduced that facilitates 'following the actors' to reveal hidden vulnerabilities and criticalities within a system.

References

1. Aanestad M, Hanseth O (2000) Implementing open network technologies in complex work practices: a case from telemedicine. In Proceedings from the IFIP WG 8.2 conference IS2000, 10–12 June. Aalborg, Denmark: Organizational and Social Perspectives on Information Technology. Kluwer Academic Publishers, 355–369
2. Akrich M (1992) The de-scription of technical objects' In: Bijker W, Law J (eds) Shaping technology, building society: studies in sociotechnical change, Mass: MIT Press, Cambridge, 205–224
3. Barabasi A-L (2003) Linked. Plume, New York Penguin Group
4. Barabasi A-L (2013) Network science. Phil Trans R Soc A 371:20120375 published 18 February 2013
5. Braha D, Bar-Yam Y (2006) From centrality to temporary fame: dynamic centrality in complex networks. Complexity 12(2):59–63
6. Brummitt CD, Hines PDH, Dobson I, Moore C, D'Souza RM (2013) Transdisciplinary electric power grid science. Proc Natl Acad Sci USA. 110 (30):12159
7. Callon M, Law J (1995) Agency and the hybrid collectif. The S Atlantic Q 94(2):481–507
8. Callon M (1999) Actor-Network theory: the market test in. Law J, Hassard J (eds) Actor network and after, Oxford and Keele: Blackwell and the Sociological Review, 181–195
9. Chapman J (2005) Predicting technological disasters: mission impossible? Disaster Prevention and Management 14(3):343–352
10. Chapman JA, Ferfolja T (2001) Fatal flaws: the acquisition of imperfect mental models and their use in hazardous situations, J Intell Capital 2(4):398–409
11. Coakes E (2003) 'Socio-technical thinking- an holistic viewpoint In: Clarke S, Coakes E, Hunter MG, Wenn A (eds) Socio-technical and human cognition elements of information systems, Information Science Publishing, Hershey, 1–4
12. Comfort LK (2006) Cities at risk: Hurricane Katrina and the drowning of New Orleans. Urban Aff Rev 41:501–516

13. Comfort LK, Hauskrecht M, Lin J-S (2004) Dynamic networks: modeling change in environments exposed to risk. In: Fiedric F. Van de Walle B (eds) Proceedings of the 5th International ISCRAM Conference – Washington, DC, USA, May 2008, 576–585
14. Dekker S (2011) Drift into failure: from hunting broken components to understanding complex systems. Ashgate Publishing, Aldershot
15. DHS (2013) Homeland security—what is critical infrastructure. http://www.dhs.gov/what-critical-infrastructure
16. Dolwick JS (2009) The social and beyond: introducing actor-network theory, J Maritime Archaeology 4(1):21–49
17. Endsley MR (1999) Situation awareness in aviation systems. In: Garland DJ, Wise, JA Hopkin VD (eds) Handbook of aviation human factors, Lawrence Erlbaum Associates, Mahwah, NJ, 257–276
18. Eurocontrol (2010) Ash-cloud of April and May 2010: impact on air traffic EUROCONTROL/CND/STATFOR STATFOR/Doc394 v1.0 28/6/10. https://www.eurocontrol.int/sites/default/files/attachments/201004-ash-impact-on-traffic.pdf. Accessed 1 July 2013
19. FEMA (2013) Hurricane sandy after action report, 1 July 2013. https://s3-us-gov-west-1.amazonaws.com/dam-production/uploads/20130726-1923-25045-7442/sandy_fema_aar.pdf. Accessed 14 Aug 2013
20. Flood RL (1999) Rethinking the fifth discipline: learning within the unknowable. Routledge Publishing, London
21. Harris AJL, Gurioli L, Hughes EE, Lagreulet S (2012) Impact of the Eyjafjallajökull ash cloud: a newspaper perspective. J Geophys Res 117(B00C08):1–35
22. Helbing D (2013) Globally networked risks and how to respond. Nature 497:51–59
23. IAEA (2012) Protection against extreme earthquakes and tsunamis in the light of the accident at the Fukushima Daiichi Nuclear Power Plant. International experts meeting, Vienna, Austria, 4–7 September 2012 http://www.iaea.org/newscenter/focus/actionplan/reports/protection040912.pdf
24. Johnson CW (2008) Understanding failures in international safety infrastructure: a comparison of European and North American power failures, In: Proceedings of the 26th international conference on system safety, Vancouver, BC, 25–29 August
25. Kroger W, Zio E (2011) Vulnerable systems. Springer Publishing, Dordrecht
26. Latour B (1987) Science in action: how to follow scientists and engineers through society. Open University Press, Milton Keynes
27. Latour B (1991) Technology is society made durable. In: Law J (ed) A sociology of monsters: essays on power, technology and domination, Routledge, London, pp 103–131
28. Latour B (1994) Pragmatogonies: a mythical account of how humans and nonhumans swap properties. Behav Sci 37(6):791–808
29. Latour B (1996) On actor-network theory. A few clarifications. Soziale Welt 47(4):369–381
30. Latour B (2005) Reassembling the social: an introduction to actor network theory. Oxford University Press, Oxford
31. Law J, Callon M (1988) Engineering and sociology in a military aircraft project: a network analysis of technological change, Social Problems, 35(3):284–297
32. Masys AJ (2005) A systemic perspective of situation awareness: an analysis of the 2002 mid-air collision over Uberlingen, Germany. Int J Disaster Prev Manag 14(4):548–557
33. Masys AJ (2010) Fratricide in air operations: opening the black box- revealing the social. PhD Dissertation, June 2010, University of Leicester, UK
34. Masys AJ (2011) The emergent nature of risk as a product of heterogeneous engineering. In: Bennett S (ed) Innovative thinking in risk, crisis and disaster management. Gower Publishing, London
35. Masys AJ (2012) Black swans to grey swans-revealing the uncertainty. Int J Disaster Prev Manag 21(3):320–325
36. Monteiro E (2000) Actor network theory and information infrastructure. In: Ciborra C (ed) From control to drift. The dynamics of corporate information infrastructures, Oxford

University Press, Oxford, 71–83.www.idi.ntnu.no/~ericm/ant.FINAL.htm, (accessed 11 July, 2005)
37. Oxford Economics (2011) UK Economic losses due to volcanic ash air travel restrictions, http://www.visitbritain.org/Images/Volcano%20Economic%20Impact%20Study%20-%20Oxford%20Economics_tcm29-15226.pdf. Accessed 1 July 2013
38. Powell JL, Owen T (2011) Actor network theory and social science: possibilities and implications. J public adm gov. 1(2):140–157. www.macrothink.org/jpag
39. Public Safety Canada (2009) National strategy for critical infrastructure. http://www.publicsafety.gc.ca/cnt/rsrcs/pblctns/srtg-crtcl-nfrstrctr/srtg-crtcl-nfrstrctr-eng.pdf
40. Reason J (1990) Human error, Cambridge University Press, New York
41. Schoemaker PJH, Day GS (2009) How to make sense of weak signals. MIT sloan management review
42. Senge P (1990) The fifth discipline: The art and practice of the learning organization, Doubleday Currency, New York
43. Somerville I (1997) Actor network theory: a useful paradigm for the analysis of the UK cable/on-line socio-technical ensemble? http://hsb/baylor.edu/eamsower/ais.ac.97/papers/somervil.html, (accessed 10 August, 2004)
44. Urry J (2002) The global complexities of September 11th. Theor Cult Soc 19(4):57–69
45. Urry J (2005) The complexities of the global. Theor Cult Soc 22(5):235–254
46. U.S.-Canada Power System Outage Task Force Final Report on the August (2003) Blackout in the United States and Canada: causes and Recommendations April 2004. http://www.nrcan.gc.ca/energy/sources/electricity/1378. Accessed on 1 June 2013
47. Vaughan D (1996) Challenger launch decision: risky technology, culture and deviance at NASA, University of Chicago Press, Chicago
48. Vespignani A (2009) Predicting the behavior of techno-social systems. Science 325:425–428
49. Vespignani A (2010) The fragility of interdependency. Nature, 464:984–985, 15 April, 2010
50. Weick KE, Sutcliffe KM (2007) Managing the unexpected: resilient performance in an age of uncertainty, 2nd edn. Wiley, San Francisco
51. Yeung HWC (2002) Towards a relational economic geography: old wine in new bottles? Paper presented at the 98th annual meeting of the association of American geographers, Los Angeles, CA. 19–23 March 2003. http://courses.nus.edu.sg/course/geoywc/publication/Yeung_AAG.pdf
52. Zolli A, Healy A-M (2007) Resilience book: why things bounce back. Free press, New York division of Simon and Schuster inc

Dealing with Complexity: Thinking About Networks and the Comprehensive Approach

Anthony J. Masys

Abstract Within the 'hyper-connected world' Helbing [9: 51], networked risks emerge that challenge our understanding and management of the defence, security and safety domain. In this 'hyper-connected world' with interconnected social/technical/political/economic systems, shocks to regional, national and global systems stemming from natural hazards, acts of armed violence, terrorism and transnational crime have significant defence and security implications. Helbing [9: 53] argues that '...Individual risks may rightly have been viewed as small, but the risk to the system as a whole was vast.'' Risk assessments thereby require appreciation for the externalities and non-linear cause-effect relationships that reside within the problem space. In light of this, we are moving from crisis management to complexity management. This chapter describes the relevance of the comprehensive approach to the defence, security and safety domain within the paradigm of network thinking

Keywords Hyper-risks · Comprehensive approach · Network thinking

1 Introduction

> I think the next century will be the century of complexity.
>
> Stephen Hawking

Stephen Hawking rightly captures the essence of the underlying theme that emerges from the chapters of this book: namely dealing with complexity. The complexity and dilemmas that characterize the defence and security landscape

A. J. Masys (✉)
University of Leicester, Leicester, UK
e-mail: Anthony.masys@gmail.com

A. J. Masys (ed.), *Networks and Network Analysis for Defence and Security*,
Lecture Notes in Social Networks, DOI: 10.1007/978-3-319-04147-6_12,

requires a paradigm shift: a shift in the way we see; a shift in the way we conceptualize; a shift that recognizes the inherent interconnectivity and interdependencies. Such a shift challenges a reductionist approach and embraces a systems thinking paradigm. The paradigm of systems thinking permits a view of the world as a complex system in which as noted by Sterman [18: 10] we come to the understanding that 'you can't do just one thing' and that 'everything is connected to everything else'. This is supported by Senge [16: 73] who is of the opinion that the discipline of the systems approach lies in a shift of mind: in seeing interrelationships rather than linear cause-effect chains and seeing processes of change rather than snapshots. System thinking paradigm that is realized through the 'network mindset' [2, 19] thereby is an appropriate approach for unearthing and communicating the complexities and interdependencies resident within the defence and security space.

According to Helbing [9: 51] we are increasingly living in a 'hyper-connected world' which creates 'hyper-risks' because of numerous networks and interdependencies. When we consider this 'hyper-connected world', networked risks emerge thereby challenging our understanding regarding the defence, security and safety domain. In this 'hyper-connected world' with interconnected social/technical/political/economic systems, shocks to regional, national and global systems stemming from natural hazards, acts of armed violence, terrorism and transnational crime have significant defence and security implications. Helbing [9: 53] argues that '…Individual risks may rightly have been viewed as small, but the risk to the system as a whole was vast.''

When dealing with terrorist organizations, serious and organized crime syndicates or critical infrastructure vulnerabilities, the 'network mindset' reveals the inherent interdependencies within the system and through this shapes intervention strategies. In this sense the 'network mindset' described in this book resonates with the 'comprehensive approach'. As described by de Coning and Friis [5: 2]

> The comprehensive approach concept should be understood in the context of an increasingly complex and interdependent international conflict management system. The scope of the crises faced by the international community is often of such a scale that no single agency, government or international organisation can manage them alone. In response, a wide range of agencies, governmental and non-governmental, and regional and international organizations have each developed specialised capacities to manage various aspects of these complex crisis systems, and together they have been able to respond with a broad range of interlinked activities.

Problems associated with defence and security are increasingly complex and interdependent. This complex problem space is value-laden, open-ended, multidimensional, ambiguous and unstable and can be labeled as 'wicked and messy'. In this sense the defence and security domain resists being tamed, bounded or managed by classical problem solving approaches.

Helbing [9: 51] poignantly argues that 'Globalization and technological revolutions are changing our planet. Today we have a worldwide exchange of people, goods, money, information, and ideas, which has produced many new opportunities, services and benefits for humanity. At the same time, however, the underlying networks have created pathways along which dangerous and damaging events can spread rapidly and globally'. This is particularly relevant with the affect of transnational crime, terrorism and natural disasters on defence and security.

Addressing the unique challenges associated with such transnational threats as terrorism and organized crime requires collaborative efforts among key defence and security stakeholders that facilitate questioning judgments and underlying assumptions, and employing critical and creative thinking in order to explore the possibility space.

2 Crisis Management and the Comprehensive Approach

Major and Schöndorf [11: 1] argue that 'today's crisis brings together social, economic and political security dimensions'. As described in Masys [12], between 50,000 and 100,000 people died, more than half of them children under five during the 2011 Horn of Africa crisis that affected Somalia, Ethiopia and Kenya. This represents a human security problem that crosses the defence, security and safety domains. The accompanying destruction of livelihoods, livestock and local market systems affected 13 million people overall. This highlights threats to human security as multiple, complex and interrelated and often mutually reinforcing.

The multidimensional nature of human security is highlighted in the 1994 Human Development Report by the United Nations Development Programme (UNDP). The report broadly defines human security as 'freedom from fear and freedom from want' recognizing seven key interrelated components (economic, food, health, environmental, personal, community and political security). The Commission on Human Security [3] argues that a new paradigm of security is required. It is recognized that:

> First, human security is needed in response to the complexity and the interrelatedness of both old and new security threats – from chronic and persistent poverty to ethnic violence, human trafficking, climate change, health pandemics, international terrorism, and sudden economic and financial downturns. Such threats tend to acquire transnational dimensions and move beyond traditional notions of security that focus on external military aggressions alone.
>
> Second, human security is required as a comprehensive approach that utilizes the wide range of new opportunities to tackle such threats in an integrated manner. Human security threats cannot be tackled through conventional mechanisms alone. Instead, they require a new consensus that acknowledges the linkages and the interdependencies between development, human rights and national security.

The interdependencies, uncertainty and complexity associated with human security, challenge traditional linear approaches to problem framing/solving and policy formulation. Systems thinking, characterized by seeing wholes and interconnections is critical to understanding these complex systems. Multiple lenses afford richer interpretation and thereby results in holistic vantage point in which to examine the problem and solution space. Houghton and Tuffley [10] argue that '…instead of structuring agreement, problem solvers should explore disagreement and deliberately highlight areas of dynamic tension so that new solution pathways can emerge. The process of synthesis means exploring tension and conflict to find places where the creative solution can be found'.

As described by the CHS [3], 'new ways of thinking' are required (if not essential) to manage the complex problems associated with human security. Major and Schöndorf [11] argue that '…to achieve successful outcomes, governments and other actors involved need to coordinate their aims, activities and instruments at the earliest possible stage and ensure these are tailored to need. This is what comprehensive approaches are all about. New concepts and structures should be introduced to guarantee the coordination and cooperation of those involved at national and international levels'. The concept of a comprehensive approach is based on the assumption and requirement for some level of coherence amongst the actors/stakeholders regarding shared goals and objective and to create a dialogue to address the various dimensions of the problem space (political, security, safety, socio-economic, humanitarian and human rights).

Mapping out conflicting values, principles and priorities, organizational and operational challenges, capability management, capacity management, challenges in leadership and management is a key 'network mindset' exercise that is essential to enabling the comprehensive approach. Considering these factors, an Activity Focus Network (AFN) model [4: 35] was developed that highlights the interdependencies resident within the comprehensive approach (Fig. 1). AFN are used to represent a complex activity system of an organization. The macro representation afforded by the AFN is built from higher-order groupings of activities/concepts reflecting the organizations social cognitions about how activities are organized [4: 38]. Through the network mindset, the interdependencies and interactions among the elements (Fig. 1), as well as the unity of the system itself will provide critical insights for understanding an organization and its systems properties [1].

Such a perspective reveals how changes to the socio-political-economic-technical system become enrolled into complex and subtle blendings, shaping collaboration, cooperation and coordination. It is recognized that 'connectivity and interdependence propagates the effects of actions, decisions and behaviours …, but that propagation or influence is not uniform as it depends on the degree of connectedness' [13].

In realizing the comprehensive approach (Fig. 1), crisis management essentially becomes complexity management [11]. Lesson from Afghanistan or the Balkans highlight that if '…one aspect of crisis management is neglected or measures are not inter-linked, there will be an impact on related efforts elsewhere' [11: 2]. Complex problems are, by their very nature, difficult to predict. Complex issues pertaining to defence, security, risk, crisis and disaster management are not amenable to detailed forecasting. Rather than fixing the shape of policy responses in advance, responses need the flexibility to adapt to emerging insights.

The network mindset thereby arises from the necessity for a complexity view of the 'wicked' problem space associated with human security. Hence managers, leaders and those involved in humanitarian or relief operations should recognize the complexity inherent in human security operations and be responsive to the

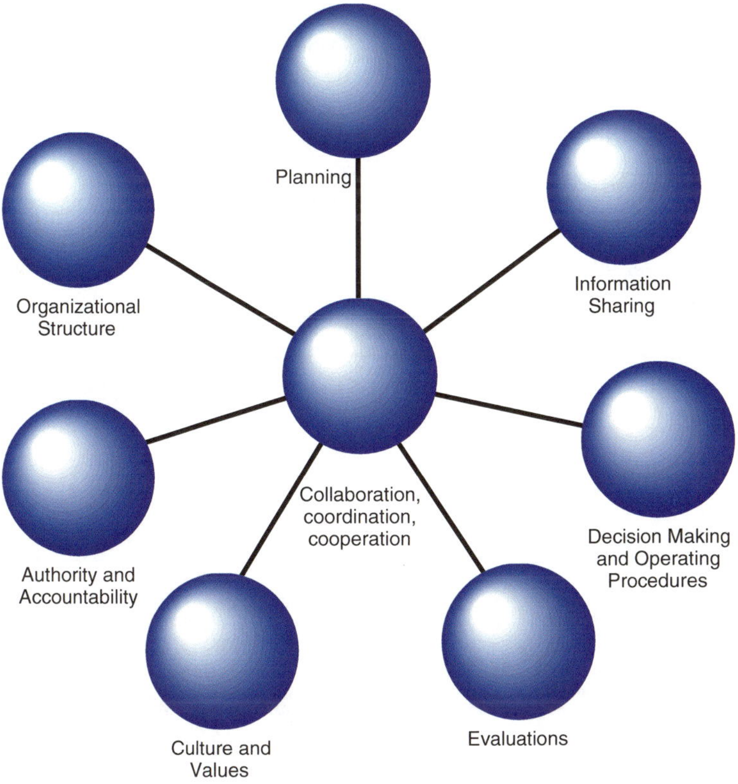

Fig. 1 Activity focus network of comprehensive approach (leveraging [20: 11–12]

emergence of weak signals and recognize potential opportunities for constructive action through an integrated comprehensive approach.

> The comprehensive approach not only makes sense – it is necessary, says NATO Secretary General Rasmussen [15].

The effective implementation of a comprehensive approach requires all actors to contribute in a concerted effort, based on a shared sense of responsibility, openness and determination, taking into account their respective strengths, mandates and roles, as well as their decision-making autonomy. This requires the network mindset (see in particular Chaps. Decision Support Through Strongest Path Method Risk Analysis, Critical Infrastructure and Vulnerability: A Relational Analysis Through Actor Network Theory, Dealing with Complexity: Thinking About Networks and the Comprehensive Approach). Peace and stability operations are managed more effectively when the interdependency and interconnectedness of the political, security, governance and development dimensions of these operations are recognized. All these initiatives have a similar aim: to achieve greater harmonization and synchronization among the activities of the different international and local actors, and across the analysis, planning, implementation, management and evaluation aspects of the programme cycle' [6: 12].

Collective endeavors are therefore borne out of a realization that an organization cannot achieve all its goals without cooperation with other organizations operating in the same domain [20: 7]. Network thinking affords the opportunity to map complex tactical and strategic interoperabilities. In this way it becomes a thinking tool and 'worldview' facilitating tool that can be applied to all phases of conflict and crisis to all the actors involved and at all operational levels.

As described in Chap. 11, inconsistencies in policies can have implications that can resonate through the system. It is therefore not feasible for organizations operating at the level of collaboration to have inconsistent policies. The requirement for improved coherence and coordination among the international community engaged in Afghanistan Friis [7: 5] or disaster response [12] calls for a 'network mindset' that supports the comprehensive approach.

3 Conclusion

Crisis management is complexity management. The network mindset that emerges as a cross-cutting theme from the chapters resonates as a supporting 'worldview' and selection of methodologies that supports a greater application of the comprehensive approach. As described in Friis and Jarmyr [8: 20]

> …a comprehensive approach does not require the full integration of all actors in a crisis area into one neat hierarchical structure. Horizontal collaboration in networks and clusters of networks is already taking place in some sectors (like humanitarian relief), and this is the model on which a comprehensive approach can be developed further in other sectors as well. That said, some degree of strategic coherence is a precondition also in this model.

That inherent coherence lies within the model described in Fig. 1. The high impact, low probability events that so characterize recent history and described in various chapters of this book highlight the necessity for complexity management: enabled through insights from embracing the 'network mindset' as the lens to support the comprehensive approach. This chapter introduces the connection between network mindset and the comprehensive approach and acts as a catalyst for further development at the strategic, operational and tactical level.

References

1. Anderson R, Crabtree B, Steele D, McDaniel R (2005) Case study research: the view from complexity science. Qual Health Res 15(5):669–685
2. Barabasi A-L (2003) Linked. Plume. Penguin Group, New York
3. CHS D (2009) United Nations Trust Fund for Human Security. Human Security in Theory and Practice: Applications of the Human Security Concept and the United Nations Trust Fund for Human Security. United Nations, New York 1–79

4. Corman SR (2006) Using activity focus networks to pressure terrorist organizations. Comput math organiz theory 12:35–49
5. De Coning C, Friis K (2008) Introduction: how to conceptualise 'comprehensive approach'? In: Friis K, Jarmyr P (eds) Comprehensive approach challenges and opportunities in complex crisis management. NUPI Report
6. De Coning C, Lurås N, Schia NN, Ulriksen S (2009) Norway's whole-of-government approach and its engagement with Afghanistan. NUPI Report Security in Practice-8 2009. Available at http://www.oecd.org/development/evaluation/dcdndep/47107380.pdf (accessed 1 Nov 2013)
7. Friis K (2010) The Politics of the Comprehensive Approach: The Military, Humanitarian and State-building Discourses in Afghanistan. NUPI Working Paper 773. Norwegian Institute of International Affairs 2010.https://www.cimicweb.org/cmo/compapp/Documents/Research%20and%20Analysis/Friis%20(2010)%20Comprehensive%20Approach%20in%20AGZ[1].pdf
8. Friis K, Jarmyr P (eds) (2008) Comprehensive Approach Challenges and opportunities in complex crisis management. NUPI Report, Security in Practice no. 11. Norsk Utenrikspolitisk Institutt 2008
9. Helbing D (2013) Globally networked risks and how to respond. Nature 497:51–59
10. Houghton L, Tuffley D (2013) Towards a methodology of wicked problem exploration through concept shifting and tension point analysis. Syst Res Behav Sci
11. Major C, Schondorf E (2011) Comprehensive approaches to crisis management. SWP Comments, 23 Sept 2011. http://www.swp-berlin.org/en/publications/swp-comments-en/swp-aktuelledetails/article/crisis_management_comprehensive_approaches.html
12. Masys AJ (2013) Human security—a view through the lens of complexity. In: Gilbert T, Kirkilionis M, Nicolis G (eds) Proceedings of the European conference on complex systems 2012., Springer proceedings in complexity Springer, Berlin, pp 325–335
13. Mitleton-Kelly E, Papaefthimiou MC (2000) Co-evolution of diverse elements interacting within a social ecosystem. In: The proceedings of feast 2000 international workshop on 'feedback and evolution in software and business processes. Imperial College, London, UK, 10–12 July 2000
14. National Crime Agency (2011) A plan for the creation of a national crime-fighting capability. Presented to Parliament by the Secretary of State for the Home Department by Command of her Majesty
15. NATO (2012) A comprehensive approach to crisis management. Available at http://www.nato.int/cps/en/natolive/topics_51633.htm. Accessed 14 Nov 2013
16. Senge P (1990) The fifth discipline: the art and practice of the learning organization. Doubleday Currency, New York
17. Senge P (2006) The fifth discipline: the art and practice of the learning organization. Doubleday Currency, New York
18. Sterman JD (2000) Business dynamics: systems thinking and modeling for a complex world. McGraw-Hill Publishing, Boston
19. Vespignani A (2009) Predicting the behavior of techno-social systems. Science 325(24):425–428
20. Williams AP (2010) Implications of operationalizing a comprehensive approach: defining what interagency interoperability really means. Int C2 J 4:1

Printed by Printforce, the Netherlands